Recovery of Non-ferrous Metal from Metallurgical Residues

Recovery of Non-ferrous Metal from Metallurgical Residues

Editor

Guo Chen

Basel • Beijing • Wuhan • Barcelona • Belgrade • Novi Sad • Cluj • Manchester

Editor
Guo Chen
School of Chemistry
and Environment
Yunnan Minzu University
Kunming
China

Editorial Office
MDPI
St. Alban-Anlage 66
4052 Basel, Switzerland

This is a reprint of articles from the Special Issue published online in the open access journal *Materials* (ISSN 1996-1944) (available at: www.mdpi.com/journal/materials/special_issues/Recovery_Metal).

For citation purposes, cite each article independently as indicated on the article page online and as indicated below:

Lastname, A.A.; Lastname, B.B. Article Title. *Journal Name* **Year**, *Volume Number*, Page Range.

ISBN 978-3-0365-9627-3 (Hbk)
ISBN 978-3-0365-9626-6 (PDF)
doi.org/10.3390/books978-3-0365-9626-6

© 2024 by the authors. Articles in this book are Open Access and distributed under the Creative Commons Attribution (CC BY) license. The book as a whole is distributed by MDPI under the terms and conditions of the Creative Commons Attribution-NonCommercial-NoDerivs (CC BY-NC-ND) license.

Contents

About the Editor vii

Guo Chen
Recovery of Non-Ferrous Metal from Metallurgical Residues
Reprinted from: *Materials* **2023**, *16*, 6943, doi:10.3390/ma16216943 1

Godfrey Dzinomwa, Benjamin Mapani, Titus Nghipulile, Kasonde Maweja, Jaquiline Tatenda Kurasha and Martha Amwaama et al.
Mineralogical Characterization of Historic Copper Slag to Guide the Recovery of Valuable Metals: A Namibian Case Study
Reprinted from: *Materials* **2023**, *16*, 6126, doi:10.3390/ma16186126 4

Szymon Orda, Michał Drzazga, Katarzyna Leszczyńska-Sejda, Mateusz Ciszewski, Alicja Kocur and Pola Branecka et al.
Investigations of the Density and Solubility of Ammonium Perrhenate and Potassium Perrhenate Aqueous Solutions
Reprinted from: *Materials* **2023**, *16*, 5481, doi:10.3390/ma16155481 26

Almagul Ultarakova, Zaure Karshyga, Nina Lokhova, Azamat Yessengaziyev, Kaisar Kassymzhanov and Arailym Mukangaliyeva
Studies on the Processing of Fine Dusts from the Electric Smelting of Ilmenite Concentrates to Obtain Titanium Dioxide
Reprinted from: *Materials* **2022**, *15*, 8314, doi:10.3390/ma15238314 37

Zhibo Zhang, Weiwei Huang, Weidong Zhao, Xiaoyuan Sun, Haohang Ji and Shubiao Yin et al.
Hot Deformation Behavior of TA1 Prepared by Electron Beam Cold Hearth Melting with a Single Pass
Reprinted from: *Materials* **2022**, *16*, 369, doi:10.3390/ma16010369 52

Pengchao Li, Guifang Zhang, Peng Yan, Peipei Zhang, Nan Tian and Zhenhua Feng
Numerical and Experimental Study on Carbon Segregation in Square Billet Continuous Casting with M-EMS
Reprinted from: *Materials* **2023**, *16*, 5531, doi:10.3390/ma16165531 66

Chunlan Tian, Ju Zhou, Chunxiao Ren, Mamdouh Omran, Fan Zhang and Ju Tang
Drying Kinetics of Microwave-Assisted Drying of Leaching Residues from Hydrometallurgy of Zinc
Reprinted from: *Materials* **2023**, *16*, 5546, doi:10.3390/ma16165546 80

Jinsong Du, Aiyuan Ma, Xingan Wang and Xuemei Zheng
Review of the Preparation and Application of Porous Materials for Typical Coal-Based Solid Waste
Reprinted from: *Materials* **2023**, *16*, 5434, doi:10.3390/ma16155434 99

Wen Jiang and Kunyu Zhao
Effect of Cu on the Formation of Reversed Austenite in Super Martensitic Stainless Steel
Reprinted from: *Materials* **2023**, *16*, 1302, doi:10.3390/ma16031302 123

Zilong Zhao, Kaiyu Wang, Guoyuan Wu, Dengbang Jiang and Yaozhong Lan
Adsorption of Sc on the Surface of Kaolinite (001): A Density Functional Theory Study
Reprinted from: *Materials* **2023**, *16*, 5349, doi:10.3390/ma16155349 137

Bing Yi, Guifang Zhang, Qi Jiang, Peipei Zhang, Zhenhua Feng and Nan Tian
The Removal of Inclusions with Different Diameters in Tundish by Channel Induction Heating:
A Numerical Simulation Study
Reprinted from: *Materials* **2023**, *16*, 5254, doi:10.3390/ma16155254 . . . . **153**

Haoli Yan, Xiaolei Zhou, Lei Gao, Haoyu Fang, Yunpeng Wang and Haohang Ji et al.
Prediction of Compressive Strength of Biomass–Humic Acid Limonite Pellets Using Artificial
Neural Network Model
Reprinted from: *Materials* **2023**, *16*, 5184, doi:10.3390/ma16145184 . . . . **166**

Xuemei Zheng, Jinjing Li, Aiyuan Ma and Bingguo Liu
Recovery of Zinc from Metallurgical Slag and Dust by Ammonium Acetate Using Response
Surface Methodology
Reprinted from: *Materials* **2023**, *16*, 5132, doi:10.3390/ma16145132 . . . . **180**

Xiaoxi Wan, Jun Li, Na Li, Jingxi Zhang, Yongwan Gu and Guo Chen et al.
Preparation of Spherical Ultrafine Silver Particles Using Y-Type Microjet Reactor
Reprinted from: *Materials* **2023**, *16*, 2217, doi:10.3390/ma16062217 . . . . **198**

Na Li, Jun Li, Xiaoxi Wan, Yifan Niu, Yongwan Gu and Guo Chen et al.
Preparation of Micro-Size Spherical Silver Particles and Their Application in Conductive Silver
Paste
Reprinted from: *Materials* **2023**, *16*, 1733, doi:10.3390/ma16041733 . . . . **212**

Chuang Zhang, Xiaolei Zhou, Lei Gao and Haoyu Fang
Study on the Roasting Process of Guisha Limonite Pellets
Reprinted from: *Materials* **2022**, *15*, 8845, doi:10.3390/ma15248845 . . . . **224**

Xuemei Zheng, Shiwei Li, Bingguo Liu, Libo Zhang and Aiyuan Ma
A Study on the Mechanism and Kinetics of Ultrasound-Enhanced Sulfuric Acid Leaching for
Zinc Extraction from Zinc Oxide Dust
Reprinted from: *Materials* **2022**, *15*, 5969, doi:10.3390/ma15175969 . . . . **243**

About the Editor

Guo Chen

Dr. Guo Chen, Professor at Yunnan Minzu University, PhD (Joint-Supervision by the Ministry of Education of the People's Republic of China and Scientific and Industrial Research Organization, Australia), Vice Director of the National and Local Joint Engineering Research Center for Green Preparation Technology of Bio-Based Materials, Vice President of the first Council of Platinum Group Metals Branch of the China Nonferrous Metals Industry Association, and committee member of the 7th Academic Committee of Heavy Nonferrous Metallurgy of the China Nonferrous Metals Society, is a member of associations including the Association for Microwave Power in Europe for research and education (AMPERE) and the American Institute of Chemical Engineers (AIChE).

Editorial

Recovery of Non-Ferrous Metal from Metallurgical Residues

Guo Chen

Kunming Key Laboratory of Energy Materials Chemistry, Yunnan Minzu University, Kunming 650500, China; guochen@kust.edu.cn; Tel.: +86-871-6591-0017

Non-ferrous metals and alloys are essential resources for the development of modern industries. With the depletion of natural minerals, the recovery of non-ferrous metals from metallurgical residues attracts researchers from multidisciplinary areas. Ideas for new recovery routes reduce the pressure on natural resources and the environment, thus enabling better manufacturing sustainability. This Special Issue primarily considers papers focused on the theoretical and engineering aspects of processing metal recovery from metallurgical residues. The purpose of the current editorial is to briefly summarize the publications included in this Special Issue.

Dzinomwa et al. examined the historic slag produced from a smelter in Namibia, which accumulated over decades of its operating life [1]. Based on the results, approximate conditions under which the different slag phases were formed were estimated, and the recovery routes for the various metals were proposed. Zheng et al. summarize the physicochemical characteristics and general processing methods of coal gangue and fly ash and review the progress in the application of porous materials prepared from these two solid wastes in the fields of energy and environmental protection, including the following: the adsorption treatment of heavy metal ions, ionic dyes, and organic pollutants in wastewater [2]. Tian et al. studied the application of microwave technology in recovering valuable metals from the leaching residue produced from zinc production [3]. Zheng et al. studied the process conditions of recycling Zn from metallurgical slag and dust material leaching using ammonium acetate (NH_3-CH_3COONH_4-H_2O) [4]. The influences of the liquid/solid ratio, stirring speed, leaching time, and total ammonia concentration as well as the interactions between these variables on the Zn effective extraction rate during the ammonium acetate leaching process were investigated. They also proposed an experimental study on ultrasound-enhanced sulfuric acid leaching for zinc extraction from zinc oxide dust [5]. Ultarakova et al. present studies on the ammonium fluoride processing of dust from the reduction smelting of ilmenite concentrate with silicon separation to obtain titanium dioxide [6]. Optimal conditions for pyrohydrolysis of titanium fluorides were determined. The effects of temperature and duration on the process were studied.

Regarding the preparation of advanced materials, the following articles are included in this Special Issue. Wang et al. prepared micron-sized silver particles using the chemical reduction method by employing a Y-type microjet reactor, silver nitrate as the precursor, ascorbic acid as the reducing agent, and gelatin as the dispersion at room temperature [7]. Using a microjet reactor, the two reaction solutions collide and combine outside the reactor, thereby avoiding microchannel obstruction issues and facilitating a quicker and more convenient synthesis process. The resistivity of conductive silver paste prepared with the as-synthesized spherical silver particles was also discussed in detail [8]. Orda et al. contribute to the technique development of purification commercial rhenium salts [9]. The adsorption behavior of Sc on the surface of kaolinite (001) was investigated using the density functional theory via the generalized gradient approximation plane-wave pseudopotential method by Zhao et al. [10]. Zhang et al. researched the thermal deformation behavior of titanium ingots prepared using EB furnaces, which can reduce the cost of titanium production [11].

Some of the attempts at optimizing the traditional metallurgical process are also presented in this Special Issue, the research method of which can be referenced for the slag treatment. Research on limonite pellet technology is crucial for iron making as high-grade iron ore resources decline. However, pellets undergo rigorous mechanical actions during production and use. Yan et al. prepared a series of limonite pellet samples with varying ratios and measured their compressive strength. Artificial neural networks (ANN) predicted the compressive strength of humic acid and bentonite-based pellets, establishing the relationship between input variables (binder content, pellet diameter, and weight) and the output response (compressive strength). Integrating pellet technology and machine learning drives limonite pellet advancement, contributing to emission reduction and environmental preservation [12]. Zhang et al. used a pelletizing method to enhance the subsequent iron-making process by applying Guisha limonite, with advantages including large reserves and low price [13]. The purpose is to provide an alternative for the sinter, thus reducing the greenhouse gas emissions during the iron-making process. A multivariate regression model for estimating the compressive strength of pellets was developed using the Box–Behnken experimental methodology, where the relevant factors were the roasting temperature, pellet diameter, and bentonite content.

For steel making, electromagnetic stirring (M-EMS) has been extensively applied in continuous casting production to reduce the quality defects of casting billets. To investigate the effect of continuously casting electromagnetic stirring on billet segregation, a 3D multiphysics coupling model was established to simulate the internal heat, momentum, and solute transfer behavior in order to identify the effect of M-EMS on the carbon segregation of a continuous casting square billet [14]. The quality of the bloom will be impacted by the non-metallic impurities in the molten steel in the tundish, which will reduce the plasticity and fatigue life of the steel. Yi et al. established a six-flow double-channel T-shaped induction heating tundish mathematical model. The effects of induction heating conditions on the removal of inclusions in the tundish were investigated, and the impact of various inclusion particle sizes on the removal effect of inclusions under induction heating was explored [15]. The effect of Cu on the formation of reversed austenite in super martensitic stainless steel was investigated by Jiang et al. using X-ray diffraction (XRD), a transmission electron microscope (TEM), and an energy-dispersive spectrometer (EDS) [16].

Funding: The authors acknowledge the financial support from the National Natural Science Foundation of China (No: 52104351), Academy of Finland (Grant No. 349833), the Science and Technology Major Project of Yunnan Province (No: 202202AG050007), and the Yunnan Fundamental Research Projects (No: 202101AU070088).

Conflicts of Interest: The authors declare no conflict of interest.

References

1. Dzinomwa, G.; Mapani, B.; Nghipulile, T.; Maweja, K.; Kurasha, J.T.; Amwaama, M.; Chigayo, K. Mineralogical Characterization of Historic Copper Slag to Guide the Recovery of Valuable Metals: A Namibian Case Study. *Materials* **2023**, *16*, 6126. [CrossRef] [PubMed]
2. Du, J.; Ma, A.; Wang, X.; Zheng, X. Review of the Preparation and Application of Porous Materials for Typical Coal-Based Solid Waste. *Materials* **2023**, *16*, 5434. [CrossRef] [PubMed]
3. Tian, C.; Zhou, J.; Ren, C.; Omran, M.; Zhang, F.; Tang, J. Drying Kinetics of Microwave-Assisted Drying of Leaching Residues from Hydrometallurgy of Zinc. *Materials* **2023**, *16*, 5546. [CrossRef] [PubMed]
4. Zheng, X.; Li, J.; Ma, A.; Liu, B. Recovery of Zinc from Metallurgical Slag and Dust by Ammonium Acetate Using Response Surface Methodology. *Materials* **2023**, *16*, 5132. [CrossRef]
5. Zheng, X.; Li, S.; Liu, B.; Zhang, L.; Ma, A. A Study on the Mechanism and Kinetics of Ultrasound-Enhanced Sulfuric Acid Leaching for Zinc Extraction from Zinc Oxide Dust. *Materials* **2022**, *15*, 5969. [CrossRef]
6. Ultarakova, A.; Karshyga, Z.; Lokhova, N.; Yessengaziyev, A.; Kassymzhanov, K.; Mukangaliyeva, A. Studies on the Processing of Fine Dusts from the Electric Smelting of Ilmenite Concentrates to Obtain Titanium Dioxide. *Materials* **2022**, *15*, 8314. [CrossRef]
7. Wan, X.; Li, J.; Li, N.; Zhang, J.; Gu, Y.; Chen, G.; Ju, S. Preparation of Spherical Ultrafine Silver Particles Using Y-Type Microjet Reactor. *Materials* **2023**, *16*, 2217. [CrossRef] [PubMed]
8. Li, N.; Li, J.; Wan, X.; Niu, Y.; Gu, Y.; Chen, G.; Ju, S. Preparation of Micro-Size Spherical Silver Particles and Their Application in Conductive Silver Paste. *Materials* **2023**, *16*, 1733. [CrossRef] [PubMed]

9. Orda, S.; Drzazga, M.; Leszczyńska-Sejda, K.; Ciszewski, M.; Kocur, A.; Branecka, P.; Gall, K.; Słaboń, M.; Lemanowicz, M. Investigations of the Density and Solubility of Ammonium Perrhenate and Potassium Perrhenate Aqueous Solutions. *Materials* **2023**, *16*, 5481. [CrossRef] [PubMed]
10. Zhao, Z.; Wang, K.; Wu, G.; Jiang, D.; Lan, Y. Adsorption of Sc on the Surface of Kaolinite (001): A Density Functional Theory Study. *Materials* **2023**, *16*, 5349. [CrossRef] [PubMed]
11. Zhang, Z.; Huang, W.; Zhao, W.; Sun, X.; Ji, H.; Yin, S.; Chen, J.; Gao, L. Hot Deformation Behavior of TA1 Prepared by Electron Beam Cold Hearth Melting with a Single Pass. *Materials* **2023**, *16*, 369. [CrossRef] [PubMed]
12. Yan, H.; Zhou, X.; Gao, L.; Fang, H.; Wang, Y.; Ji, H.; Liu, S. Prediction of Compressive Strength of Biomass–Humic Acid Limonite Pellets Using Artificial Neural Network Model. *Materials* **2023**, *16*, 5184. [CrossRef] [PubMed]
13. Zhang, C.; Zhou, X.; Gao, L.; Fang, H. Study on the Roasting Process of Guisha Limonite Pellets. *Materials* **2022**, *15*, 8845. [CrossRef]
14. Li, P.; Zhang, G.; Yan, P.; Zhang, P.; Tian, N.; Feng, Z. Numerical and Experimental Study on Carbon Segregation in Square Billet Continuous Casting with M-EMS. *Materials* **2023**, *16*, 5531. [CrossRef]
15. Yi, B.; Zhang, G.; Jiang, Q.; Zhang, P.; Feng, Z.; Tian, N. The Removal of Inclusions with Different Diameters in Tundish by Channel Induction Heating: A Numerical Simulation Study. *Materials* **2023**, *16*, 5254. [CrossRef] [PubMed]
16. Jiang, W.; Zhao, K. Effect of Cu on the Formation of Reversed Austenite in Super Martensitic Stainless Steel. *Materials* **2023**, *16*, 1302. [CrossRef]

Disclaimer/Publisher's Note: The statements, opinions and data contained in all publications are solely those of the individual author(s) and contributor(s) and not of MDPI and/or the editor(s). MDPI and/or the editor(s) disclaim responsibility for any injury to people or property resulting from any ideas, methods, instructions or products referred to in the content.

Article

Mineralogical Characterization of Historic Copper Slag to Guide the Recovery of Valuable Metals: A Namibian Case Study

Godfrey Dzinomwa [1,*], Benjamin Mapani [1], Titus Nghipulile [2], Kasonde Maweja [1], Jaquiline Tatenda Kurasha [1], Martha Amwaama [1] and Kayini Chigayo [1]

[1] Department of Civil, Mining, and Process Engineering, Faculty of Engineering and the Built Environment, Namibia University of Science and Technology, Private Bag, Windhoek 13388, Namibia; bmapani@nust.na (B.M.); mkasonde@nust.na (K.M.); jkurasha@nust.na (J.T.K.); mamwaama@nust.na (M.A.); kchigayo@nust.na (K.C.)
[2] Minerals Processing Division, Mintek, Private Bag X3015, Randburg 2125, South Africa; titusn@mintek.co.za
* Correspondence: gdzinomwa@nust.na

Citation: Dzinomwa, G.; Mapani, B.; Nghipulile, T.; Maweja, K.; Kurasha, J.T.; Amwaama, M.; Chigayo, K. Mineralogical Characterization of Historic Copper Slag to Guide the Recovery of Valuable Metals: A Namibian Case Study. *Materials* **2023**, *16*, 6126. https://doi.org/10.3390/ma16186126

Academic Editor: Guo Chen

Received: 26 July 2023
Revised: 25 August 2023
Accepted: 26 August 2023
Published: 8 September 2023

Copyright: © 2023 by the authors. Licensee MDPI, Basel, Switzerland. This article is an open access article distributed under the terms and conditions of the Creative Commons Attribution (CC BY) license (https:// creativecommons.org/licenses/by/ 4.0/).

Abstract: The depletion of the ore reserves in the world necessitates the search for secondary sources such as waste products (tailings and slag). The treatment and cleaning up of such secondary sources also has a positive impact on the environment. A smelter in Namibia we examined had historic slag which accumulated over decades of its operating life, thus posing the challenge of how best to collect representative samples to evaluate and propose viable methods of recovering contained metals. In this study, analytical and mineralogical characterization of the slag was performed using X-ray fluorescence (XRF) analysis, atomic absorption spectrometer (AAS), ICP-OES, scanning electron microscopy energy dispersive spectroscopy (SEM-EDS) analysis, and optical microscopy analysis. The chemical analyses showed that the metal values contained in the slag were mainly copper, lead, and zinc whose average contents were approximately 0.35% Cu, 3% Pb, and 5.5% Zn. About 10.5% Fe was also contained in the slag. Germanium was detected by scanning electron microscopy, but was however below detection limits of the chemical analysis equipment used. Based on the results, approximate conditions under which the different slag phases were formed were estimated and the recovery routes for the various metals were proposed. Analysis by both optical and scanning electron microscopy revealed that Zn and Fe occurred mainly in association with O as oxides, while Cu and Pb were mainly associated with S as sulphides. The slag consisted of three different phases, namely the silicate phase (slag), metallic phase and the sulphide phases. The phases in the slag were mainly silicate phases as well as metallic and sulphide phases. It was observed that the metallic and sulphide phases were dominant in the finer size fractions (-75 μm) whereas the sulphide phase was also present in the coarser size fractions ($+300$ μm). An important finding from the microscopy examination was that the sulphide phases were interstitial and could be liberated from the slag. This finding meant that liberation and subsequent concentration of the sulphide phases was feasible using conventional processing techniques.

Keywords: copper slag; mineralogy; slag re-processing; pyrometallurgy; hydrometallurgy

1. Introduction

Ore reserves around the world are currently being depleted, with very few new discoveries occurring. This necessitates the search for the secondary sources of metals such as tailings [1–4], slag [5–9], and electronic waste printed circuit boards [10–16]. Copper is one of the metals whose ore reserves have been depleted in many areas, with slag re-processing being common as a way of supplementing the fresh concentrates in smelters [17]. Approximately 2 tons of slag per tonne of copper produced is generated in a typical pyrometallurgical process [6,7,18]. Environmental and socio-economic impacts of such slags are some compelling considerations in favour of further recovery of valuable metals [19,20]. The composition of the slag is dependent on the mineralogy of the ore, but it typically contains oxides of gangue elements such as iron, silicon, magnesium, calcium, phosphorus, and copper [21–23]. The slag also contains some proportions of copper matte that is mechanically dragged when tapping from the settling zone of the smelter crucible [24].

Valuable metals such as gold, silver, germanium, and copper contained in sulphide concentrates are mainly dissolved in the copper matte phase. However, the operation of the smelters leads to incomplete settling of the molten metals and sulphides from the liquid slag. Maweja et al. [22] have shown that more than 80% of copper in the matte smelting slags was in sulphide form, which is inherent to mechanical drawing of the matte during slag tapping. Converter slags have high copper contents and are not disposed to waste, therefore they are not considered in this study. Table 1 gives the compositions of some copper slags obtained in plants using different processes and equipment. Typical copper slag composition ranges are Fe (as FeO and Fe_3O_4) 20–40%; SiO_2 25–40%; Al_2O_3 up to 10% and CaO up to 10%.

Table 1. Compositions of copper smelter slags in wt%, Refs. [7,22,25,26].

Copper Smelting Method	FeO + Fe_3O_4	SiO_2	CaO	MgO	Al_2O_3	Cu	S	Co	ZnO	Au + Ge + Ag
Water jacket furnace (DR Congo)	29	31	10	5	7	1.4	0.6	2	10	10 g/ton
DC electric furnace (DR Congo)	21	38	12	11	6	0.3	0.1	0.5	4	-
Mitsubishi	54	32	7	-	4	2	0.6	-	-	-
ISASMELT method	49	33	2.3	2	0.2	1	2.8	-	-	-
Converter slag	76	22	2	1	8	2	4	-	-	-
Inco flash smelter	<60	<33	1.7	1.6	4.7	0.9	1.1	-	-	-
Vanukov smelting	<58	<24	2	1.5	3	2.5	0.6	-	-	-
Airtight blast furnace	<43	<35	<12	4	8	1.4	0.3	-	-	-
Flash smelting	<59	<33	<10	2	7	0.25	0.6	-	-	-

The variety of slag compositions of Table 1, is inherent to both the concentrate composition and the extraction method applied. Each slag composition requires an appropriate treatment method for further metal recovery or for utilization as construction material [27,28]. Thermal reduction of copper smelter slags with solid carbon is commonly applied in industry to recover valuable metals [21,29–31]. This method requires temperatures above the melting temperature of slags, which is 1300–1500 °C. For example, the ST plant in the Democratic Republic of Congo processes 22,000 tons/month of matte smelter slag by direct reduction with coal in a DC current furnace to recover copper, cobalt, zinc, and germanium from a white alloy. The furnace operates at about 1400 °C. Above this temperature, silica reduction becomes significant and this can result in low grade white alloy product. In Turkey, high-pressure oxidative leaching is rather considered for the simultaneous recovery of 90% of copper, cobalt and zinc from Küre-Kastamonu historical slag containing 0.84% Cu, 0.34% Co, 0.23% Zn, 1.3% S, and about 56% (FeO + Fe_3O_4) as shown in the work by [32,33] have demonstrated the effects of heat removal (cooling rate) during slag casting and post casting heat treatment on the leaching of copper, cobalt, and zinc from the slag in acidic media. The formation of silica gels in acidic leaching of amorphous slags will hinder separation processes such as solvent extraction of metal ions from the leaching solutions. Bioleaching treatment of smelter slags has been investigated, but the yields of this method remains moderate for copper, zinc and nickel, with 62%, 35%, and 44% dissolution, respectively, after 29 days [34]. The authors have proceeded by further precipitation of the metals under controlled pH conditions. Flotation is extensively used to recover copper and other metals from smelter slags [35,36]. Flotation with xanthate collectors enhanced by chemical and mechanical activation is applied to recover copper from slags with high efficiency above 90% [37]. A comprehensive review of the methods of copper recovery from slags is presented by Kundu et al. [38], where the authors compare the efficiency and environmental impact of methods of flotation, hydrometallurgy and pyrometallurgy to recover copper from the slags. The variety of methods used in industry and the extensive investigation in new processes show the importance of slag characterization as it enables the selection of a technically and economically viable combination of unit operations for the recovery of metals from slags.

Namibia Custom Smelters in the Republic of Namibia is a standalone operation with no mining activities on site (i.e., they buy the concentrates from producers). There is always a continuous need to supplement the primary source with secondary materials such as the smelter slag. Their smelter can treat complex copper concentrates such as those containing arsenic (As). The smelter consists of an Ausmelt top submerged lance furnace, two Peirce Smith converters, a sulphuric acid plant, a slag milling and flotation plant, and an arsenic trioxide production plant. That process allows the smelter to produce blister copper, sulphuric acid and arsenic trioxide. The blister copper (98.5% Cu) is delivered to refineries in Asia and Europe where it is further refined to copper metal. The Arsenic trioxide is sold worldwide, and the sulphuric acid is sold to Namibian mining companies in the uranium and copper industry for leaching purposes. Since the 1900s when the smelter commenced with a blast furnace operation, historic slag has been stockpiled. Acknowledging the improvement in the technology over the decades, the historic slag can possibly be reprocessed to recover copper and other metal values. Previous studies on the slag samples collected on the dumps in the smelter area showed that there were some recoverable metal values including the high-tech elements such as germanium [39,40]. The objective of this study was to characterize that historic slag chemically and mineralogically, explain the conditions under which the phases were formed, and propose the processing routes for various commercially extractable metals.

2. Materials and Methods

The slag heaps at the smelter were in two categories; the pre-1964 slag that accumulated from 1901 to 1964 from the Pb-blast furnace; and the slag that was produced with a relatively new smelter that was installed in 1964. It was deemed that the historic pre-1964 slag heap was more likely to contain higher amounts of valuable metals due to the less developed extractive technology used at the time. The slag heap typically measured 3–5 m in height and about 5 hectares in surface area. A total of 84 holes were drilled on the historic slag heap. Each drill hole was sampled at 1 m intervals with a 2 kg sample being collected. Each sample was logged and described. Similar samples were then composited in groups of 7–10 samples from 12–20 drillholes. These samples were homogenized in groups of four (4), where each sample represented between 12 and 20 drillholes, depending on the log sheet similarity. The four samples weighed about 20–24 kg. The four samples were each conned and quartered, from which a representative sample of 20 kg was obtained. The representative sample ensured that the materials from all depth sections of the slag profile were represented and sampled. The composited sample was then crushed to −5 mm size so as to reduce the grain size variation for analytical purposes.

The representative 20 kg sample was further blended using the coning and quartering method prior to splitting. The rotary splitter was used to split the blended sample into 1 kg aliquots that were used for particle size analysis, chemical analyses, and mineralogical characterization as shown in Figure 1.

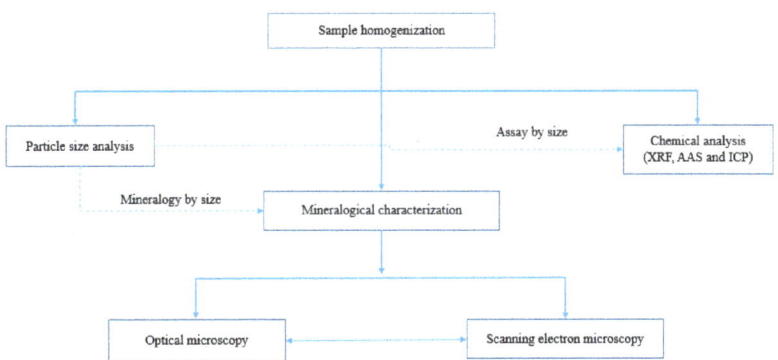

Figure 1. Process diagram for the methodology employed to characterize the copper slag.

Particle size analysis was performed on two of the 1 kg sub-samples using different techniques (i.e., one sample was subjected to dry screening using the vibrating sieve shaker for 15 min while the other sample was manually wet screened). Sieves with the aperture sizes ranging between 38 and 3350 µm were used. Wet screening products were dried overnight in the oven set at a temperature of 80 °C. The size-by-size products were reserved for chemical analysis and mineralogical characterization.

Chemical analysis was determined using three different techniques and thus providing validation of the assays. AAS analysis of Fe, Pb, Ca, Zn, and Cu were performed on a Thermo Scientific ICE 3000 series Atomic Absorption Spectrometer (Basel, Switzerland). The ICP-OES analyses were conducted using the Perkin Elmer Optima 8000 (Johannesburg, South Africa). XRF analyses were performed using the benchtop X-ray fluorescence (XRF) spectrometer model NEX CG supplied by Applied Rigaku Technologies from Austin, TX, USA. This benchtop XRF machine uses an in-built calibration procedure referred to as the multi-channel analyser (MCA) Analysis was carried out on a composite head sample as well as samples in the five particle size ranges +3350 µm, +850, +300, +75, and −75 µm. The analysis of the classified slag materials was aimed at identifying the possibility of a preferential partition or concentration of valuable metals in slag particles in certain granulometric sizes. The AAS and ICP-OES samples were digested using aqua-regia (hydrochloric acid: nitric acid [1:3]) solution. For each sample, 1.0 g was added to 25 mL of aqua regia solution in a 100 mL beaker and heated for 2 h. Following digestion, the sample was filtered and transferred with the help of 2% nitric acid in a 100 mL volumetric flask. Appropriate dilution factors were applied to allow for measurements of the concentration of the analytes.

The microstructures and phase distributions of the granulometric ranges were analyzed in backscattered and in secondary electron modes in a scanning electron microscope (SEM) machine. The samples for microscopy consisted of two types:

1. Loose powders for observation of the particle morphologies and phases in the outer layers of as received condition.
2. Slag material mounted on resin, ground on grit papers then polished to metallographic surface condition using diamond slurries up to the 3 µm for observation inside the cross sections of the particles.

The samples of the materials obtained after granulometric separation in six particle size ranges of +3350, −3350 + 850, −850 + 300, −300 + 75, −75 + 38, and −38 µm were hot mounted on resin and polished for observation using the JEOL JSM-IT300 scanning electron microscope coupled with the Thermo Scientific NORAN System 7 (NS7) energy dispersive spectroscopy (EDS) software (Advancedlab, Basel, Switzerland). The samples were lightly carbon coated using the Quorum Q150T sputter coater (Advancedlab, Switzerland). SEM analysis was carried out in a high vacuum in the backscattered mode (BSE) at an acceleration voltage of 10–15 kV, a probe current of 50 nA, and a working distance of 15 mm. Other samples taken from these classified granulometric size ranges were attached on carbon tapes, lightly carbon coated and observed in the SEM secondary electron (SE) mode without polishing. These samples were also mounted on a polished thin section to allow for observation in a BX51 Olympus optical microscope in both reflection and transmission modes for phase identification.

3. Results and Discussion

3.1. Particle Size Analysis of the Received Slag Material

Figure 2 presents the particle size distributions of the received smelter slag after wet and dry screening. Since the size distributions of wet and dry sieving matched, it can be concluded that the slag material was not contaminated with soluble materials or the fine airborne materials from the surrounding fields of the dump. It can be observed that the 'as received' slag sample was 95% passing 3.35 mm while the particles that are coarser than 300 µm and those finer than 75 µm constitute 92% (w/w) and 1.2% (w/w), respectively, of the slag sample. The 80% passing size of the as-received smelter slag material is 2.5 mm.

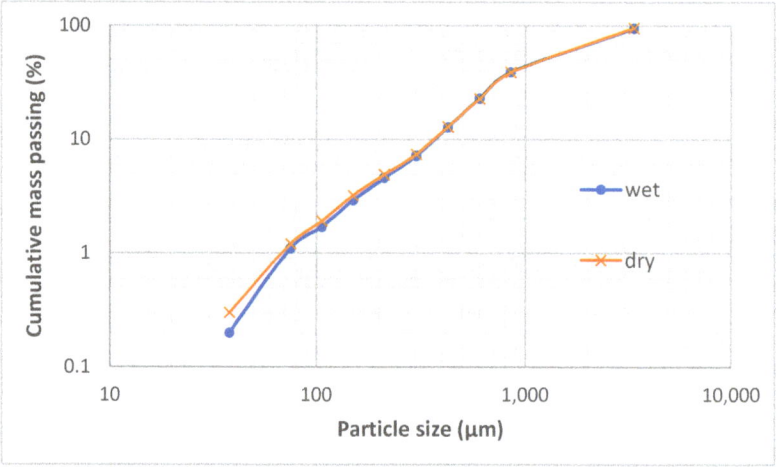

Figure 2. Particle size distribution of unwashed (dry sieved) and water washed (wet sieved) historic smelter slag materials.

3.2. Chemical Analysis

The chemical compositions of the composite smelter slag samples are listed in Table 2. The copper assays range between 0.3 and 0.4% in the composite sample while that of iron ranges between 9 and 12%. The assay-by-size (ABS) results are also listed in Table 2 to show the metal distributions in different particle size fraction. It is acknowledged that in Table 2, only the results of the coarsest size fraction (+3.35 mm) and finest size fraction (−75 µm) are presented, with the rest of the ABS results included in Appendix A. In general, it can be observed that the contents of metal values (with the exception of Fe, Zn, and Mn) are higher in the −75 µm (liberated particles) and lowest in the +3.35 mm (unliberated size fraction). Since −75 µm constitutes 1.2% (w/w) as shown in Figure 2, the assay of 1.2% Cu suggests that about 5% of the total copper in the slag is contained in the −75 µm size fraction while the 95% of the total copper content is still locked in the slag particles coarser than +75 µm. The slag contents in copper and zinc are comparable to those of the DC electric furnace smelter in the Democratic Republic of Congo. Maweja et al. [33] reported the effects of annealing and leaching conditions on the recovery of copper and zinc from the DC electric furnace slag.

Table 2. Elemental compositions of the composite historic smelter slag samples and assay-by size (+3.35 mm and −75 μm).

Sample Type	Element (% w/w)											
	Fe	Si	Ca	Mg	Al	Mn	Cu	Zn	Pb	As	Mo	Precious Metals
Composite samples	9–11	13–19	11	4	2	0.1	0.3–0.4	4–7	2–4	0.2	0.25	
Sieved at +3350 μm	10		11			0.13	0.3	3.2	3.2	0.2	0.2	
Sieved at −75 μm	11		8			0.12	1.2	3.1	3.5	0.8	0.2	

3.3. Morphology and Phase Distribution Characterisation of the Slag Materials

3.3.1. Scanning Electron Microscopy

The compositional contrast observed in backscattered (BSE) mode images (Figures 3a–c, 4a,b and 5a) indicate that higher proportions of particles of material phases containing the heavy elements were found in the fine particles (−38 μm and −75 + 38 μm). The proportion of the phases of heavy elements is lower in the particles coarser than 300 μm. SEM BSE also revealed that the particles of the heavy metal phases present in the fine granulometric size ranges were separated from the slag matrix particles (i.e., higher degree of liberation between particles of metallic and slag phases). The separation of these particles from the slag matrix is attributed to the non-miscibility between the molten slag and the metal compounds contained in the matte that were mechanically dragged when tapping the slag from the hearth zone of the smelting furnace.

The morphologies and sizes of the particles of the matte or metal phases and the slag matrix particles were investigated under secondary electrons (SE) mode observation in the SEM machine. The images of Figures 3–5 illustrate the actual shapes. The SEM SE images indicate higher proportion of the high aspect ratio (Length/Diameter) needle-like particles present in the finer materials of −38 μm (Figure 3c). These needle-like particles are relatively coarser within the fine size material fraction, but their relative volumetric proportion in the slag material decreases as the granulometric size ranges increase in the −75 + 38 μm (Figure 3d) and then in the −300 + 75 μm (Figure 3f). Finally, no needle-like particles are observed in the coarse size ranges of slag material fractions above 300 μm (Figures 4c,d and 5b). The nature and formation mechanism of the needle-like particles is discussed in the next section.

The scanning electron microscopy images show that the small particles of matte or metal phases (high reflectance) found in the coarse granulometric ranges, above 300 μm (Figures 4 and 5), were rather trapped by occlusion inside the smelting liquid slag and solidified upon quenching in the high pressure water jet. Large slag particles have cracks that formed during the fast quenching in the water jet. The formation of the cracks is attributed to the low thermal conductivity of the slag [41–44] and the difference in coefficients of thermal expansion, thus shrinkage, between the slag and the matte (metal) phases, which result in thermal stresses inside the solidified material. The particle size distribution in Figure 2 suggested that the materials of granulometric size larger than 300 μm represent about 90% (w/w) of the sample of the received historic slag material. It, therefore, implies that the major quantities of matte and metal phases, that were mechanically dragged with the slag, are trapped inside the slag particles larger than 300 μm but their concentration values in these large particles are lower than in the finer materials (<300 μm). The fine particle size ranges (<300 μm) represent only about 8% (w/w) of the historic smelter slag but with higher contents of matte and metal phases.

3.3.2. Morphology and Composition of Fine Particle Materials

The SEM SE observation of particles of the received slag sample material collected in granulometric ranges smaller than 300 μm revealed four types of particle shapes as shown in Figures 6 and 7 (i.e., the flakes, the spheres with smooth surfaces, the needle-like particles, and the ovoid particles with rough surfaces). The SEM SE images of Figures 6 and 7, infer that the flaky particles represent the main constituent of the slag materials in all particle size ranges obtained after sieving the historic slag material. The ovoid particles with rough surfaces are present in a wide size range from 30 to 200 μm. The SEM SE images suggest the ovoid particles are the second important phase present in the material passing the 300 μm sieves. The needle-like particles have aspect ratios (Length/Diameter) higher than 10. Such particles of length bigger than 500 μm have passed through the 38 μm sieve square apertures since their diameter is as small as 25–30 μm. The spherical particles compose the least proportion of phase found in the slag material that passed through the 300 μm sieves. They have a diameter range of 30–70 μm and are distinguishable from the ovoid particles by their smooth surface appearance under SEM SE mode.

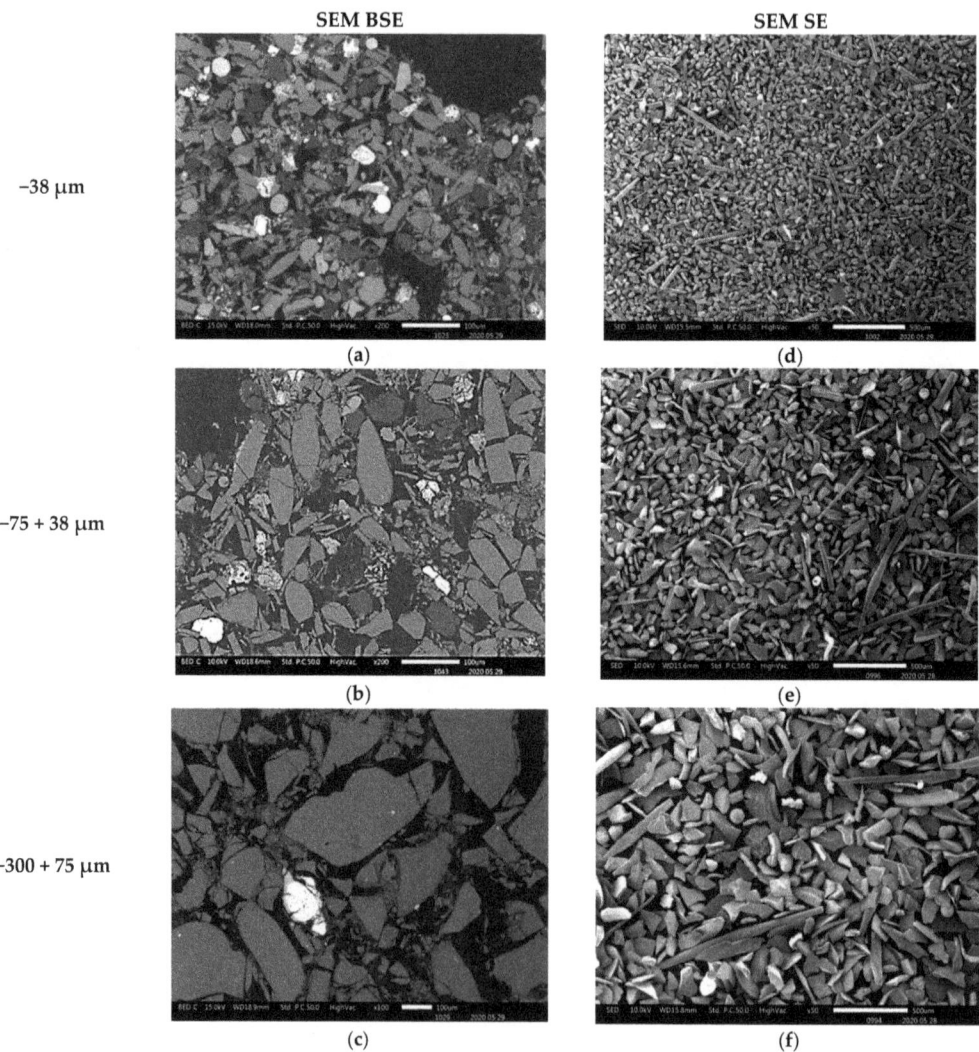

Figure 3. SEM BSE of polished slag particles (**a**–**c**) and SE of received slag material (**d**–**f**) micrographs showing the shapes of the particles. The high reflectance phases are metallic sulphides, whereas the rest of the material are silicates.

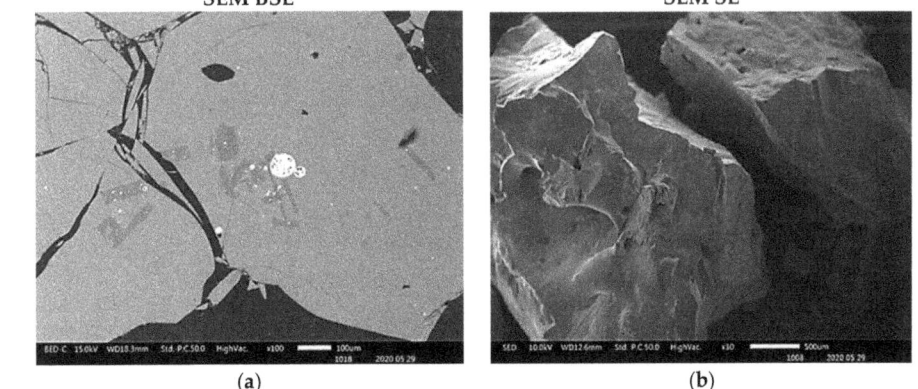

Figure 4. SEM BSE of polished slag particles (**a**,**b**) and SE of received slag material (**c**,**d**) micrographs showing the shapes of the particles. The high reflectance phases are metal sulphides of Cu, Zn, Pb and Fe.

Figure 5. SEM BSE of polished slag particles (**a**) and SE of received slag material (**b**) micrographs showing the shapes of the particles. The high reflectance phases are metal sulphides of Cu, Zn, Pb and Fe.

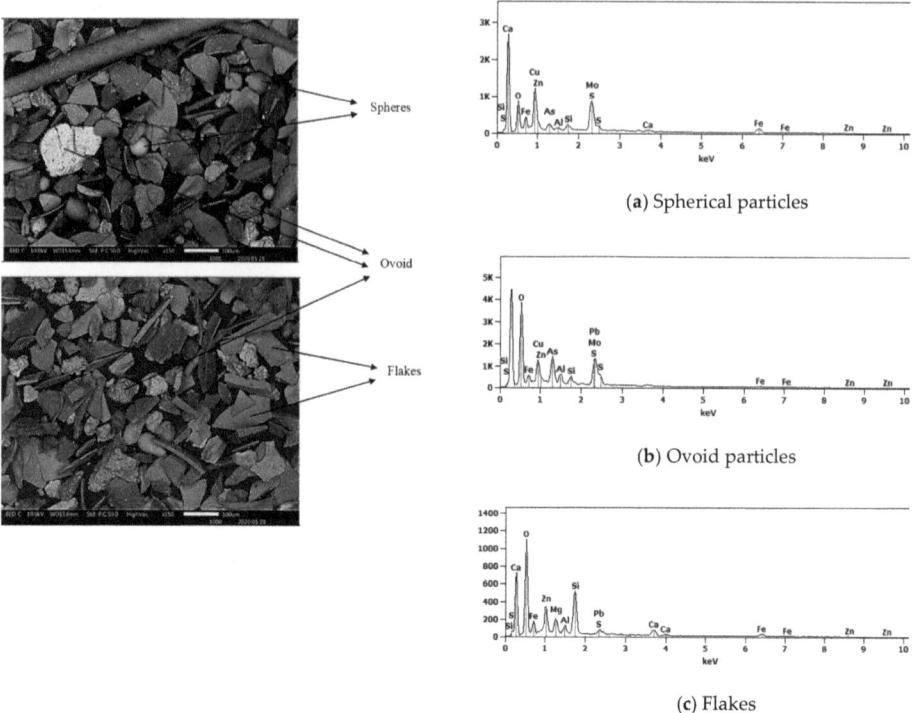

(a) Spherical particles

(b) Ovoid particles

(c) Flakes

Figure 6. SEM SE images showing particles of different morphologies and sizes in the slag material of granulometric range between 75 and 300 μm and SEM EDS spectra (a) spherical particle, (b) ovoid particle and (c) flakes.

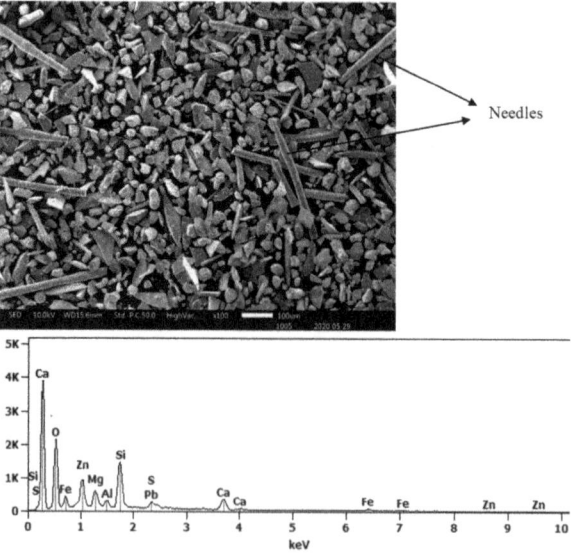

Figure 7. SEM SE image and EDS spectrum showing needle-like slag particles in material received with particle sizes smaller than 38 μm.

Energy dispersive X-ray spectroscopy (EDS) analysis of these four types of particles in the slag sample indicates that they are different phases in compositions. Their corresponding composition ranges are given in Table 3. The EDX method is considered as semi-quantitative, but has the advantage of evaluating the composition of selected grains or regions within the sample of materials under consideration. The method is not recommended for the quantification of light elements, and thus the oxygen contents only serve as qualitative indicators of the oxide phases present in the slag material. Similarly, the sulphur contents reveal the presence of the sulphide phases.

Table 3. EDS semi-quantitative area analysis composition ranges in % (w/w) of particles in received slag material passing the 300 μm sieves. (The oxygen and sulphur contents are only indicative of the oxide and sulfide phases present).

Element	Flakes	Needles	Ovoid	Spheres
Si	13–19	14–37	<0.4	0.3–9.0
Ca	12–20	<9.0	-	<1.8
Fe	4–21	12–20	1.0–39	<9.0
Zn	11–15	8.7–17	0.3–1.4	1.7–3.0
Pb	0.7–2.6	3.4–7.0	15–39	48–79
Cu	0.0–1.2	0.0–0.93	12–44	3.7–4.4
As	-	-	2.5–30	2.5–15
Ge	-	-	<0.16	<0.63
Mg	3.8–5.3	<5.0	-	<0.07
Al	1.1–2.8	1.9–2.8	-	-
S	<0.6	<0.4	1.3–55	<4.0
O	29–37	27–34	<7.0	9.0–20

The results of the EDS area analysis of the particles of the slag materials infer the following:

The flaky particles (i.e., the main component of the solid slag) consist of metal oxide compounds or [(Fe, Ca, Mg, Zn, Al, Si) O]. The flake type of particles contain the highest zinc content of 11–15% (w/w) of all the particles present in the slag. The elemental mapping in Figure 8 shows the preferential location of zinc inside the regions of the flake and needle-like shape particles. The flake type of particles have the lowest contents of sulphide formers Pb and Cu. The EDS analysis showed no peaks for arsenic and germanium in these particles.

The needle-like particles are also composed of the [(Fe, Ca, Mg, Zn, Al, Si) O] compounds, however with higher silicon and lead contents, and lower calcium content than those found the flakes. An acidity index (1) of the slag is introduced to compare the chemistry of the flake shape and needle-like particles components of the historic slag material as follows:

$$i = \frac{\frac{Si\%}{28}}{\frac{Ca\%}{40} + \frac{Fe\%}{56} + \frac{Mg\%}{24} + \frac{Zn\%}{65} + \frac{Pb\%}{207}} \quad (1)$$

The calculation yields values of the acidity index equal to 0.5 for the flakes and 1.2 for the needle-like particles of the slag material (i.e., the acidity index of the needles is more than double that of the flakes). This difference suggests that the molten slag in the smelting furnace contained at least two different liquid phases; one main liquid phase of the composition of the flakes and a second liquid phase, in smaller quantity, of the composition as the needle-like particles. The needles phases therefore are much higher in silica-rich phases than other shapes. This second liquid might have consisted of droplets or veins that formed the elongated needles (L > 500 μm) with fine cross section diameters (D < 75 μm) upon tapping and quenching of the slag in the high-pressure water jet. The higher acid index of the needle-like shape particles indicates the tendency of formation of polymeric silicates with tridimensional structures, which results in high viscosity of the molten slag phase. The coexistence of significant numbers of the needle-like particles with the flakes in the fine granulometric size ranges may be ascribed to the separation, due to lack of miscibility, of the two phases in liquid slag inside the hearth of the furnace. Some micron and sub-micron size particles, of similar appearance to the ovoid and spherical shape phase are trapped onto the surface of the needle-like particles, not onto the flake-shaped particles. It is inferred, therefore, that the high viscosity of the liquid phase of high acidity index (needle-like) hindered the decantation of the droplets of liquid sulphide (metal) phases through the hearth zone into the matte or metal phase.

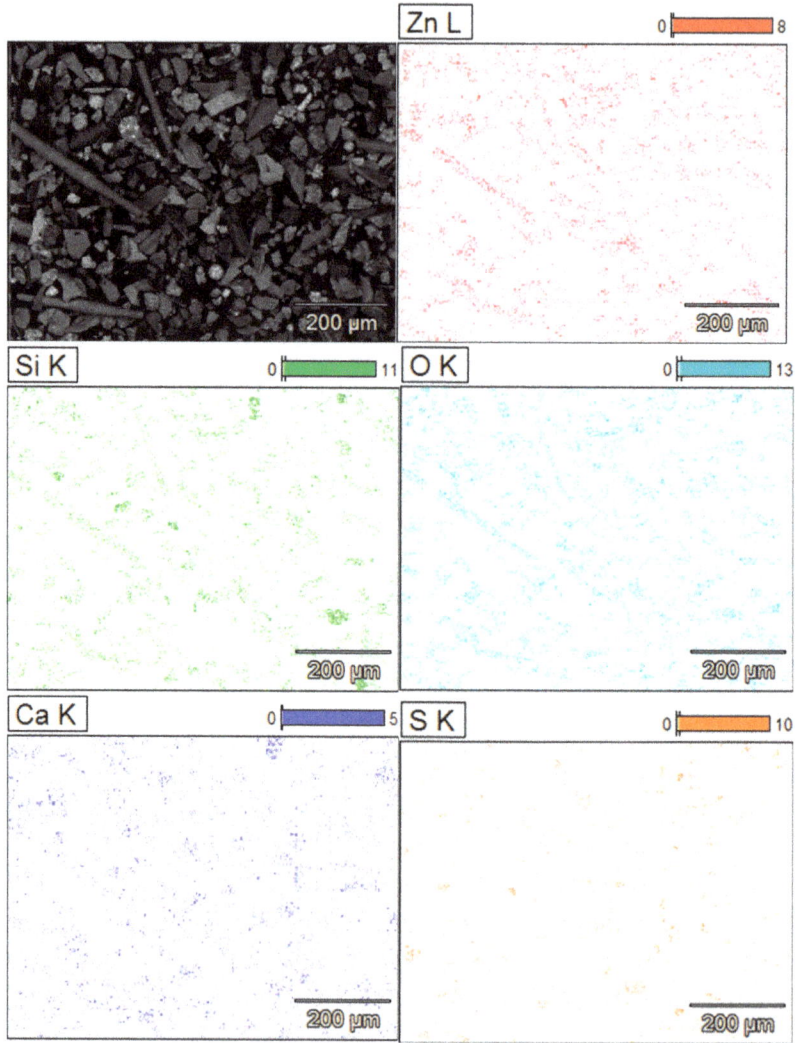

Figure 8. Elemental mapping of slag material showing the presence of Zinc along with silicon, and calcium in the flakes and the needle-like particles of oxide phase particles. Needles have less calcium compared to other shapes.

On the other hand, the ovoid particles are distinguishable by their rough surfaces and have very low contents in oxide formers Si, Ca, Fe Mg, Al, and Zn. They rather have the highest concentrations of Cu, Fe, As, and S along with the second highest content of Pb of all the four types of particles encountered in the historic slag material. It is concluded from the results of EDS analysis in Table 3 that the ovoid particles consist of the sulphide compounds such as Cu_2S, FeS, PbS, and As_2S_3 that constituted the matte product in the smelter. The wide range of variation of the four elements Cu, Fe, Pb and As is ascribed to the heterogeneity within the ovoid shape particles due to the separation of non-miscible sulphides phases upon cooling. The calculated equilibrium phase diagrams in Figure 9, show the lack of miscibility between PbS and the other sulphides such as FeS, Cu_2S or ZnS below 600 °C. The lack of miscibility with other sulphides and metals is also exploited as a means of producing high grade mattes and Pb metal in smelting furnace.

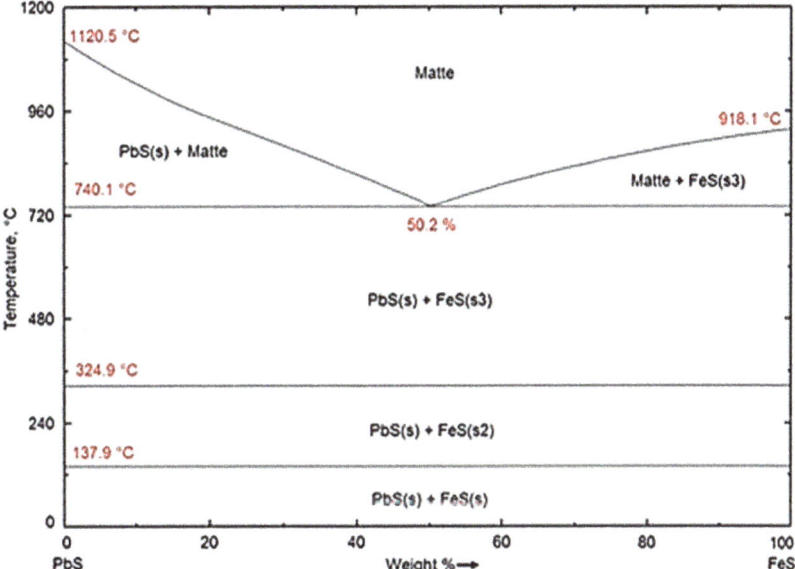

Figure 9. Calculated equilibrium phase diagrams showing the lack of miscibility between PbS and other metal sulphides below 600 °C.

Finally, the spherical particles with smooth surfaces have the highest Pb content of all the four types of particles, which reaches 79% (w/w) in some areas, with relatively low sulphur. The strong oxygen peak infers that these particles consist of a mixture of PbO and Pb metal formed in the melt upon the conversion reactions (2) and (3). The spherical particles have dissolved some amounts of Si, Fe, As, and Zn. The appearance of these free particles in the slag material shows that their material is not miscible either with the liquid slag (flakes and needle-like particles materials) or with the sulphide material of the ovoid shape particles.

$$2PbS + 3O_2 \rightarrow 2PbO + 2SO_2 \tag{2}$$

$$PbS + 2PbO \rightarrow 3Pb + SO_2 \tag{3}$$

The EDS analysis suggested the presence of arsenic and germanium was in the sulphide constituents of the ovoid shape particles and in the high Pb content spherical particles. The L_α X-ray energies of germanium and arsenic are close to each other at 1.188 keV and 1.282 keV, respectively. The EDS method showed little selectivity between these two elements especially in the presence of low germanium concentration and low $\frac{Ge\%}{As\%}$ as expected in this case. This method of analysis suggests that arsenic and germanium are collected in the ovoid shape particles of the sulphide phases and more in the spherical particles of the (Pb, PbO, PbS) phase. The sulphide ovoid shape particles and the lead spherical particles would contain up to 1600 ppm and 6300 ppm germanium respectively. It should however be recalled here that according to the results of particle size classification, these particles and all the slag material passing the sieve at 300 µm represent about 8% (w/w) of the total mass of slag sample received. A conservative approach based on the estimation that those particles only represent 2% (w/w) of the slag material yield a germanium content within the range 32–126 ppm in the historic smelter slag received.

3.3.3. Morphology and Composition of Coarse Size (>300 µm) Slag Particles

The morphologies of the slag particles within the size ranges +300–600 µm, +600–850 µm, and +850–3350 µm are illustrated in Figures 4 and 5. Unlike the fine particle size ranges (<300 µm) where flakes, needles, ovoid and spherical shape particles were found, the coarser particles are bulky and irregular multifaceted with multiple cracks. These coarse particles contain some amount of small particles of entrapped matte or metals as seen in the micrographs of the polished particles of sizes larger than 850 µm and 3350 µm which are shown in Figure 10. The EDS chemical analysis of the phases present in these coarse slag particles (in Table 4) indicates the presence of two slag phases and the small particles of matte or metal phases. Table 4 shows that the matrix and the secondary phases of the bulky slag particles (>300 µm) have chemical composition ranges similar to that of the flaky particles found in the fine materials (<300 µm). It therefore infers that the formation of the small flakes resulted from grinding of the molten slag by the high-velocity flow water jet, whereas the coarse bulky slag particles were formed in the low-velocity flow regions of the water jet. Pressure and flow rate distribution inside the water jet determined the granulometric distribution of the solidified slag.

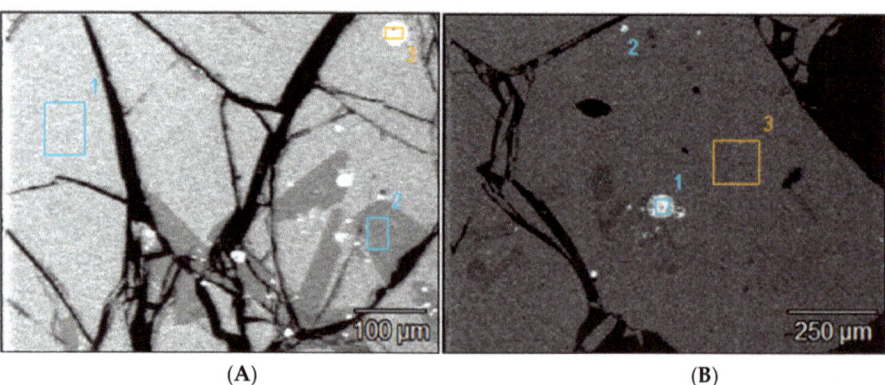

Figure 10. SEM BSE image showing the matrix, a secondary phase and matte fine particles areas in slag particles larger than 850 µm, (**A**) +850–3350 µm and (**B**) +3350 µm, the numbers 1, 2, 3 in the SEM micrographs represent the selected grains for SEM EDS analysis, not relevant.

The secondary phase found in the coarse slag particles has slightly higher silicon and calcium contents, but lower Pb content than the main phase of the slag matrix. The matrix and the secondary phases of the coarse particles have acidity index 0.4 and 0.7 respectively, which are close to the acidity index $i = 0.5$ of the flake shape particles found in the fine material of the received sample of historic smelter slag. It was noticed earlier that the needle-like shape particles only formed at high acidity index 1.2, under high silicon content (37 wt%), which leads to the polymerisation of silicate structures. These silicates have higher viscosity at the slag melting temperature.

Table 4. EDS semi-quantitative area analysis composition ranges in % (w/w) of the phases observed in coarse slag particles (+300–3350 μm). The oxygen and sulphur contents are only indicative of the oxide or sulphide compounds present.

Element	Slag Matrix	Secondary Phase of the Slag	Trapped Matte/Metal Particles
Si	12–13	19	1.2–8.7
Ca	12–16	21	0.5–1.8
Fe	19–23	3	4.3–8.6
Zn	13–17	13	2.3–3.0
Pb	<6.9	<0.7	48–59
Cu	-	-	15–16
As	-	-	2.7–3.2
Ge	-	-	0–0.63
Mg	3.2–4.1	4.7	-
Al	2.0–2.1	1.2	0.2–0.4
S	<0.3	-	0.4–0.9
O	25–29	38	9.7–12

3.3.4. Optical Microscopy Analysis of the Slag Constituents

As a way of simplifying the process of identification, all phases identified are given names of natural equivalents. The process of slag formation leads also to the formation of glass and several silicate phases with specific shapes that have been identified both under the SEM and in optical microscopy. Glass is mainly silicate material that is formed when silicate melt is cooled rapidly. The petrography is better understood by realizing that in the slag development, free metal particles become droplets in the slag, and then cool down and tend to form spherical and ovoid shapes. The following sulphide phases have been identified: galena (PbS); wurtzite (ZnS); sphalerite (ZnS); chalcopyrite ($CuFeS_2$); pyrrhotite ($Fe_{1-x}S$); pyrite (FeS_2); minor cubanite ($CuFe_2S_3$) and covellite (CuS). The silicates present include fayalite (Fe_2SiO_4), monticellite ($CaMgSiO_4$), mellilite ($Ca(Mg,Fe,Zn)Si_2O_7$), anorthite ($CaAl_2Si_2O_8$) and Pb-Plagioclase ($PbAl_2Si_2O_8$). Zn, Mg, Fe, and Ca spinels and wuestite (FeO) are also present.

Six size fractions (−38, −75 + 38, −300 + 75, −600 + 300, −3350 + 850, and +3350 μm) of polished mounted samples were analyzed with the optical microscope, with results shown in Figures 11 and 12. In comparing photographs from the SEM and from the optical microscopy, it is worth noting that brightness in SEM is a function of atomic number whereas in microscopy it is a function of the bonding structure. The spinels (Zn, Mg, Fe, and Ca spinels) appear to be the first ones to form in the scorification process. The metallic phases are dominant in the −38, −75 + 38, −300 + 75 μm fractions. The coarse-grained samples contain fewer grains of metallic phases. Three main phases were observed, namely the silicate phase (slag), metallic phase and the sulphide phases as shown in Figure 11.

The analysis of Figure 12 also reveals that the metallic and sulphide phases are predominantly contained in the slag finer size fractions (−38, −75 + 38 and −300 + 75 μm) whereas the sulphide phases are also present in the coarser size fractions (300, 850 and 3350 μm).

The SEM EDS was used to determine the elemental composition of the phases. The metallic phase consists of elements such as copper (Cu), lead (Pb), germanium (Ge), and some zinc (Zn). Metallic phases in finer size fraction occur as interstitial grains within the slag matrix and this shows that the metals can easily be separated from the slag (Figure 13).

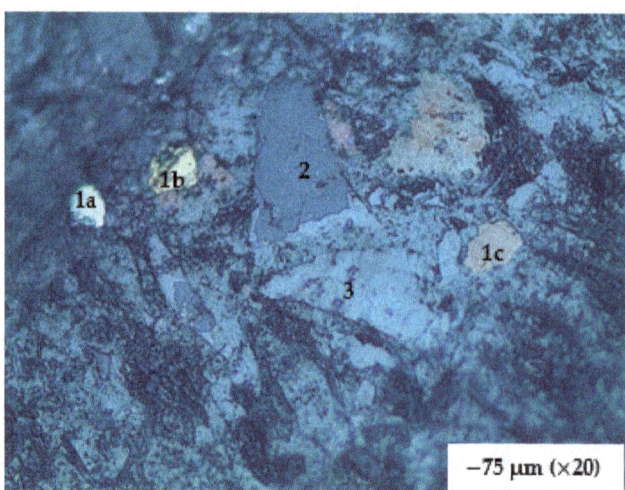

Figure 11. A classical occurrence of the slag material showing the different phases present. Three phases in a −75 μm size fraction. 1a (pyrite), b (chalcopyrite), and c (pyrrhotite) = metallic sulphide phases; 2 = sulphide phase (ZnS); 3 = silicate phase (spinel in slag). Length of bottom of photograph = 2 mm. The metallic phases are in the order of 0.15–0.25 mm in size.

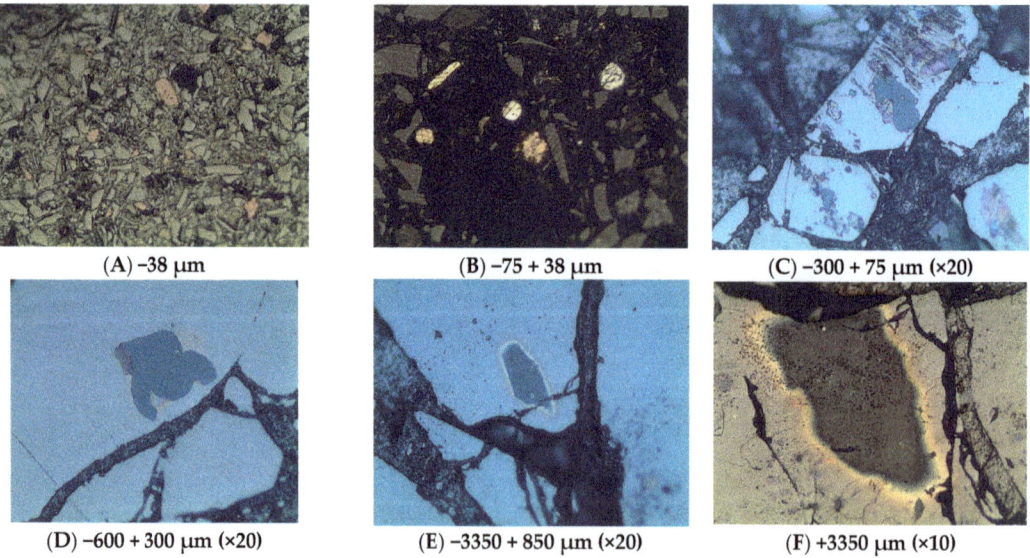

Figure 12. Phases (metallic, oxide and sulphide phases) within various size fractions indicated at the bottom of each photograph: (**A**) Mostly slag with minor pyrrhotite; (**B**) Grains of chalcopyrite (bright yellow) with pyrrhotite, dull yellow; (**C**) Large galena phases (PbS) with enclosed sphalerite (ZnS); (**D**,**E**) mainly galena phases enclosing low reflectance probable ZnS; (**F**) A metalloid of probably spinel with wurtzite (ZnS). Length of bottom photograph = 0.4 mm.

The silicate phases consist mostly of olivine and pyroxenes rich in zinc whereas the sulphide phases are predominantly made up of lead sulphide (PbS) grains which are being rimmed by copper and zinc sulphides; occasionally chalcopyrite occurs as a single phase. This proposes that PbS crystallized first as droplets in the slag, followed by ZnS attaching itself, together with CuS, on the nucleated PbS phases (see Figure 14). The sulphide phase occurs within the slag matrix.

−38 μm (×10)
(A)

−75 μm (×10)
(B)

75 μm (×10)
(C)

Figure 13. Metallic phases occurring as individual grains within the slag in the finer size fractions. In (**A**) the morphology of the slag material is observed as being of various sizes and shapes. In (**B**) (center the sulphide phases occur as spherical phases, indicating a high temperature of formation as droplets in the slag. In (**C**) (far right) is an example showing the very bright ovoid phase (sulphide); intermediate brightness, spinels and very low brightness, silicates.

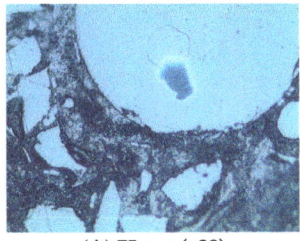
(A) 75 μm (×20)

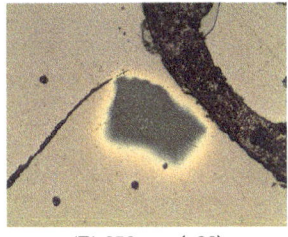
(B) 850 μm (×20)

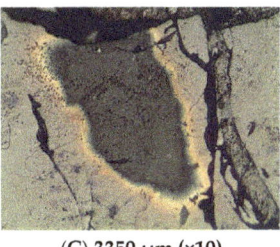
(C) 3350 μm (×10)

Figure 14. In (**A**), a nearly perfect sphere of PbS. A PbS phase occurring in (**B**) within the slag with bright edges being rimmed by CuS and ZnS. The grey phase in (**B**,**C**) has a high amount of ZnS.

The composition of glass was verified by the SEM EDS, and it is variable and enriched in Ca and Fe. There are several angular grains of feldspars (Figures 12 and 13). These attest to the fact that the duration of melting of the historic slag in the furnace was not high enough to allow for good formation of euhedral crystals or the temperature of slag formation was not sufficiently high [45] to completely melt the furnace charge. Spinels in the slag represent the formation of oxide phases. Some of these spinels will form if oxidation conditions are not high enough in the slag, verging towards reducing conditions where sulphide droplets are stable and form minerals. The most commonly observed sulphides are galena (PbS), wurtzite and sphalerite (ZnS), pyrrhotite (Fe_{1-x} S); and various sulphides of Cu and Fe such as pyrite (FeS_2), cubanite, covellite, and chalcocite. Wurtzite is a high temperature phase that forms at about 1020 °C [46] and contains some Fe as observed in the SEM. This suggests that temperatures of that order were at least reached in the furnace.

The important finding here is that the sulphide phases are interstitial and can be liberated from the slag. The scanning microscopy analysis of the slag material shows the presence of the free particles of matte phase and metal (ovoid and spherical) phases in material of size ranges below 300 μm. Slag particles coarser than 300 μm have trapped the matte and metal particles inside. The coarse particles of slag material have large cracks, which indicate that they are brittle and friable. The interfaces between the matte or metal particles and the slag material are incoherent with very little or no bonding strength, this results in the complete separation between them at particle sizes below 300 μm. It may therefore be necessary to mill the slag material to 300 μm to achieve the liberation of all the matte and metal particles trapped inside the coarse slag particles.

4. Recommended Processing Routes for the Recovery of Valuable Metals in the Slag

A combination of mineralogical and chemical methods indicated that the historic slag from Tsumeb smelter contains extractable metals such as copper, zinc, lead and arsenic whose average compositions are 0.35% Cu, 5.5% Zn, 3% Pb, and 0.2% As, respectively. Ge is a trace element, with its presence only observed using the SEM. Germanium as a strategic metal is widely used in advanced technology applications such as thermal solar panels and optic fibres. Germanium is currently extracted from zinc and lead sulphide ores [47,48] or coal deposits as well as metallurgical residues from the processing of the sulphide ores and coals (e.g., smelting flue dust and coal fly ashes) by volatization concentration from lignite coal [49]. Typical concentrations of Ge in these resources range from 30 to 200 ppm [50–52]. The estimated content range of 32–126 ppm germanium in the historic Tsumeb slag is well comparable to that in resources from which it is obtained as a by-product of metal sulphide treatment worldwide by hydrometallurgy and pyrometallurgy processes [40,49,51]. The world's germanium production was only 300 metric tons coming from China (about 60%), Canada, Finland, Russia, and the United States [53]. The demand forecast of germanium estimates an annual growth rate of 3.5% during the period time 2023–2030. Germanium ingots trade price in London was about U$1,300,000/ton in July 2023, which is about 500 times the prices of zinc and lead, which were priced at about U$2300 and U$2100/ton respectively, and copper traded at U$8300/ton during the same period. The higher price justifies the extraction of germanium from low-grade resources. The simultaneous extraction of zinc, lead, copper, and germanium from the Tsumeb historic slag can hence be justified economically. The suggested processing routes for the metals in the Tsumeb slag are discussed below.

a. Copper, lead and arsenic

Energy dispersive of X-ray spectroscopy analysis shows that copper, arsenic and most of lead are contained in round shaped sulphide particles which can easily be detached from the slag matrix. These particles can be liberated by low-energy milling of the slag to below 106 μm. Flotation, with xanthate collector, can be applied to concentrate these sulphide minerals into a feed recycle to the smelter [8]. Alternatively, the sulphide concentrate can be roasted to produce the calcine which can be advanced to the leaching tank to recover copper [38].

b. Zinc and lead

The SEM EDS analysis shows that most of zinc and fraction of lead are found in dissolved form in the slag matrix material. Recovery of zinc and lead from the complex silicate compound requires chemical reactions such as reduction with solid coal and fuming at temperature above the melting temperature of the slag [54]. Alternatively lead and zinc can be leached from the slag matrix in alkaline medium. This is possible since Pb and Zn can both form complexes with hydroxyl ions (OH^-) [55]. Alkaline leaching presents the advantage of lower dissolution of iron and contamination of the leaching solution [56]. It also avoids the risk of formation of gel due to the collapse of silicate structures in acidic medium [57]. Purification methods such as cementation can then be applied to recover Pb and Zn metals from the pregnant leach solution.

c. Germanium

SEM EDS analysis have revealed the presence of germanium in the sulphide particles with lead content. Germanium can be recovered simultaneously with lead, copper and zinc sulphides. Germanium can also be recovered by acid leaching of the white alloy produced by direct reduction of the slag with solid coal, followed by solvent extraction and precipitation. Such process is implemented in the Democratic Republic of Congo for the simultaneous recovery of copper, cobalt, zinc, and germanium from copper matte smelting slag [22,33].

5. Conclusions

The depletion of the ore reserves in the world necessitates the search for secondary sources such as slags for the continued production of metals. Environmental and socio-economic impacts of treating such slags and further recovering valuable metals from them cannot be overlooked. The smelter in Namibia accumulated a significant amount of blast furnace slag during its operating life which can be traced back to 1900's. A representative slag sample was prepared for mineralogical characterization using a combination of analytical and mineralogical techniques (AAS, XRF, ICP OES, SEM EDS and optical microscopy). The chemical analyses showed that the metal values contained in the slag were mainly copper, lead, and zinc whose average contents were approximately up to 0.4% Cu, 4% Pb and 7% Zn. About 11% Fe was also contained in the slag. Germanium is a trace element, with its concentration estimated from the SEM EDS analysis to be in the range of 32–126 ppm. Optical and scanning electron microscopy have shown the mineral phases for both metals and slag. This information is useful in developing the extraction methods for various metals in the slag.

The granulometric classification shows high volume fractions of sulphide particles present in the slag material of the size ranges smaller than 300 µm, and are readily separable from the slag material. The presence of liberated sulphide particles shows that there is no interface bonding between the sulphide phases and the slag matrix, thus mechanical separation between these phases is feasible.

Most sulphide and metal particles are trapped inside the slag particles larger than 300 µm, which represent about 90% of the slag material. The large cracks observed in the SEM and optical images suggest that these coarse slag particles are brittle and friable. Low energy milling will therefore suffice to fracture the large slag particles as a means to liberate the trapped sulphide particles.

Based on the elemental analysis and mineralogical characterisation work undertaken, it is recommended that metallurgical test work be carried out to determine an economically viable and safe process route for recovery of valuable metals contained in the slag. The target metals would include Zn, Pb, Cu, Ge, Mo, and Ga. The metallurgical test work proposed includes pre-concentration of valuable minerals using gravity separation of the sulphide particles from the slag matrix material. Zinc can be extracted by coal reduction and fuming or by leaching. Germanium can be extracted by leaching the white alloy obtained by direct reduction of the slag with coal or from the concentrate of sulphide minerals.

Author Contributions: Conceptualization, G.D. and B.M.; methodology, B.M., G.D., K.M., T.N., J.T.K., M.A. and K.C.; formal analysis, B.M., G.D., K.M., T.N., J.T.K., M.A. and K.C.; investigation, B.M., G.D., K.M., T.N., J.T.K., M.A. and K.C.; data curation, B.M., G.D., K.M., T.N., J.T.K., M.A. and K.C.; writing—original draft preparation, T.N., B.M., G.D., K.M., J.T.K., M.A. and K.C.; writing—review and editing, G.D., B.M., K.M., T.N., J.T.K. and M.A.; supervision, B.M., G.D., B.M. and K.M.; project administration, G.D. All authors have read and agreed to the published version of the manuscript.

Funding: This research received no external funding.

Informed Consent Statement: Not applicable.

Data Availability Statement: Data is contained within the article.

Acknowledgments: The authors would like to thank the Namibian Custom Smelter management for granting us the permission to use some of the project findings in this paper. The laboratory staff in the Department of Civil, Mining and Process Engineering at the Namibia University of Science and Technology (NUST) are acknowledged for their assistance during the test work, microscopy and logistics.

Conflicts of Interest: The authors declare no conflict of interest.

Appendix A. Assay-by Size

Table A1. Assay by size from AAS.

Elements	Elemental Composition (% w/w)				
	+3350 μm	−3350 + 850 μm	−850 + 300 μm	−300 + 75 μm	−75 μm
Si					
Ca	11.07	11.81		10.81	8.34
Mn					
Fe	9.01	11.97		10.7	11.05
Cd					
Ni					
Cu	0.33	0.33		0.51	1.51
Zn	0.8	0.86		0.86	1.05
As					
Mg					
Co					
Pb	3.33	3.97		4.37	5.53
Ag					
Ge					
Mo					
Other (%)	75.46	71.06		72.75	72.52

Table A2. Assay by size from ICP.

Elements	Elemental Composition (% w/w)				
	+3350 μm	−3350 + 850 μm	−850 + 300 μm	−300 + 75 μm	−75 μm
Si					
Ca					
Mn	0.12	0.14	0.15	0.13	0.09
Fe	10.45	14.29	17.81	14.53	9.75
Cd	0.01	0.01	0.01	0.02	0.05
Ni	0.001	0.001	0.002	0.001	0.001
Cu	0.28	0.33	0.35	0.45	1.08
Zn	3.08	3.12	3.16	2.98	2.1
As	0.23	0.3	0.32	0.38	0.71
Mg	2.45	2.66	2.9	2.67	0
Co	0.01	0.012	0.012	0.011	0.008
Pb	2.03	2.23	2.28	2.23	2.04
Ag	0.0013	nd	0.004	nd	nd
Ge	nd	nd	nd	nd	nd
Mo					
Other (%)	81.3377	76.907	73.002	76.598	84.171

Table A3. Assay by size from XRF.

Elements	Elemental Composition (% w/w)				
	+3350 μm	−3350 + 850 μm	−850 + 300 μm	−300 + 75 μm	−75 μm
Si					
Ca					
Mn	0.13	0.13	0.08	0.15	0.14
Fe	10.02	8.96	5.35	8.22	11.91
Cd					
Ni					
Cu	0.26	0.21	0.11	0.2	1.06
Zn	7.13	5.7	2.93	4.72	7.37
As	0.18	0.14	0.08	0.2	0.88
Mg					
Co					
Pb	2	1.55	0.74	1.07	2.92
Ag					
Ge					
Mo	0.19	0.12	0.06	0.11	0.19
Other (%)	80.09	83.19	90.65	85.33	75.53

References

1. Álvarez, M.L.; Méndez, A.; Rodríguez-Pacheco, R.; Paz-Ferreiro, J.; Gascó, G. Recovery of zinc and copper from mine tailings by acid leaching solutions combined with carbon-based materials. *Appl. Sci.* 2021, *11*, 5166. [CrossRef]
2. Vardanyan, N.; Sevoyan, G.; Navasardyan, T.; Vardanyan, A. Recovery of valuable metals from polymetallic mine tailings by natural microbial consortium. *Environ. Technol.* 2019, *40*, 3467–3472. [CrossRef] [PubMed]
3. Sarker, S.K.; Haque, N.; Bhuiyan, M.; Bruckard, W.; Pramanik, B.K. Recovery of strategically important critical minerals from mine tailings. *J. Environ. Chem. Eng.* 2022, *10*, 107622. [CrossRef]
4. Mulenshi, J. *Reprocessing Historical Tailings for Possible Remediation and Recovery of Critical Metals and Minerals—The Yxsjöberg Case*; Lulea University of Technology: Luleå, Sweden, 2021.
5. Gabasiane, T.S.; Danha, G.; Mamvura, T.A.; Mashifana, T.; Dzinomwa, G. Characterization of copper slag for beneficiation of iron and copper. *Heliyon* 2021, *7*, e06757. [CrossRef]
6. Piatak, N.M.; Parsons, M.B.; Seal, R.R. Characteristics and environmental aspects of slag: A review. In *Applied Geochemistry*; Elsevier Ltd.: Amsterdam, The Netherlands, 2015; Volume 57. [CrossRef]
7. Gorai, B.; Jana, R.K.; Premchand. Characteristics and utilisation of copper slag—A review. *Resour. Conserv. Recycl.* 2003, *39*, 299–313. [CrossRef]
8. Sipunga, E. *Optimization of the Flotation of Copper Smelter Slags from Namibia Custom Smelters' Slag Mill Plant*; University of Witwatersrand: Johannesburg, South Africa, 2015.
9. Lohmeier, S.; Lottermoser, B.G.; Schirmer, T.; Gallhofer, D. Copper slag as a potential source of critical elements—A case study from Tsumeb, Namibia. *J. South. Afr. Inst. Min. Metall.* 2021, *121*, 129–142. [CrossRef]
10. Franke, D.; Suponik, T.; Nuckowski, P.M.; Golombek, K.; Hyra, K. Recovery of metals from printed circuit boards by means of electrostatic separation. *Manag. Syst. Prod. Eng.* 2020, *28*, 213–219. [CrossRef]
11. Mori de Oliveira, C.; Bellopede, R.; Tori, A.; Marini, P. Study of Metal Recovery from Printed Circuit Boards by Physical-Mechanical Treatment Processes. *Mater. Proc.* 2022, *5*, 121. [CrossRef]
12. Pietrelli, L.; Francolini, I.; Piozzi, A.; Vocciante, M. Metals recovery from printed circuit boards: The pursuit of environmental and economic sustainability. *Chem. Eng. Trans.* 2018, *70*, 271–276. [CrossRef]
13. Jadhav, U.; Hocheng, H. Hydrometallurgical Recovery of Metals from Large Printed Circuit Board Pieces. *Sci. Rep.* 2015, *5*, 14574. [CrossRef]
14. Trinh, H.B.; Lee, J.; Kim, S.; Lee, J.C.; Aceituno, J.C. F.; Oh, S. Selective recovery of copper from industrial sludge by integrated sulfuric leaching and electrodeposition. *Metals* 2021, *11*, 22. [CrossRef]
15. Park, I.; Yoo, K.; Alorro, R.D.; Kim, M.S.; Kim, S.K. Leaching of copper from cuprous oxide in aerated sulfuric acid. *Mater. Trans.* 2017, *58*, 1500–1504. [CrossRef]

16. Łukomska, A.; Wiśniewska, A.; Dąbrowski, Z.; Lach, J.; Wróbel, K.; Kolasa, D.; Domańska, U. Recovery of Metals from Electronic Waste-Printed Circuit Boards by Ionic Liquids, DESs and Organophosphorous-Based Acid Extraction. *Molecules* **2022**, *27*, 4984. [CrossRef]
17. Derkowska, K.; Świerk, M.; Nowak, K. Reconstruction of copper smelting technology based on 18–20th-century slag remains from the old copper basin, Poland. *Minerals* **2021**, *11*, 926. [CrossRef]
18. Echeverry-Vargas, L.; Rojas-Reyes, N.R.; Estupiñán, E. Characterization of copper smelter slag and recovery of residual metals from these residues. *Rev. Fac. De Ing.* **2017**, *26*, 61–71. [CrossRef]
19. Tian, H.; Guo, Z.; Pan, J.; Zhu, D.; Yang, C.; Xue, Y.; Li, S.; Wang, D. Comprehensive review on metallurgical recycling and cleaning of copper slag. *Resour. Conserv. Recycl.* **2021**, *168*, 105366. [CrossRef]
20. Gabasiane, T.S.; Danha, G.; Mamvura, T.A.; Mashifana, T.; Dzinomwa, G. Environmental and socioeconomic impact of copper slag—A review. *Crystals* **2021**, *11*, 1–16. [CrossRef]
21. Zuo, Z.; Feng, Y.; Dong, X.; Luo, S.; Ren, D.; Wang, W.; Wu, Y.; Yu, Q.; Lin, H.; Lin, X. Advances in recovery of valuable metals and waste heat from copper slag. *Fuel Process. Technol.* **2022**, *235*, 107361. [CrossRef]
22. Maweja, K.; Mukongo, T.; Mbaya, R.K.; Mochubele, E.A. Effect of annealing treatment on the crystallisation and leaching of dumped base metal smelter slags. *J. Hazard. Mater.* **2010**, *183*, 294–300. [CrossRef]
23. Nadirov, R.K. Recovery of Valuable Metals from Copper Smelter Slag by Sulfation Roasting. *Trans. Indian Inst. Met.* **2019**, *72*, 603–607. [CrossRef]
24. Isaksson, J.; Vikström, T.; Lennartsson, A.; Samuelsson, C. Influence of Process Parameters on Copper Content in Reduced Iron Silicate Slag in a Settling Furnace. *Metals* **2021**, *11*, 992. [CrossRef]
25. Zhu, Z.; He, J. *Modern Copper Metallurgy*; Science Press: Beijing, China, 2003.
26. Li, I.; Wang, H.; Hu, J.H. Study on development of the comprehensive utilization of copper slag. *Energy Metall. Ind.* **2009**, *28*, 44–48.
27. Ahmad, J.; Majdi, A.; Deifalla, A.F.; Isleem, H.F.; Rahmawati, C. Concrete Made with Partially Substitutions of Copper Slag (CPS): A State Art of Review. *Materials* **2022**, *15*, 5196. [CrossRef]
28. Shi, C.; Meyer, C.; Behnood, A. Utilization of copper slag in cement and concrete. *Resour. Conserv. Recycl.* **2008**, *52*, 1115–1120. [CrossRef]
29. Reddy, R.G.; Prabhu, V.L.; Mantha, D. Kinetics of Reduction of Copper Oxide from Liquid Slag Using Carbon. *High Temp. Mater. Process.* **2003**, *22*, 25–34. [CrossRef]
30. Jerzy, Ł.; Blacha, L.; Jodkowski, M.; Smalcerz, A. The use of waste, fine-grained carbonaceous material in the process of copper slag reduction. *J. Clean. Prod.* **2021**, *288*, 125640. [CrossRef]
31. Prince, S.; Young, J.; Ma, G.; Young, C. Characterization and recovery of valuables from waste copper smelting slag. In Proceedings of the 10th International Conference on Molten Slags, Fluxes and Salts (MOLTEN16), Seattle, DC, USA, 22–25 May 2016; Reddy, R.G., Chaubal, P., Pistorius, P.C., Pal, U., Eds.; The Minerals, Metals and Materials Society: Pittsburgh, PA, USA, 2016; pp. 889–898.
32. Seyrankaya, A.; Canbazoglu, M. Recovery of cobalt, copper and zinc from Kure-Kastamonu historical copper slag by high pressure oxidative acid leaching. *Russ. J. Non-Ferr. Met.* **2021**, *62*, 390–402. [CrossRef]
33. Maweja, K.; Mukongo, T.; Mutombo, I. Cleaning of a copper matte smelting slag from a water-jacket furnace by direct reduction of heavy metals. *J. Hazard. Mater.* **2009**, *164*, 856–862. [CrossRef]
34. Kaksonen, A.H.; Lavonen, L.; Kuusenaho, M.; Kolli, A.; Närhi, H.; Vestola, E.; Puhakka, J.A.; Tuovinen, O.H. Bioleaching and recovery of metals from final slag waste of the copper smelting industry. *Miner. Eng.* **2011**, *24*, 1113–1121. [CrossRef]
35. Sarrafi, A. Recovery of copper from reverberatory furnace slag by flotation. *Miner. Eng.* **2004**, *17*, 457–459. [CrossRef]
36. Karimi, N.; Vaghar, R.; Mohammadi, M.R. T.; Hashemi, A.A. Recovery of Copper from the Slag of Khatoonabad Flash Smelting Furnace by Flotation Method. *J. Inst. Eng. India Seri. D* **2013**, *94*, 43–50. [CrossRef]
37. Linsong, W.; Zhiyong, G.; Honghu, T.; Li, W.; Haisheng, H.; Wei, S. Copper recovery from copper slags through flotation enhanced by sodium carbonate synergistic mechanical activation. *J. Environ. Chem. Eng.* **2022**, *10*, 107671. [CrossRef]
38. Kundu, T.; Senapati, S.; Kanta, S.; Angadi, S.I.; Rath, S.S. A comprehensive review on the recovery of copper values from copper slag. *Powder Technol.* **2023**, *426*, 118693. [CrossRef]
39. Jarosikova, A.; Ettler, V.; Mihaljevic, M.; Kribek, B.; Mapani, B. The pH-dependent leaching behavior of slags from various stages of a copper smelting process: Environmental implications. *J. Environ. Manag.* **2017**, *187*, 178–186. [CrossRef]
40. Ettler, V.; Mihaljevič, M.; Strnad, L.; Křibek, B.; Hrstka, T.; Kamona, F.; Mapani, B. Gallium and germanium extraction and potential recovery from metallurgical slags. *J. Clean. Prod.* **2022**, *379*, 134677. [CrossRef]
41. Sibarani, D.; Hamuyuni, J.; Luomala, M.; Lindgren, M.; Jokilaakso, A. Thermal Conductivity of Solidified Industrial Copper Matte and Fayalite Slag. *Jom* **2020**, *72*, 1927–1934. [CrossRef]
42. Kang, Y.; Lee, J.; Morita, K. Thermal conductivity of molten slags: A review of measurement techniques and discussion based on microstructural analysis. *ISIJ Int.* **2014**, *54*, 2008–2016. [CrossRef]
43. Derin, B.; Cinar, F.; Yiicel, O. The Electrical Characteristics of Copper Slags in a 270 Kva Dc Arc Furnace. In Proceedings of the 3rd Balkan Metallurgical Conference, Ohrid, Macedonia, 24–27 September 2003; pp. 265–270.
44. Matsushita, T.; Watanabe, T.; Hayashi, M.; Mukai, K. Thermal, optical and surface/interfacial properties of molten slag systems. *Int. Mater. Rev.* **2011**, *56*, 287–323. [CrossRef]
45. Hauptmann, A. *The Archaeometallurgy of Copper*, 1st ed.; Springer: Berlin/Heidelberg, Germany, 2007. [CrossRef]

46. Ettler, V.; Johan, Z. Minéralogie des phases métalliques des mattes sulfurées de la métallurgie du plomb. *Comptes Rendus Geosci.* **2003**, *335*, 1005–1012. [CrossRef]
47. Paradis, S. Indium, germanium and gallium in volcanic-and sediment-hosted base-metal sulphide deposits. In *Symposium on Strategic and Critical Materials Proceedings*; Simandl, G.J., Neetz, M., Eds.; British Columbia Geological Survey: Victoria, BC, Canada, 2015; pp. 23–29.
48. Höll, R.; Kling, M.; Schroll, E. Metallogenesis of germanium-A review. *Ore Geol. Rev.* **2007**, *30*, 145–180. [CrossRef]
49. Ruiz, A.G.; Sola, P.C.; Palmerola, N.M. Germanium: Current and Novel Recovery Processes. In *Advanced Material and Device Applications with Germanium*; Lee, S., Ed.; IntecOpen: London, UK, 2018. [CrossRef]
50. Berger, J.A.; Schmidt, M.E.; Gellert, R.; Boyd, N.I.; Desouza, E.D.; Flemming, R.L.; Izawa, M.R.M.; Ming, D.W.; Perrett, G.M.; Rampe, E.B.; et al. Zinc and germanium in the sedimentary rocks of Gale Crater on Mars indicate hydrothermal enrichment followed by diagenetic fractionation. *J. Geophys. Res. Planets* **2017**, *122*, 1747–1772. [CrossRef]
51. Avarmaa, K.; Klemettinen, L.; O'brien, H.; Taskinen, P.; Jokilaakso, A. Critical metals Ga, Ge and In: Experimental evidence for smelter recovery improvements. *Minerals* **2019**, *9*, 367. [CrossRef]
52. Melcher, F.; Oberthür, T.; Rammlmair, D. Geochemical and mineralogical distribution of germanium in the Khusib Springs Cu-Zn-Pb-Ag sulfide deposit, Otavi Mountain Land, Namibia. *Ore Geol. Rev.* **2006**, *28*, 32–56. [CrossRef]
53. CRMA. Germanium. *European Critical Raw Materials*. 2023. Available online: https://www.crmalliance.eu/germanium (accessed on 24 August 2023).
54. Reddy, R.G.; Prabhu, V.L.; Mantha, D. Zinc Fuming from Lead Blast Furnace Slag. *High Temp. Mater. Process.* **2002**, *21*, 377–386. [CrossRef]
55. Liu, Q.; Zhao, Y.; Zhao, G. Production of zinc and lead concentrates from lean oxidized zinc ores by alkaline leaching followed by two-step precipitation using sulfides. *Hydrometallurgy* **2011**, *110*, 79–84. [CrossRef]
56. Silwamba, M.; Ito, M.; Hiroyoshi, N.; Tabelin, C.B.; Hashizume, R.; Nakata, H.; Nakayama, S.; Ishizuka, M. Alkaline Leaching and Concurrent Cementation of Dissolved Pb and Zn from Zinc Plant Leach Residues. *Minerals* **2022**, *12*, 393. [CrossRef]
57. Kazadi, D.M.; Groot, D.R.; Steenkamp, J.D.; Pöllmannb, H. Control of silica polymerization during ferromanganese slag sulphuric acid digestion and water leaching. *Hydrometallurgy* **2016**, *166*, 214–221. [CrossRef]

Disclaimer/Publisher's Note: The statements, opinions and data contained in all publications are solely those of the individual author(s) and contributor(s) and not of MDPI and/or the editor(s). MDPI and/or the editor(s) disclaim responsibility for any injury to people or property resulting from any ideas, methods, instructions or products referred to in the content.

Article

Investigations of the Density and Solubility of Ammonium Perrhenate and Potassium Perrhenate Aqueous Solutions

Szymon Orda [1,*], Michał Drzazga [1], Katarzyna Leszczyńska-Sejda [1], Mateusz Ciszewski [1], Alicja Kocur [2], Pola Branecka [2], Kacper Gall [2], Mateusz Słaboń [2] and Marcin Lemanowicz [2]

1. Łukasiewicz Research Network—Institute of Non-Ferrous Metals, Centre of Hydroelectrometallurgy, ul. Sowińskiego 5, 44-100 Gliwice, Poland; michal.drzazga@imn.lukasiewicz.gov.pl (M.D.); katarzyna.leszczynska-sejda@imn.lukasiewicz.gov.pl (K.L.-S.); mateusz.ciszewski@imn.lukasiewicz.gov.pl (M.C.)
2. Department of Chemical Engineering and Process Design, Faculty of Chemistry, Silesian University of Technology, ul. ks. M. Strzody 7, 44-100 Gliwice, Poland; alicja.kocur@polsl.pl (A.K.); kacper.gall@student.polsl.pl (K.G.); mateusz.slabon@student.polsl.pl (M.S.); marcin.lemanowicz@polsl.pl (M.L.)
* Correspondence: szymon.orda@imn.lukasiewicz.gov.pl

Abstract: Rhenium is largely used as an additive to nickel- and cobalt-based superalloys. Their resistance to temperature and corrosion makes them suitable for the production of turbines in civil and military aviation, safety valves in drilling platforms, and tools working at temperatures exceeding 1000 °C. The purity of commercial rhenium salts is highly important. Potassium, which is a particularly undesirable element, can be removed by recrystallization. Therefore, it is crucial to possess detailed knowledge concerning process parameters including the dissolved solid concentration and the resulting saturation temperature. This can be achieved using simple densimetric methods. Due to the fact that data concerning the physicochemical properties of ammonium perrhenate (APR) NH_4ReO_4 and potassium perrhenate (PPR) $KReO_4$ are imprecise or unavailable in the scientific literature, the goal of this study is to present experimental data including the solubility and density of water solutions of both salts. In the experiments, a densimeter with a vibrating cell was used to precisely determine the densities. Although the investigated solutions did not fit into the earlier proposed mathematical model, some crucial conclusions could still be made based on the results.

Keywords: rhenium; ammonium perrhenate; potassium perrhenate; densimetric method; solubility curve

1. Introduction

Metallic rhenium is a valuable metal characterized by parameters such as a density of 21.02 g/cm^{-3}, a melting point of 3181 °C, a heat capacity of 0.137 J·g^{-1}·K^{-1}, a Young's modulus of 463 GPa, and a hardness (Mohs scale) of 7.0 [1–3]. The application of rhenium includes different branches of industry: aviation, defense, petrochemicals, medicine and electricity. An average of 83% of rhenium is used for the production of superalloys [2,4,5]. The addition of metallic Re to nickel-based alloys increases thermal and corrosion resistance and thus the durability of materials used for jet engines [6]. The second area of application is the petrochemical industry. About 12% of rhenium is used in reforming high-octane gasoline [2,5]. Bi-metallic catalysts (Re-Pt) are characterized by significantly higher operation stability at high temperatures and under low pressure [7,8]. The lifetime of Re-Pt catalysts during the deposition of coal on surfaces is longer compared to traditional ones [7]. In medical applications, rhenium is used for the production of stents and X-ray lamps. Metal is also used in therapeutic medicine [5]. Radioactive isotope rhenium [187]Re decays, emitting radiation, which is not dangerous to the human body [3,9]. Other applications of this metal include the production of thermocouples, heating elements, electrodes, electric contact points, flashbulbs, and electromagnets [4,10,11]. The development of the aerospace industry has had an influence on rhenium demand and thus metal prices.

Molybdenite concentrates from copper mines are the most important primary sources of rhenium [1,6,10,12,13]. Rhenium occurs there as ReS_2. During the roasting of molybdenite concentrates, fumes and dust containing the volatile form of rhenium, Re_2O_7, are generated [14]. Fumes containing rhenium heptoxide in contact with water create scrubber liquor containing ReO_4^-. Then, after concentration, the ammonium perrhenate, the main commercial rhenium compound, is produced. Limited access to primary sources leads to the recycling of rhenium from spent catalysts and superalloy waste [15–17]. Nowadays, there exist numerous technologies including recovery for the rhenium from deactivated catalysts [7,8,18].

Ammonium perrhenate is the main commercial rhenium compound. The most important step in the production of its crystalline form is the separation of rhenium from solutions created during technological processes. The reported methods of rhenium separation involve ion exchange, adsorption, solvent extraction, and precipitation. In the first case, different types of resins have been investigated [4,6,10,12,19–29]. Rhenium from scrubber liquor may be adsorbed using active carbon or nano particles of iron(III) oxide coated with active carbon [30,31]. In the case of solutions enriched with rhenium a suitable approach is solvent extraction for which many different extractants have been studied [32–35]. Finally, apart from the above-mentioned methods, another form of rhenium recovery is precipitation [36,37].

Crude ammonium perrhenate (CAP) usually requires additional treatment, i.e., purification. One of the most undesirable pollutants is potassium. Potassium ions form a sparingly soluble compound—potassium perrhenate (PPR). Researchers have tested electrodialysis membranes and ion exchange to remove potassium from CAP [15,38–40]. However, CAP recrystallization easily removes impurities such as potassium. One can find just a few papers dedicated to the investigations of APR crystallization. There is one publication about the multistage recrystallization of CAP to remove potassium [41]. Apart from that, the influence of stirring on the recrystallization of ammonium perrhenate has been examined [42]. Other properties, the thermal expansion coefficient, the standard enthalpies formation, and the heat capacity single crystals APR and PPR, have also been determined [43–46]. It was found that the thermal expansion coefficient of ammonium perrhenate indicates an anomaly compared to potassium perrhenate [47]. The mutual solubility of sodium perrhenate in a water–ethanol system was also tested [48].

In order to properly carry out the crystallization process, it is obligatory to know the mechanism and kinetics of the APR and PPR crystallization process [49]. Due to limited or obsolete data concerning the physicochemical properties of APR and PPR water solutions, we decided to deeply investigate the solubilities and densities of these salts. Potassium perrhenate is the main impurity of ammonium perrhenate—the main commercial rhenium compound used for the production of metallic rhenium. This salt crystallizes from a ternary solution and since its solubility is significantly lower than the solubility of the main product, a recrystallization process has to be employed. Moreover, the performed research allowed us to make an attempt to fit the experimental data to the mathematical models available in the literature. In the future, this work will be a solid foundation for investigations into APR and PPR crystallization, like the process design and control or its numerical simulations.

2. Materials and Methods

Two rhenium salts were used for density measurements: ammonium perrhenate (analytical grade, INNOVATOR sp. z.o.o., Gliwice, Poland) and potassium perrhenate. The potassium salt was prepared via the neutralization of perrhenic acid by potassium hydroxide (analytical grade, Chempur, Piekary Slaskie, Poland). Perrhenic acid was obtained from the APR by the ion exchange method [50]. The saturated solutions of each salt were prepared using reverse osmosis (RO) water (0.06 µS cm^{-1} conductivity, Hydrolab, Straszyn, Poland). For both reagents' phase compositions, determination was performed by X-ray diffraction. XRD analysis was performed using a Rigaku MiniFlex600 (Rigaku Co.,

Tokyo, Japan) diffractometer. The obtained diffractograms are included as Supplementary Materials (Figures S1 and S2).

Excess amounts of salt were dissolved in RO water in a thermostated vessel reactor of 0.5×10^{-3} m^3. The saturated solution was stirred for 2 h using a magnetic stirrer at a constant temperature. The temperature was controlled by a thermostat Huber Ministats CC-K6 (temperature accuracy +/−0.02 K). After stabilization, a sample of saturated solution was collected by a syringe equipped with a microporous filter (0.22 µm pore size). To prevent crystallization during sampling, the syringe, filter, and needle were heated to 10 K above the saturation temperature each time before use. A small portion of solution (approximately 3 mL) was taken from the vessel and transferred to digital densimeter DMA 4500M (Anton Paar, Singapore), which determined the density with an accuracy of $+/-5 \times 10^{-5}$ g·cm^{-3} at atmospheric pressure. The temperature was controlled automatically by the densimeter within ±0.02 K accuracy. The densities of the samples were determined based on the oscillating U-tube principle. The glass cell was excited and oscillated at a given frequency depending on the mass of the investigated fluid. The density was calculated based on the corresponding frequency. In contrast with classical methods, the precise measurements may be completed using small volumes of fluids. After 30 min, another measurement was made. If the two subsequently measured densities were similar, then it was assumed that the equilibrium state was achieved. The saturated solutions of APR and PPR salts were prepared in the same way.

The samples for the solubility and density measurements were collected from saturated solutions. The density was measured in the range between the saturation temperature and a 10 K higher temperature with 1 K steps. This process was performed automatically by the densimeter. Moreover, an additional sample was taken in order to determine the solubility of the salts using the gravimetric method. Multiple beakers were filled with samples and placed in a laboratory dryer at 105 °C. They were dried until a constant mass was achieved. The solubility was calculated on the basis of the mass difference. Experiments were carried out at six saturation temperatures: 283.15 K; 293.15 K; 303.15 K; 313.15 K; 323.15 K; and 333.15 K. All measurements were repeated three times.

3. Results and Discussion

3.1. Solubility Curve

The solubility data are presented in Tables S1 and S2, attached as Supplementary Materials. The experimental data can be described by the following mathematical model [51]:

$$\log_{10} S = A + \frac{B}{T} + C \cdot \log_{10} T \quad (1)$$

Based on the experimental data coefficients, A, B, and C in Equation (1) were obtained using the least squares method (LSM). The values are presented in Table 1.

Table 1. Regression and correlation coefficients of Equation (1).

Coefficient	Ammonium Perrhenate	Potassium Perrhenate
A	13.77561	13.99451
B	−1590.72975	−2009.10125
C	−3.07759	−2.90039
R^2	0.99959	0.99951
A	13.77561	13.99451

Figure 1 presents the experimental data, the calculated values, and the data taken from the literature for ammonium and potassium perrhenate. The solubilities of the two salts were compared.

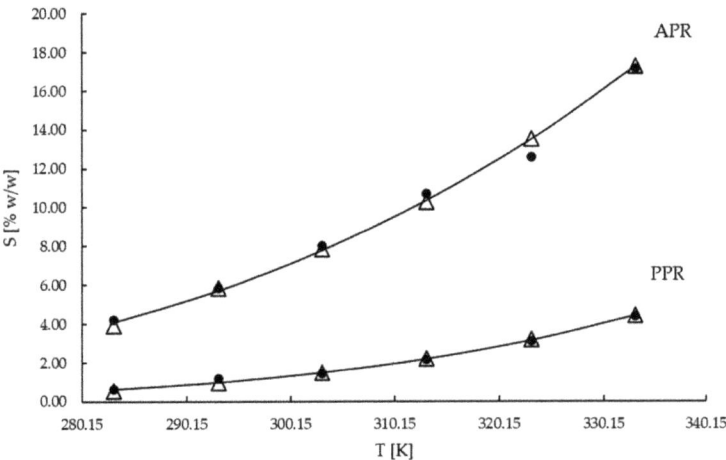

Figure 1. Solubility of NH_4ReO_4 and $KReO_4$: △ experimental; − calculated; • literature [52].

The standard deviation bars are not presented in the figure due to their small values. The highest standard deviation was equal to 0.33 % w/w. As one may notice, the points obtained on the basis of our experiments are in better agreement with the mathematical model than the data available in the literature. The solubility of ammonium perrhenate changes noticeably with temperature, which would seem to make the isohydric crystallization by cooling a suitable process for its production. However, Figure 1 perfectly illustrates the challenge concerning the removal of potassium perrhenate. As one may notice, the solubility of this salt is very low compared with ammonium perrhenate. Only at 60 °C does the solubility of $KReO_4$ match the solubility of NH_4ReO_4 at 10 °C.

3.2. Density Mesurements
3.2.1. Ammonium Perrhenate

Experimental data on the density measurements for the undersaturated region are presented in Table S3 in the Supplementary Materials. For the average values of density, standard deviations were calculated, and they are presented in the Supplementary Materials. Based on the collected data, Figure 2 was plotted. In the limited range, close to the saturation line, the density of the solution can be described by the linear function of temperature [53,54]:

$$\rho = a \cdot T + b \qquad (2)$$

The linear equation coefficients (a and b) and the coefficient of determination (R^2) for the undersaturated region are given in Table 2 for all investigated saturation temperatures. The same type of linear correlation was obtained by Frej et al. [53], Marciniak [54,55], and Bogacz et al. [55,56].

Table 2. Coefficients and the determination coefficients for undersaturated NH_4ReO_4 solutions (automatic measurement). The second column represents the density of the saturated solution.

T [K]	ρ [g·cm^{-3}]	a [g·cm^{-3}·K^{-1}]	b [g·cm^{-3}]	R^2 [-]
283.15	1.03009	−0.00019	1.08392	0.99320
293.15	1.04246	−0.00031	1.13205	0.99845
303.15	1.05765	−0.00041	1.18075	0.99933
313.15	1.07816	−0.00056	1.25512	0.99677
323.15	1.10233	−0.00139	1.55029	0.99369
333.15	1.12943	−0.00205	1.81105	0.99234

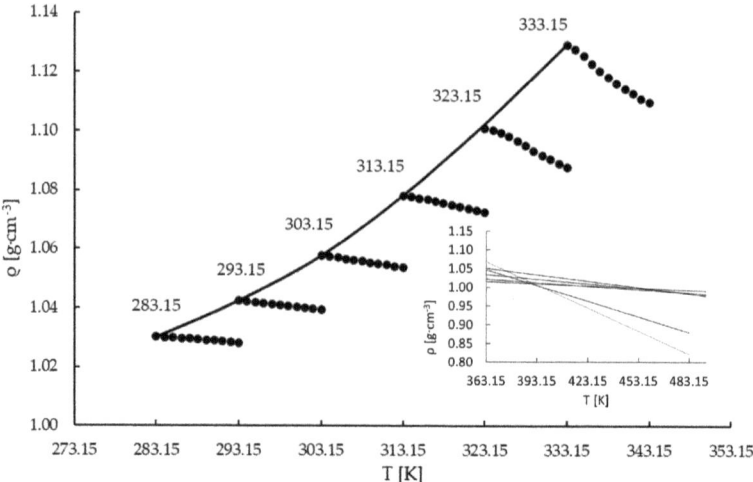

Figure 2. Density vs. temperature of concentrated NH$_4$ReO$_4$ solutions (•) and density curve in temperature saturation (−).

The density curve at the saturation temperature (Figure 2) is described by the following second-order polynomial equation:

$$\rho = a \cdot T^2 + b \cdot T + c \qquad (3)$$

In Equation (3), the correlations coefficients (a, b, and c) could be obtained by using the LSM and the experimental density data. In Table 3, the coefficients and determination coefficients are presented.

Table 3. Coefficients and determination coefficient of polynomial Equation (3) for NH$_4$ReO$_4$.

a × 10^5	b	c	R^2
1.85031	−0.00916	2.10943	0.98980

It can be observed that the correlation slope decreases with the increase in the saturation temperature. Using the slope coefficients of Equation (2) from Table 2 for the extrapolated temperature range, an attempt was made to determine a construction point (Figure 2). It can be seen that for temperatures 323.15 and 333.15 K, the a coefficient is significantly lower than for lower temperatures. Thus, the extrapolated density lines did not intersect at the same point, and determination of construction points for the investigated temperature range was not possible.

Additional density measurements for ammonium perrhenate solutions at 323.15 K and 333.15 K in the unsaturated region were performed by measuring a fresh sample of saturated solution. The density data are presented in the Supplementary Materials in Table S3. For each measurement, a fresh sample of saturated solution was injected into the device, its temperature was adjusted to the target value, and the density was measured. Figure 3 presents the comparison of the density data for the solution saturation temperatures of 323.15 K and 333.15 K for automatic measurement (temperature scan of density—a whole data series was made based on one sample) and manual measurements (each point in the series was determined based on the fresh sample). It may be noticed that in the first case the slope coefficient is visibly higher compared to the second case. This is consistent with the data presented in Figure 3 where the trend of these two data series deviated from the rest of the samples. Also, the correlation coefficients for Equation (2) and the determination coefficients for the undersaturated regions were calculated, and they are presented in Table 4. In our opinion, this behavior may be explained by the formation of gas (ammonia) microbubbles during the measurements [53]. This observation is also supported by the thermodynamic calculations performed using HSC Chemistry 9.6.1 software (Outotec, module Gem—Equilibrium Composition), according to which ammonia evolution is possible for temperatures exceeding 320 K (Figure S3 in the Supplementary Materials). It is worth noting that the first points for both measurement methods are identical. This simply means that the properties of the sample changed during the measurement. The temperature scan took about 1 h. For such a long period of time and at elevated temperatures microbubbles of ammonia could be created within the sample. Since the density measurements are based on cell vibration analysis, the inertia of bubbles resulted in a decrease in the measured value. The longer the sample was measured, the higher the deviation that was achieved due to the higher accumulation of bubbles.

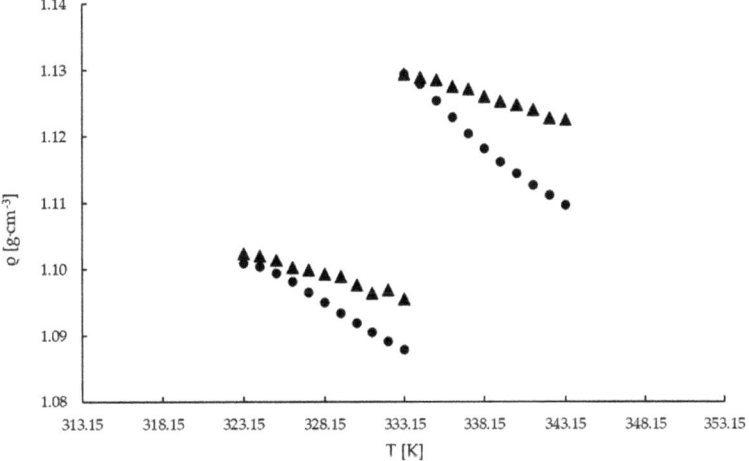

Figure 3. Density measurements of NH_4ReO_4 solutions: • automatic; ▲ manual.

Table 4. Coefficients and the determination coefficients for undersaturated NH_4ReO_4 solutions (manual measurements).

T [K]	ρ [g·cm^{-3}]	a [g·cm^{-3}·K^{-1}]	b [g·cm^{-3}]	R^2 [-]
323.15	1.02334	−0.00069	1.32514	0.98140
333.15	1.12943	−0.00073	1.37276	0.99253

Due to the relatively low solubility of the salt, its impact on the solution density was limited. Yet, the polynomial nature of the saturation line was preserved. These data are an interesting starting point for future investigations on crystallization and solubility kinetics. Moreover, the phenomenon of ammonium formation and its impact on multicomponent crystallization should be investigated more deeply in the future.

3.2.2. Potassium Perrhenate

The density data on potassium perrhenate and the standard deviations are presented in the Supplementary Materials in Table S5. Based on the density values and the same procedure as for ammonium perrhenate, Figure 4 was plotted. The coefficients (a, b, and R^2) describing the slope of the density lines are presented in Table 5. Moreover, the coefficients in Equation (3) for the potassium perrhenate solution are presented in Table 6.

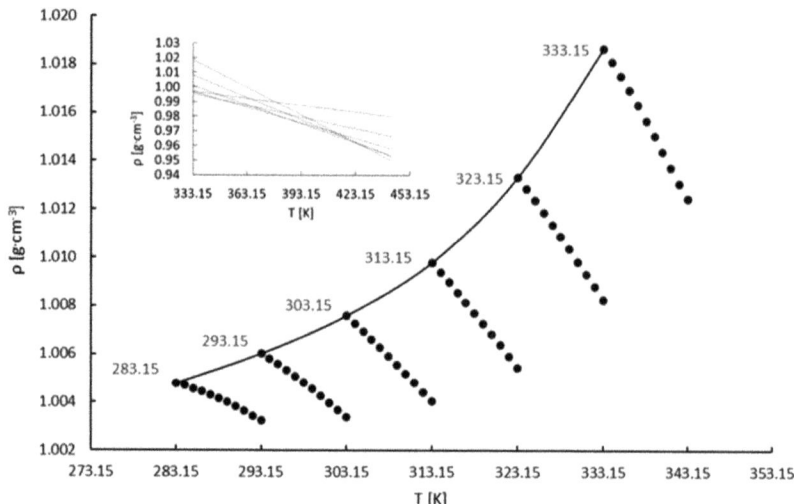

Figure 4. Density vs. temperature of the concentrated KReO$_4$ solutions.

Table 5. Coefficients and the square of the Pearson correlation coefficients for undersaturated KReO$_4$ solutions.

T [K]	P [g·cm^{-3}]	a [g·cm^{-3}·K^{-1}]	b [g·cm^{-3}]	R^2 [-]
283.15	1.00479	−0.00016	1.04893	0.98903
293.15	1.00601	−0.00026	1.08350	0.99751
303.15	1.00759	−0.00035	1.11526	0.99908
313.15	1.00978	−0.00043	1.14579	0.99949
323.15	1.01332	−0.00051	1.17722	0.99971
333.15	1.01865	−0.00063	1.22731	0.99939

Table 6. Coefficients and determination coefficient of polynomial Equation (3) for KReO$_4$.

a × 10^6	b	c	R^2
3.93827	−0.00221	1.29264	0.96180

Additional PPR density measurements using fresh solutions were not made. At low saturation temperatures, i.e., 283.15, 293.15, and 303.15 K, it was not possible to determine the construction point as the extrapolated lines did not intersect at a single point. It is shown in Figure 4. This may be related to the low solubility of PPR compared with APR (Figure 1). When the saturation temperatures decrease the change in density with temperature is lower. This can be observed in Figure 4. On the other hand, by using data for saturation temperatures of 313.15, 323.15, and 333.15 K, the construction point was obtained and plotted in Figure 5. All extrapolated lines intersect at one pole point, whose coordinates are (424.50 K, 0.96168 g·cm^{-3}) (Figure 5). Thus, the methodology for the determination of the construction point for potassium perrhenate solutions is appropriate at temperatures higher than 303.15K. It is important to emphasize that the obtained pole point has no physical meaning and it is used only for calculations [56].

In most cases, the application of the densimetric method to aqueous solutions of salts allows one to construct the pole point. Then, based on the density measurements, the pole point may be used for the determination of the saturation temperature [55,56]. In our opinion, the reason why the experimental data did not fit to the mathematical model proposed by Bogacz et al. [56] is associated with the solubility of the salts. Bogacz et al. investigated potassium chloride and potassium sulphate, whose solubilities significantly surpasses the solubilities of APR and PPR. It is also likely that the ions' interaction plays a crucial role in the analysis of the behavior of such saturated systems.

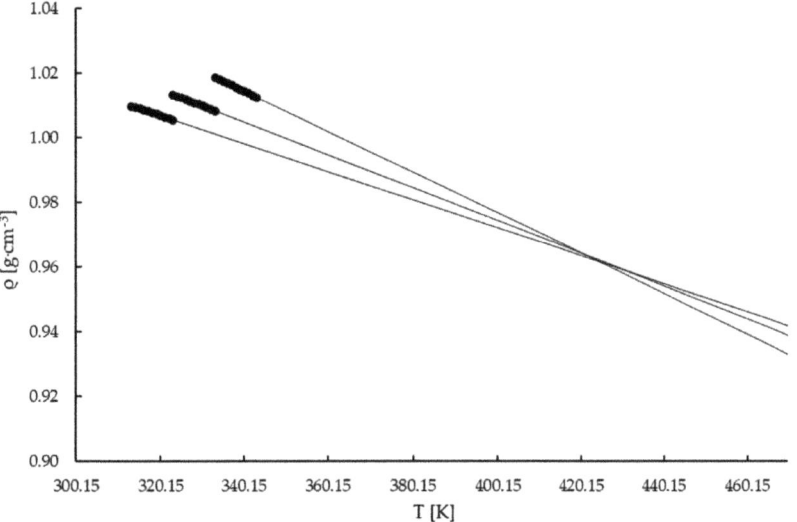

Figure 5. Construction point of the concentrated KReO$_4$ solution.

As in the case of ammonium perrhenate, the presence of salt had a small impact on the densities of the solutions, which were slightly higher than for the base liquid, i.e., water. Due to the significant difference between the solubilities of both salts the densities of the ammonium perrhenate solutions were always higher than the solutions of potassium perrhenate. Once again, the second-order polynomial of the saturation curve was preserved.

4. Conclusions

The densities and solubilities of aqueous solutions of APR and PPR were examined. In the case of solubility, the experimental data were correlated as a function of temperature using empirical equations, and good agreement was found (even better than when compared to the data already available in the literature). Based on the density data, an attempt was made to fit a mathematical model proposed in the literature. Unfortunately, it was impossible to find the pole points for both salts for the whole investigated temperature range. Surprisingly, it was noticed that for the ammonia salt, microbubbles were created during the density measurements for higher temperatures, which had a significant impact on the readings. For these cases, manual measurements had to be performed. The reason why the experimental data did not fit the mathematical model is, in our opinion, the low solubility of the investigated salts in relation to the systems for which the model was developed. Still, this work will be a solid foundation for investigations on APR and PPR crystallization, like the process design and control or its numerical simulation.

Supplementary Materials: The following supporting information can be downloaded at: https://www.mdpi.com/article/10.3390/ma16155481/s1. Figure S1: XRD analysis of NH$_4$ReO$_4$; Figure S2: XRD analysis of KReO$_4$; Figure S3: Equilibrium amount of ammonia evolved from saturated NH$_4$ReO$_4$ aqueous solution vs. temperature (HSC Chemistry 9.6.1, Gem module); Table S1: Solubility of ammonium perrhenate; Table S2: Solubility of potassium perrhenate; Table S3: APR density versus temperature (automatic measurements); Table S4: APR density versus temperature (manual measurements); Table S5: PPR density versus temperature.

Author Contributions: Conceptualization: S.O.; formal analysis: S.O., M.L. and M.D.; supervision: M.L. and M.D.; validation: S.O.; investigation: S.O., M.L., A.K., P.B., K.G. and M.S.; visualization: S.O.; writing—original draft: S.O.; writing—review and editing: M.L., K.L.-S., M.D. and M.C. All authors have read and agreed to the published version of the manuscript.

Funding: This work was financially supported by a grant from the Ministry of Science and Higher Education in Poland (grant number DWD/5/0572/2021 "Implementation doctorate").

Institutional Review Board Statement: Not applicable.

Informed Consent Statement: Not applicable.

Data Availability Statement: Not applicable.

Conflicts of Interest: The authors declare no conflict of interest.

References

1. Lunk, H.-J.; Drobot, D.V.; Hartl, H. Discovery, Properties and Applications of Rhenium and Its Compounds. *ChemTexts* **2021**, *7*, 6. [CrossRef]
2. Leszczyńska-Sejda, K.; Benke, G.; Kopyto, D.; Chmielarz, A.; Drzazga, M.; Ciszewski, M.; Kowalik, P.; Goc, K.; Bugla, K.; Grabowski, T.; et al. Development of Rhenium Technologies in Poland. In Proceedings of the Conference Paper Proceeding Congress European Metallurgical Conference EMCGDMB, Salzburg, Austria, 27–30 June 2021.
3. Millensifer, A.T.; Sinclair, D.; Jonasson, I.; Lipmann, A. Rhenium. In *Critical Metals Handbook*; Gunn, G., Ed.; John Wiley & Sons: Oxford, UK, 2013; pp. 340–360. ISBN 978-1-118-75534-1.
4. Lutskiy, D.S.; Ignatovich, A.S.; Sulimova, M.A. Determination of the Sorption Characteristics of Ammonium Perrenate Ions on Anion Exchange Resin AV-17-8. *J. Phys. Conf. Ser.* **2019**, *1399*, 055069. [CrossRef]
5. *Rhenium: Outlook to 2029*; Roskill Information Services Ltd.: London, UK, 2019; ISBN 978-1-910-92279-8.
6. Zhang, B.; Liu, H.-Z.; Wang, W.; Gao, Z.-G.; Cao, Y.-H. Recovery of Rhenium from Copper Leach Solutions Using Ion Exchange with Weak Base Resins. *Hydrometallurgy* **2017**, *173*, 50–56. [CrossRef]
7. Kasikov, L.G.; Petrova, A.M. Processing of Deactivated Platinum-Rhenium Catalysts. *Theor. Found. Chem. Eng.* **2009**, *43*, 544–552. [CrossRef]
8. Angelidis, T.N.; Rosopoulou, D.; Tzitzios, V. Selective Rhenium Recovery from Spent Reforming Catalysts. *Ind. Eng. Chem. Res.* **1999**, *38*, 1830–1836. [CrossRef]
9. Leddicotte, G.W. *The Radiochemistry of Rhenium*; National Academy of Sciences: Washington, DC, USA, 1981.
10. Lan, X.; Liang, S.; Song, Y. Recovery of Rhenium from Molybdenite Calcine by a Resin-in-Pulp Process. *Hydrometallurgy* **2006**, *82*, 133–136. [CrossRef]
11. John, D.A.; Seal, R.R.; Polyak, D.E. *Critical Mineral Resources of the United States—Economic and Environmental Geology and Prospects for Future Supply*; Professional Paper 1802; U.S. Geological Survey: Reston, VA, USA, 2017; Chapter P1.
12. Maltseva, E.E.; Blokhin, A.A.; Murashkin, Y.V.; Mikhaylenko, M.A. An Increase in Purity of Ammonium Perrhenate Solutions with Respect to Molybdenum(IV) with the Sorption Recovery of Rhenium(VII) from Mo-Containing Solutions. *Russ. J. Non-Ferr. Met.* **2017**, *58*, 463–469. [CrossRef]
13. Fang, D.; Song, Z.; Zhang, S.; Li, J.; Zang, S. Solvent Extraction of Rhenium(VII) from Aqueous Solution Assisted by Hydrophobic Ionic Liquid. *J. Chem. Eng. Data* **2017**, *62*, 1094–1098. [CrossRef]
14. Ammann, P.R.; Loose, T.A. Rhenium Volatilization during Molybdenite Roasting. *Metall. Mater. Trans. B* **1972**, *3*, 1020–1022. [CrossRef]
15. Guro, V.P. Ammonium Perrhenate Purification and Rhenium Recovery from Heat-Resistant Rhenium-Nickel Superalloys. In Proceedings of the 21st International Conference on Metallurgy and Materials, Brno, Czech Republic, 23–25 May 2012.
16. Srivastava, R.R.; Kim, M.; Lee, J. Novel Aqueous Processing of the Reverted Turbine-Blade Superalloy for Rhenium Recovery. *Ind. Eng. Chem. Res.* **2016**, *55*, 8191–8199. [CrossRef]
17. Tang, J.; Sun, Y.; Hou, G.; Ding, Y.; He, F.; Zhou, Y. Studies on Influencing Factors of Ammonium Rhenate Recovery from Waste Superalloy. *Appl. Sci.* **2018**, *8*, 2016. [CrossRef]
18. Kennth, N.H.; Xinghui, M. Recovery of Platinum Group Metals and Rhenium from Materials Using Halogen Reagents. U.S. Patent 5,542,957, 6 August 1996.
19. Cyganowski, P.; Cierlik, A.; Leśniewicz, A.; Pohl, P.; Jermakowicz-Bartkowiak, D. Separation of Re(VII) from Mo(VI) by Anion Exchange Resins Synthesized Using Microwave Heat. *Hydrometallurgy* **2019**, *185*, 12–22. [CrossRef]
20. Fisher, S.A.; Meloche, V.W. Ion Exchange Separation of Rhenium from Molybdenum. *Anal. Chem.* **1952**, *24*, 1100–1106. [CrossRef]
21. Guo, X.; Ma, Z.; Li, D.; Tian, Q.; Xu, Z. Recovery of Re(VII) from Aqueous Solutions with Coated Impregnated Resins Containing Ionic Liquid Aliquat 336. *Hydrometallurgy* **2019**, *190*, 105149. [CrossRef]
22. Kholmogorov, A.G.; Kononova, O.N.; Kachin, S.V.; Ilyichev, S.N.; Kryuchkov, V.V.; Kalyakina, O.P.; Pashkov, G.L. Ion Exchange Recovery and Concentration of Rhenium from Salt Solutions. *Hydrometallurgy* **1999**, *51*, 19–35. [CrossRef]

23. Mozammel, M.; Sadrnezhaad, S.K.; Badami, E.; Ahmadi, E. Breakthrough Curves for Adsorption and Elution of Rhenium in a Column Ion Exchange System. *Hydrometallurgy* **2007**, *85*, 17–23. [CrossRef]
24. Nebeker, N.; Hiskey, J.B. Recovery of Rhenium from Copper Leach Solution by Ion Exchange. *Hydrometallurgy* **2012**, *125–126*, 64–68. [CrossRef]
25. Shu, Z.; Yang, M. Adsorption of Rhenium(VII) with Anion Exchange Resin D318. *Chin. J. Chem. Eng.* **2010**, *18*, 372–376. [CrossRef]
26. Virolainen, S.; Laatikainen, M.; Sainio, T. Ion Exchange Recovery of Rhenium from Industrially Relevant Sulfate Solutions: Single Column Separations and Modeling. *Hydrometallurgy* **2015**, *158*, 74–82. [CrossRef]
27. Xiong, C.; Yao, C.; Wu, X. Adsorption of Rhenium(VII) on 4-Amino-1,2,4-Triazole Resin. *Hydrometallurgy* **2008**, *90*, 221–226. [CrossRef]
28. Zagorodnyaya, A.N.; Abisheva, Z.S.; Sharipova, A.S.; Sadykanova, S.E.; Bochevskaya, Y.G.; Atanova, O.V. Sorption of Rhenium and Uranium by Strong Base Anion Exchange Resin from Solutions with Different Anion Compositions. *Hydrometallurgy* **2013**, *131–132*, 127–132. [CrossRef]
29. Zakhar'yan, S.V.; Gedgagov, E.I. Anion-Exchange Separation of Rhenium and Selenium in Schemes for Obtaining Ammonium Perrhenate. *Theor. Found. Chem. Eng.* **2013**, *47*, 637–643. [CrossRef]
30. Seo, S.; Choi, W.S.; Yang, T.J.; Kim, M.J.; Tran, T. Recovery of Rhenium and Molybdenum from a Roaster Fume Scrubbing Liquor by Adsorption Using Activated Carbon. *Hydrometallurgy* **2012**, *129–130*, 145–150. [CrossRef]
31. Vosough, M.; Shahtahmasebi, N.; Behdani, M. Recovery Rhenium from Roasted Dust through Super Para-Magnetic Nano-Particles. *Int. J. Refract. Met. Hard Mater.* **2016**, *60*, 125–130. [CrossRef]
32. Xiong, Y.; Lou, Z.; Yue, S.; Song, J.; Shan, W.; Han, G. Kinetics and Mechanism of Re(VII) Extraction with Mixtures of Tri-Alkylamine and Tri-n-Butylphosphate. *Hydrometallurgy* **2010**, *100*, 110–115. [CrossRef]
33. Sato, T.; Sato, K. Liquid-Liquid Extraction of Rhenium (VII) from Hydrochloric Acid Solutions by Neutral Organophosphorus Compounds and High Molecular Weight Amines. *Hydrometallurgy* **1990**, *25*, 281–291. [CrossRef]
34. Ali, M.C.; Suzuki, T.; Tachibana, Y.; Sasaki, Y.; Ikeda, Y. Selective Extraction of Perrhenate Anion in Nitric Acid Solution Using 2,2'-(Imino)Bis(N,N'-Dioctylacetamide) as an Extractant. *Sep. Purif. Technol.* **2012**, *92*, 77–82. [CrossRef]
35. Gerhardt, N.I.; Palant, A.A.; Dungan, S.R. Extraction of Tungsten (VI), Molybdenum (VI) and Rhenium (VII) by Diisododecylamine. *Hydrometallurgy* **2000**, *55*, 1–15. [CrossRef]
36. Melaven, A.D.; Bacon, J.A. Process for Recovering Rhenium 1947. U.S. Patent 2,414,965, 28 January 1947.
37. Zagorodnyaya, A.N.; Abisheva, Z.S. Rhenium Recovery from Ammonia Solutions. *Hydrometallurgy* **2002**, *65*, 69–76. [CrossRef]
38. Zagorodnyaya, A.N.; Abisheva, Z.S.; Agapova, L.Y.; Sharipova, A.S. Purification of Crude Ammonium Perrhenate from Potassium by Recrystallization, Sorption, and Membrane Electrodialysis. *Theor. Found. Chem. Eng.* **2019**, *53*, 841–847. [CrossRef]
39. Zagorodnyaya, A.N.; Abisheva, Z.S.; Sadykanova, S.E.; Sharipova, A.S.; Akcil, A. Purification of Ammonium Perrhenate Solutions from Potassium by Ion Exchange. *Miner. Process. Extr. Metall. Rev.* **2017**, *38*, 284–291. [CrossRef]
40. Leszczyńska-Sejda, K.; Majewski, T.; Benke, G.; Piętaszewski, J.; Anyszkiewicz, K.; Michałowski, J.; Chmielarz, A. Production of High-Purity Ammonium Perrhenate for W–Re–Ni–Fe Heavy Alloys. *J. Alloys Compd.* **2012**, *513*, 347–352. [CrossRef]
41. Zagorodnyaya, A.N.; Sharipova, A.S.; Linnik, K.A.; Abisheva, Z.S. Multi-Stage Recrystallization of Crude Ammonium Perrhenate. *Theor. Found. Chem. Eng.* **2018**, *52*, 717–724. [CrossRef]
42. Tang, J.; Feng, L.; Zhang, C.; Sun, Y.; Wang, L.; Zhou, Y.; Fang, D.; Liu, Y. The Influences of Stirring on the Recrystallization of Ammonium Perrhenate. *Appl. Sci.* **2020**, *10*, 656. [CrossRef]
43. Kruger, G.J.; Reynhardt, E.C. Ammonium Perrhenate at 295 and 135 K. *Acta Crystallogr. Sect. B* **1978**, *34*, 259–261. [CrossRef]
44. Brown, R.J.S.; Smeltzer, J.G.; Heyding, R.D. Nuclear Quadrupole Resonance and Thermal Expansion in Perrhenate Salts. *J. Magn. Reson.* **1976**, *24*, 269–274. [CrossRef]
45. Weir, R.D.; Staveley, L.A.K. The Heat Capacity and Thermodynamic Properties of Potassium Perrhenate and Ammonium Perrhenate from 8 to 304 K. *J. Chem. Phys.* **1980**, *73*, 1386–1392. [CrossRef]
46. Reynolds, E.M.; Yu, M.; Thorogood, G.J.; Brand, H.E.A.; Poineau, F.; Kennedy, B.J. Thermal Expansion of Ammonium Pertechnetate and Ammonium Perrhenate. *J. Solid State Chem.* **2019**, *274*, 64–68. [CrossRef]
47. Johnson, R.A.; Rogers, M.T. Anomalous Temperature Dependence of the NQR Frequency in NH_4ReO_4. *J. Magn. Reson.* **1974**, *15*, 584–589. [CrossRef]
48. Casas, J.M.; Sepúlveda, E.; Bravo, L.; Cifuentes, L. Crystallization of Sodium Perrhenate from $NaReO_4$–H_2O–C_2H_5OH Solutions at 298 K. *Hydrometallurgy* **2012**, *113–114*, 192–194. [CrossRef]
49. Lemanowicz, M.; Mielańczyk, A.; Walica, T.; Kotek, M.; Gierczycki, A. Application of Polymers as a Tool in Crystallization—A Review. *Polymers* **2021**, *13*, 2695. [CrossRef] [PubMed]
50. Leszczyńska-Sejda, K.; Benke, G.; Chmielarz, A.; Anyszkiewicz, K. Methods of Synthesis of Perrhenic Acid from Aqueous Solutions of Ammonium Perrhenate. *Hydrometallurgy* **2009**, *1*, 117–132.
51. Carl, L.Y. *The Yaws Handbook of Physical Properties for Hydrocarbons and Chemicals*; Lamar University: Beamont, TX, USA, 2012.
52. Struwe, F.; Pietsch, E. *Gmelins Handbuch der Anorganischen Chemie*, 8th ed.; Deutche Chemische Gesellschaft: Frankfurt am Main, Germany, 1972; ISBN 3-527-86901-8.
53. Ghosh, S.K. Decomposition of Ammonium Nitrite in Solution. *Z. Phys. Chem.* **1956**, *206*, 321–326. [CrossRef]
54. Marciniak, B. Density and ultrasonic velocity of undersaturated and supersaturated solutions of fluoranthene in trichloroethylene, and study of their metastable zone width. *J. Cryst. Growth.* **2012**, *236*, 347–356. [CrossRef]

55. Bogacz, W. Densimetric method for determination of potassium sulphate aqueous solutions saturation point Densymetryczna metoda wyznaczania punktu nasycenia wodnych roztworów siarczanu(VI) potasu. *Przem. Chem.* **2017**, *1*, 212–213. [CrossRef]
56. Bogacz, W.; Al-Rashed, M.H.; Lemanowicz, M.; Wójcik, J. A Simple Densimetric Method to Determine Saturation Temperature of Aqueous Potassium Chloride Solution. *J. Solut. Chem.* **2016**, *45*, 1071–1076. [CrossRef]

Disclaimer/Publisher's Note: The statements, opinions and data contained in all publications are solely those of the individual author(s) and contributor(s) and not of MDPI and/or the editor(s). MDPI and/or the editor(s) disclaim responsibility for any injury to people or property resulting from any ideas, methods, instructions or products referred to in the content.

Article

Studies on the Processing of Fine Dusts from the Electric Smelting of Ilmenite Concentrates to Obtain Titanium Dioxide

Almagul Ultarakova, Zaure Karshyga *, Nina Lokhova, Azamat Yessengaziyev, Kaisar Kassymzhanov and Arailym Mukangaliyeva

The Institute of Metallurgy and Ore Beneficiation, Satbayev University, Almaty 050013, Kazakhstan
* Correspondence: z.karshyga@satbayev.university; Tel.: +7-747-8822187

Abstract: This article presents studies on the ammonium fluoride processing of dusts from the reduction smelting of ilmenite concentrate with separation of silicon to obtain titanium dioxide. Optimal conditions for pyrohydrolysis of titanium fluorides were determined. The effects of temperature and duration on the process were studied. The optimal conditions for pyrohydrolysis of titanium fluorides were a temperature of 600 °C and duration of 240–300 min. The degree of titanium fluoride conversion to titanium oxide was 99.5% at these conditions. Titanium dioxide obtained by pyrohydrolysis of titanium fluorides was purified from iron, chromium, and manganese impurities. The effect of hydrochloric acid solution concentration, S:L ratio, and the process duration on the purification degree of titanium fluoride pyrohydrolysis was studied. The following optimum purification conditions were determined: hydrochloric acid solution concentration 12.5–15 wt%, temperature 25–30 °C, S:L = 1:6÷8, duration 20–30 min. The purified titanium dioxide consisted mainly of anatase. The pigmented titanium dioxide of rutile modification with 99.8 wt% TiO_2 was obtained after calcination at 900 °C for 120 min.

Keywords: ilmenite concentrate; dusts; waste; purification; sublimate; pyrohydrolysis; silicon; titanium dioxide

Citation: Ultarakova, A.; Karshyga, Z.; Lokhova, N.; Yessengaziyev, A.; Kassymzhanov, K.; Mukangaliyeva, A. Studies on the Processing of Fine Dusts from the Electric Smelting of Ilmenite Concentrates to Obtain Titanium Dioxide. *Materials* **2022**, *15*, 8314. https://doi.org/10.3390/ma15238314

Academic Editors: Zihang Liu and Guo Chen

Received: 13 October 2022
Accepted: 20 November 2022
Published: 23 November 2022

Publisher's Note: MDPI stays neutral with regard to jurisdictional claims in published maps and institutional affiliations.

Copyright: © 2022 by the authors. Licensee MDPI, Basel, Switzerland. This article is an open access article distributed under the terms and conditions of the Creative Commons Attribution (CC BY) license (https:// creativecommons.org/licenses/by/ 4.0/).

1. Introduction

Titanium and its compounds are important components of modern production. The properties of titanium, including high strength, good corrosion resistance, and s light weight, make it indispensable in such areas of application as aerospace, shipbuilding, chemical manufacture, military, medical care, etc. China, Japan, Russia, Kazakhstan, USA, and Ukraine are the largest producers of titanium sponge [1–4]. Ilmenite concentrate is used as raw material for titanium production during reduction electric smelting resulting in production of titanium slag and substandard pig iron. The charge for smelting is fed in a loose state, accompanied by a large dust entrainment. The dust cannot be returned to the smelting process or fed into the chlorinators due to its high silica content so it is stored together with other solid waste in specially designated landfill areas.

The silica content in the ore-thermal smelting dusts of ilmenite concentrates can be in the range of 5–20 wt%. They also contain iron oxides up to 30 wt%. Considerable amounts of titanium dioxide are lost together with the dust generated in the electric smelting process of ilmenite concentrate. Its content therein reaches 50 wt%, and its additional extraction from the dust waste will enable not only loss reduction, but also the ability to obtain additional commercial products.

There is no effective integrated processing of titanium production waste in the world. Available technologies for the production of titanium products are aimed at processing traditional mineral raw materials, the majority of which are ilmenite ores and concentrates.

Of all the mined titanium ore, 5% goes to the production of metallic titanium, and about 90–95% of the mined titanium raw materials are processed into pigment titanium dioxide TiO_2, which is in special demand and has a high degree of purity. The superiority of TiO_2 as a white pigment is mainly due to its high refractive index and consequently its light-scattering ability which provide excellent concealment and brightness. Titanium dioxide is non-toxic and harmless; it is used in solar cells [5], fuel cells, chemical current sources, protective and optical coatings, as a photocatalyst for water purification [6], in medicine as a material with antibacterial and antiviral effects [7], for air purification and disinfection in public places [8], for creation of self-cleaning surfaces [9], in photoelectrochemical water decomposition reactions to produce hydrogen [10,11], and in photocatalysts for destruction of toxic organic compounds [12–14]. Titanium dioxide nanomaterials doped with nitrogen and fluorine have a higher photocatalytic activity [15,16].

The existing industrial technologies for producing titanium dioxide are sulfuric acid and chlorine methods, which were introduced in the middle of the twentieth century in a number of countries. The titanium-containing raw material is treated with concentrated sulfuric acid in the sulfuric acid process, and a sulfate solution is obtained. Titanium dioxide is precipitated during its hydrolytic decomposition [17–21]. Under the chlorine technology, first, rutile is subjected to the action of chlorine gas, with the formation of titanium tetrachloride, which is then converted into a pigment at high temperature in a mixture of air and oxygen [22]. The main disadvantage of sulfuric acid technology is the formation of a large amount of waste solutions, the disposal of which is difficult. It is necessary to use only high-quality rutile in the chlorine technology, and its production requires additional preliminary preparation of titanium raw materials.

Recently, ammonium fluorine processing has become one of the promising methods for processing titanium-containing raw materials. The fundamentals of the ammonium fluoride method for processing titanium-containing raw materials were laid by Svendsen and described in his patents [23,24]. The methods consist of heating and fluorination of rutile or ilmenite in a mixture with ammonium fluoride or bifluoride, followed by sublimation of titanium fluoride compounds, their decomposition, and the production of titanium dioxide.

The reagent for fluorination, ammonium bifluoride, under normal conditions does not pose a significant environmental hazard, has high chemical activity, and a number of technologically favorable physical and chemical properties, such as a melting point of 126.2 °C and a boiling point of 238 °C, which are accompanied by decomposition into NH_3 and HF, and good solubility in water at 434 g/dm^3 [25].

The physicochemical basis of the fluorination process with ammonium bifluoride is that oxygen-containing compounds of transitional and many non-transitional elements, when interacting with NH_4HF_2, form ammonium fluoro- or oxofluorometallates, which are very convenient for processing [26], the physicochemical properties of which ensure the solubility of products and the possibility of separating mixtures by sublimation. The great advantage of these complex salts is that the selective tendency to sublimation or to thermal dissociation to non-volatile fluorides, which guarantees a deep separation of the components, and the stepwise elimination of NH_4F vapors makes it possible to collect the desublimate of the latter and use it in a closed cycle.

As noted above, silicon may concentrate in the dusts of electric smelting of ilmenite concentrates during their formation, and its presence does not allow them to return to the smelting process. Previous studies on ammonium fluoride treatment of cakes obtained from nitric acid leaching of titanium production sludge with ammonium bifluoride have shown the potential for separating silicon and beneficiating the residue with titanium [27,28].

In [29], the fluorination of titanium slag with ammonium hydrofluoride resulted in titanium dioxide with admixtures of other oxides left in the solid product after sublimation of silicon hexafluoride. Further separation of titanium from other components is carried out using leaching with a solution of ammonium hydrofluoride. Titanium is precipitated from the solution with ammonia water. The content of TiO_2 in the resulting product is more than 90 wt%. The disadvantage of this method is the need for repeated washing of the precipitate, consisting of $(NH_4)_2TiOF_5$ or $(NH_4)_3TiOF_5$, to shift the equilibrium to the formation of $Ti(OH)_4$.

In studies by [30,31], the processing of ilmenite concentrate includes fluorination of raw materials by sintering with a fluoride reagent, heat treatment of the fluorinated mass to separate fluorination products by sublimation, and pyrohydrolysis of the residue after sublimation to obtain iron oxide. In fluorination, ammonium fluoride, ammonium bifluoride, or a mixture thereof in an inert gas flow is used as a fluoride reagent, and the sublimation products are trapped with water to obtain an ammonium fluorotitanate solution. Titanium is also precipitated with aqueous ammonia. A coarse-grained precipitate of $(NH_4)_2TiO(OH)_3F$ is obtained, but TiO_2 is in an amorphous form after calcination up to 900 °C, and anatase (95 wt%) is obtained only after maintaining it at 900 °C for 5 h. Precipitation at a concentration of $(NH_4)_2TiF_6$, 300 g/dm^3 with ammonia water (25% NH_3) to pH = 9 results in the production of $(NH_4)_2TiO(OH)_4$; after filtering and calcining at 500 °C for 2 h, the content fluorine ion in TiO_2 is 0.3 wt%. This TiO_2, which is anatase stable at temperatures up to 950–1000 °C, is used as a catalyst in organic synthesis processes. To obtain the rutile form of TiO_2, it is necessary to subject the precipitate after the first filtration to repulpation and washing in ammonia water to pH = 11–12, which makes it possible to reduce the content of fluorine ion to 0.007 wt%. The process of rutilization of titanium dioxide is 99 wt% at 700 °C for 4 h, or at 800 °C for 2 h.

The proposed method for producing pigment titanium dioxide is multi-stage, energy intensive, and requires a significant consumption of ammonia.

There is a known method [32], according to which, metallic aluminum is added to a solution of fluorotitanic acid with a concentration of 250–500 g/dm^3 TiO_2 with stirring of the solution at the rate of 0.15–1.5% of aluminum relative to TiO_2 in solution. In this case, aluminum acts as a pigment-rutilizing additive. Then, high-temperature pyrolysis (burning) of the solution is performed at 500–700 °C in the tower nozzle pyrolysis furnace, running on natural gas. Gaseous products containing HF are captured and disposed, with production of concentrated hydrofluoric acid. The dry product of pyrolysis—titanium dioxide containing fluorine, is pulped in water at a solid to liquid ratio of 1:3 and stirred for 1.5–2.0 h at 40–50 °C to accelerate the removal of fluorine from titanium dioxide. Milk of lime is added until 6.5–7.07 pH is reached. As a result, fluorine is bound in the form of a water-insoluble CaF_2 that does not adversely affect paint coatings. The disadvantage of this method is the high total content of fluorine in the product.

The method of pyrolysis in two furnaces installed in series has been described in [33]. The process can be divided into two stages.

Stage I: the precipitate containing titanium salts is subject to the first hot hydrolysis at 340–400 °C for 1–3 h after removal of water by drying; typically, this stage is performed in a furnace under superheated steam and with continuous stirring. All bonds of fluorine and ammonia are broken under these conditions, with production of an intermediate product in the form of a powder consisting of TiO_2 (95–97 wt%) and $TiOF_2$.

Stage II: The first stage results in TiO_2 containing traces of $TiOF_2$ that is a grey and blue powder and, therefore, contaminates the final TiO_2 product that should be characterized with a high degree of whiteness, in contrast. This pollutant is removed by subsequent pyrohydrolysis at 700–900 °C. The second pyrohydrolysis is usually performed in a continuously stirred oven for 60–180 min. It is advisable to introduce air and superheated steam into the furnace to complete the reaction. The end product of the second oven consists of a powder with particles of variable sizes from about 0.1 μm to about 4 μm.

Providing for the specific fineness of the TiO_2 powder produced in this way, it can be placed on the market without further grinding.

There is a method [34] intended to obtain titanium dioxide with pyrohydrolysis of fluorammonium salts of titanium in the gas phase in the presence of water vapor. This method differs from others due to pyrohydrolysis performed by heating the reactor up to 450–500 °C at a water vapor temperature from 700 to 1200 °C, preferably 900–1000 °C, with the use of ammonium hexafluorotitanate; water vapors are obtained by burning hydrogen in oxygen in the burner, and an additional amount of water vapor, obtained by evaporation at boiling temperature, is added to their amount. However, mixtures of hydrogen and oxygen are explosive, so the method requires very strict adherence to technological regulations and safety precautions.

As the review of scientific, technical, and patent literature has shown, the use of the ammonium fluoride processing method makes it possible to successfully regenerate the used fluoride reagents. This indicates significant advantages over the sulphate method, where a large amount of dilute hydrolytic sulfuric acid is formed, contaminated with various impurities, which makes it difficult to return it back to the process. In addition, there is a danger of working with concentrated sulfuric acid during the decomposition of titanium-containing raw materials, which is accompanied by gas and reaction mass emissions. In the chlorine method, during the processing of ilmenite, at the stage of separation of titanium, silicon, aluminum, and iron chlorides, difficulties arise due to the proximity of their physicochemical properties. Additionally, from an environmental point of view, the use of chlorine in technology and the danger of phosgene formation during chlorination in the presence of carbon-containing reducing agents require strict adherence to technological regulations and safety measures. Use of the ammonium fluorine method will simplify the method for obtaining titanium dioxide from dusts of ore-thermal smelting by reducing the number of technological operations and the number of reagents with the possibility of their regeneration, improve the quality of the resulting titanium dioxide, and create the possibility of using a safer and more environmentally friendly method.

Production wastes have a complex polycomponent composition; their processing is a difficult task. The ammonium fluoride method makes it possible to separate the target components with high selectivity and obtain end products from them. Therefore, it is of interest to conduct research on ammonium fluoride processing of waste dust from the electric smelting of ilmenite concentrate, for which the process of obtaining titanium dioxide and its purification is an important stage. This work is aimed at studying titanium fluoride sublimes, the processes of their pyrohydrolysis, and purification of the resulting titanium dioxide.

2. Materials and Methods

Fine dust of electric smelting of ilmenite concentrate was provided by "Ust-Kamenogorsk Titanium and Magnesium Plant" JSC, with the following composition wt%: 20.57 Ti, 14.15 Fe, 12.85 Si, 0.47 Cr, and 3.42 Mn.

Analysis methods: X-ray diffraction analysis (XRD) was performed on diffractometer D8 ADVANCE "BRUKER AXS GmbH", (Germany, Karsruhe) radiation Cu-Kα, database PDF-2 International Center for Diffraction Data ICDD (Swarthmore, PA, USA).

X-ray fluorescence analysis was performed using an Axios PANalytical spectrometer with wave dispersion (The Netherlands, Almelo).

The chemical analysis of the samples was performed using an Optima 8300 DV inductively coupled plasma optical emission spectrometer (Waltham, MA, USA, Perkin Elmer Inc.).

Experimental procedure: the initial components were thoroughly mixed in the required ratio to perform the sublimation processes for silicon fluoride and titanium. The charge sample was placed in an alundum boat and loaded into an electric furnace. Fluorination was performed in a LOP LT-50/500–1200 tubular furnace. The argon feed rate was 1.0–1.5 dm^3/h.

A sample of titanium fluorides was placed in an alundum boat and loaded into an electric furnace during the pyrohydrolysis process. The steam feed rate was 1.5 to 2.0 dm^3/h.

Experiments on pyrohydrolysis of the titanium fluorides were performed in the laboratory set shown in Figure 1.

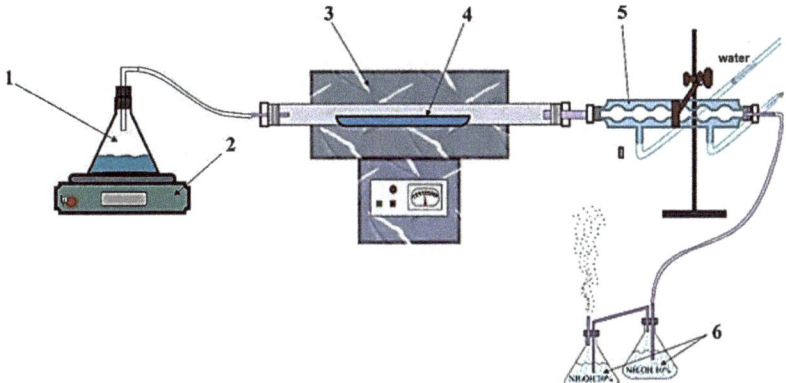

Figure 1. Laboratory setup for pyrohydrolysis. 1: flask with water; 2: electric stove; 3: oven LOIP LF/500–1200; 4: alundum boat; 5: refrigerator; 6: flasks with 10 wt% ammonia solution.

Titanium dioxide was purified with a hydrochloric acid solution in a thermostated reactor with a volume of 0.5 dm^3. The pulp was stirred with a glass stirrer. Stirrer rotation speed was 450 rpm.

The kinetic parameters of titanium fluoride pyrohydrolysis processes (activation energies (E_a) and reaction rate constants (K_e)) were calculated based on the chemical analysis data for elements in the reaction products. The degree of reaction product formation required for further calculations was determined with the following formula:

$$\alpha = m/m_{calc}, \qquad (1)$$

where m is the mass of the resulting product, and m_{calc} is its theoretically possible amount. Two methods were used for calculations: the generalized Erofeev topochemical equation and the Emanuel–Knorre method.

3. Results and Discussion
3.1. Sublimates of Titanium Fluorides

Silicon fluorides were separated in the form of sublimates at the first stage of processing of fine dust obtained from the electric smelting of ilmenite concentrates; a sample of dust was mixed with ammonium bifluoride in a 1:1 ratio. The process was performed at 260 °C for 6 h [35–37].

Titanium fluorides were produced by processing the residue after silicon sublimation from dust at 610 °C for 2 h. The composition of titanium fluorides is shown in Table 1.

Table 1. Content of controlled components in titanium fluorides, wt%.

Ti	Fe	Si	Cr	Al	F	O
58.99	0.56	0.005	0.095	0.140	20.377	19.76

The studied sample consists of two phases, according to XRD analysis (Figure 2). The main share is $(NH_4)_{0.8}TiOF_{2.8}$—73.3 rel. % and $(NH_4)_2TiF_6$—26.7 rel. %.

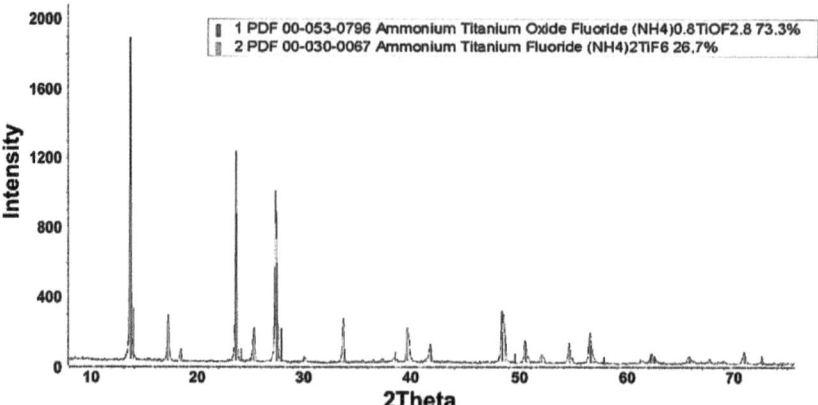

Figure 2. XRD pattern of the original sample of titanium fluorides.

In [38], the process of fluorination of natural ilmenite with ammonium bifluoride was studied. According to a thermal study of the process, the compound $(NH_4)_3Ti(OH)_{0.4}F_{6.6}$ was initially formed, which, with increasing temperature, further decomposed stepwise, with the elimination of NH_3 and HF molecules:

$$(NH_4)_3Ti(OH)_{0.4}F_{6.6} \rightarrow (NH_4)_2Ti(OH)_{0.4}F_{5.6} + NH_3 + HF \qquad (2)$$

$$(NH_4)_2Ti(OH)_{0.4}F_{5.6} \rightarrow NH_4TiO_{0.4}F_{4.2} \uparrow + NH_3 + 1.4HF \qquad (3)$$

$$NH_4TiO_{0.4}F_{4.2} \rightarrow NH_4TiO_{0.4}F_{4.2} \uparrow \qquad (4)$$

According to [38], volatile compounds of titanium TiF_4 and ammonium oxofluorotitanate (reaction 4) with the general formula $NH_4TiO_xF_{5-2x}$, were formed, which incongruently sublimated, as the authors assumed, with the formation of an adduct of titanium with ammonia.

In the sublimates obtained in this work (Figure 2), according to XRD, the main component is ammonium oxyfluorotitanate, with the composition $(NH_4)_{0.8}TiOF_{2.8}$, and $(NH_4)_2TiF_6$ comprises a little more than a quarter. The composition of sublimates somewhat differs from the results of thermal analysis in [38]. The dusts of the electric smelting of the ilmenite concentrate differ from those of the ilmenite concentrate itself; moreover, the residues of fluorination and sublimation of silicon fluorides are used to sublimate titanium fluoride compounds. Dust contains titanium-containing phases: iron titanium oxides (Fe_2TiO_5 and $Fe_{1.5}Ti_{0.5}O_3$), magnesium titanium oxide ($MgTi_2O_5$), and titanium oxide (TiO_2) [35]. The desiliconized residues are mostly represented by such titanium-containing phases as $(NH_4)_2TiF_6$, Ti_6O_{11}; there is also a $(NH_4)_{0.8}TiOF_{2.8}$ phase [37]. Taking into account the results of XRD analysis of dusts and desiliconized residues of their fluorination, the process of titanium fluoride formation can be represented for the main phases of iron titanium oxides (Fe_2TiO_5 and $Fe_{1.5}Ti_{0.5}O_3$), and magnesium titanium oxide ($MgTi_2O_5$), in dusts, by the following reactions:

$$Fe_2TiO_5 + 9NH_4HF_2 \rightarrow (NH_4)_2TiF_6 + 2(NH_4)_3FeF_6 + NH_3 + 5H_2O \qquad (5)$$

$$Fe_{1.5}Ti_{0.5}O_3 + 5NH_4HF_2 \rightarrow 0.5(NH_4)_2TiF_6 + (NH_4)_3FeF_6 + 0.5FeF_2 + NH_3 + 3H_2O \qquad (6)$$

$$MgTi_2O_5 + 7NH_4HF_2 \rightarrow 2(NH_4)_2TiF_6 + MgF_2 + 3NH_3 + 5H_2O \qquad (7)$$

$$Fe_2TiO_5 + 7.4NH_4HF_2 \rightarrow (NH_4)_{0.8}TiOF_{2.8} + 2(NH_4)_3FeF_6 + 0.6NH_3 + 4H_2O \qquad (8)$$

$$Fe_{1.5}Ti_{0.5}O_3 + 4.2NH_4HF_2 \rightarrow 0.5(NH_4)_{0.8}TiOF_{2.8} + (NH_4)_3FeF_6 + 0.5FeF_2 + \\ +0.8NH_3 + 2.5H_2O \quad (9)$$

$$MgTi_2O_5 + 3.8\,NH_4HF_2 \rightarrow 2(NH_4)_{0.8}TiOF_{2.8} + MgF_2 + 2.2\,NH_3 + 3H_2O \quad (10)$$

Trivalent iron predominates in the dust, and ferrous iron is also present in the $Fe_{1.5}Ti_{0.5}O_3$ phase. According to the data of [38], trivalent iron present in natural ilmenite was completely reduced to divalent iron, with the release of gaseous nitrogen. However, in accordance with the results of the studies [39], the samples contained an iron-containing phase $(NH_4)FeF_6$ with ferric iron; at the same time, the studies were carried out in an argon atmosphere. Based on the foregoing, in reactions (6) and (8) iron is present in the form of two phases: $(NH_4)FeF_6$ and FeF_2.

3.2. Determination of the Optimal Conditions for the Pyrohydrolysis of Titanium Fluorides

In order to obtain titanium dioxide, the sublimates obtained after fluorination and sublimation of the desiliconized residue from the ammonium fluoride treatment of dusts were subjected to pyrohydrolysis.

The pyrohydrolysis of the titanium fluorides is described by the reactions [40,41]:

$$NH_4TiOF_3 + H_2O = TiO_2 + NH_3 + 3HF \quad (11)$$

$$(NH_4)_2TiF_6 + 2H_2O = TiO_2 + 2NH_3 + 6HF \quad (12)$$

According to [38], the pyrohydrolysis of ammonium fluorotitanates with the formation of titanium dioxide can proceed sequentially, forming intermediate compounds, according to the scheme:

$$(NH_4)_2Ti(OH)_xF_{6-x} \rightarrow NH_4TiOF_3 \rightarrow (NH_4)_{0.8}TiOF_{2.8} \rightarrow (NH_4)_{0.3}TiOF_{2.3} \rightarrow TiO_2 \quad (13)$$

The effect of various parameters on the process of pyrohydrolysis has been studied. At the first stage, *the influence of the temperature of the pyrohydrolysis process* was studied. The study of the temperature effect on the degree of pyrohydrolysis of the titanium fluorides was performed in the range of 300–700 °C; the duration of the experiments was 300 min. The research results are presented in Figures 3 and 4.

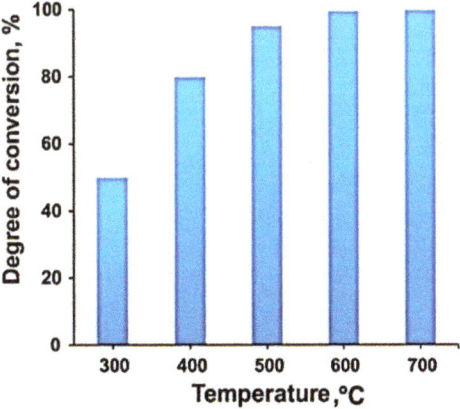

Figure 3. Dependence of the conversion degree of titanium fluorides into titanium dioxide on the pyrohydrolysis temperature.

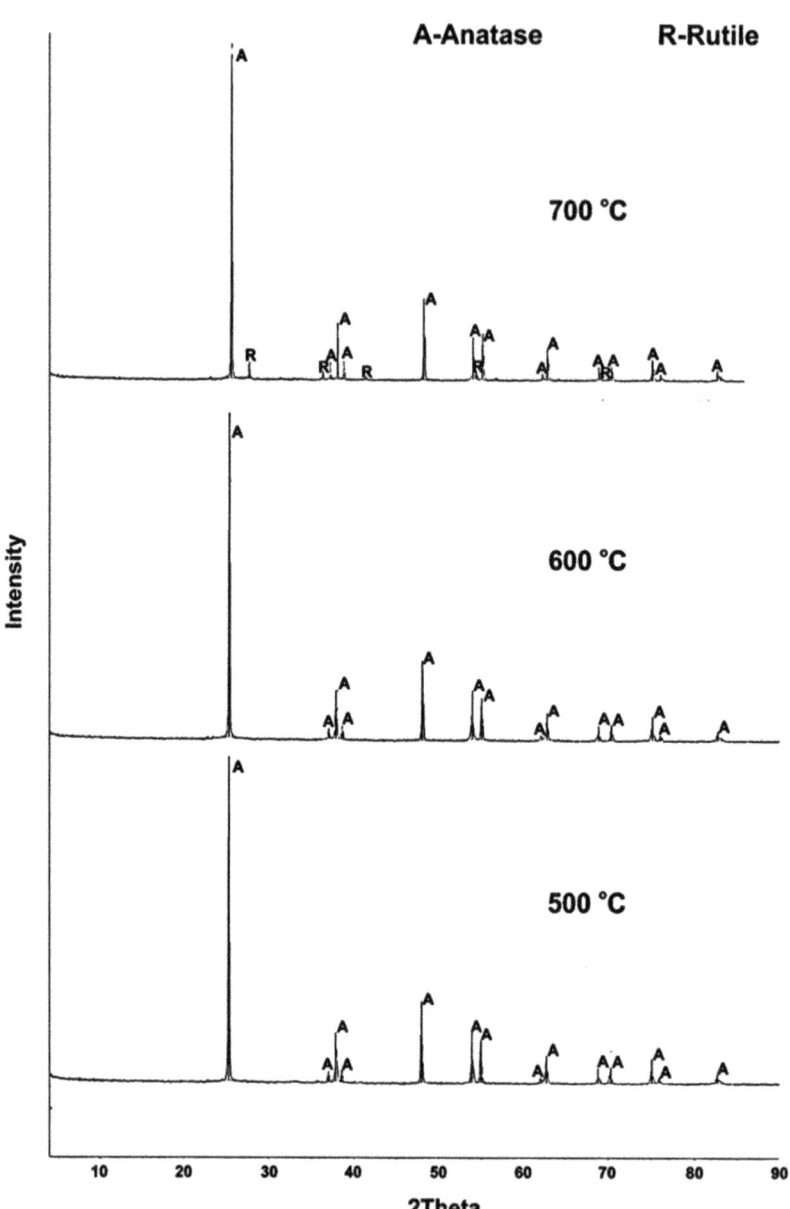

Figure 4. XRD patterns of titanium fluoride pyrolysis products (duration 300 min).

The data shown in Figure 3 indicate that the conversion degree of titanium fluorides into oxide during the process at 600 °C reached 99.5%, and a further increase in temperature to 700 °C did not have a significant effect since the pyrohydrolysis process was almost completely finished at 600 °C.

However, the structural modification of the resulting titanium dioxide was of interest. A series of experiments was conducted in the range of 400–700 °C in order to study the effect of the titanium fluoride pyrolysis temperature on titanium dioxide modification. The

XRD analysis results show (Figure 4) that the pyrolysis of the initial product at 500–600 °C for 300 min resulted in the formation of an anatase monophase.

An increase of up to 700 °C in the process temperature resulted in the onset of the anatase → rutile transition. The proportions of anatase and rutile were 94.8 and 5.2 rel. %, respectively.

The use of pyrohydrolysis of fluoroammonium compounds for the production of titanium dioxide predetermines the presence of fluorine ions in the product, which, during the condensation of titanium oxide compounds in the temperature range of 400–700 °C and higher, apparently contributes to the formation of an anatase structure [31,42]. Depending on the final consumer destination, it is possible to subject the product of pyrohydrolysis to further processing and obtain titanium dioxide of the required modification with the other corresponding characteristics.

Thus, studies have shown that titanium fluorides are almost completely converted into titanium dioxide at 600 °C.

At the next stage, the influence of the duration of the titanium fluoride pyrohydrolysis process was studied.

A series of experiments was conducted with duration of 60–300 min at 600 °C, with a steam supply rate of 1.5 dm^3/g. The results of the experiments are shown in Figure 5.

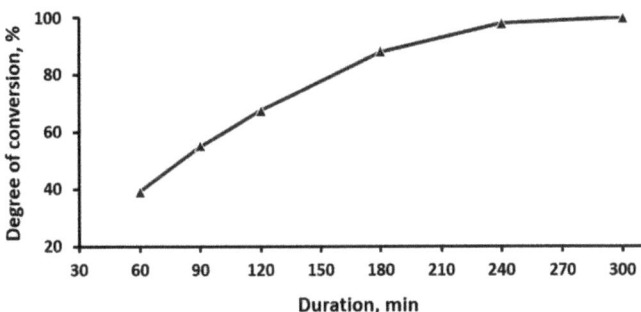

Figure 5. Dependence of the conversion degree of titanium fluorides into titanium dioxide on the process duration.

The interval 60–240 min in Figure 5 reflects a linear dependence of the degree of conversion of titanium fluorides into oxides on the duration of pyrohydrolysis. When the process was held for 240–300 min, the pyrohydrolysis was almost totally completed, and the degree of conversion of titanium fluorides into oxides reached their maximum values. The fluoride conversion rate decreased after 240 min of the process. The composition of the pyrohydrolysis product obtained at 240 min of the process is presented in Table 2; the fluorine content in the resulting product was 0.447 wt%.

Table 2. The content of the main components in the pyrohydrolysis product of titanium fluorides (600 °C, 5 g), wt%.

TiO$_2$	Fe$_2$O$_3$	Cr$_2$O$_3$	Al$_2$O$_3$	SiO$_2$	F
98.5	0.807	0.183	0.266	0.012	0.447

Table 2 shows that, in terms of the mass content of the main component, the resulting titanium dioxide could meet the requirements for obtaining pigment grades [43]; however, the content of impurities in the product was high and required additional processing.

The rate constants and activation energies were calculated for the pyrohydrolysis process, in accordance with reactions (11) and (12), which are presented in Table 3.

Table 3. The values of the rate constants and activation energies of the processes described by reactions (11) and (12).

	Reaction (11)		Reaction (12)	
K_e, s^{-1}		E_a, kJ/mol	K_e, s^{-1}	E_a, kJ/mol
16.1×10^{-5}		14.7	12.9×10^{-5}	15.3

The pyrohydrolysis of titanium fluorides proceeded in the outer diffusion region, i.e., the rate of the process was limited by the insufficiently acceptable access of water vapor to titanium fluoride compounds, which must be taken into account in the further carrying out of the process to ensure sufficient mixing of the reactants.

Thus, the optimal conditions for the pyrohydrolysis of titanium fluorides are a process temperature of 600 °C and a duration of 240–300 min.

3.3. Purification of Titanium Dioxide from Impurities and Obtaining Rutile

Pigmented titanium dioxide is used mainly in the rutile form. Titanium dioxide produced as a result of pyrohydrolysis of titanium fluorides had a grayish tint and needed to be purified from such impurities as iron, chromium, and manganese. The oxides of these metals dissolve well in hydrochloric acid solutions; therefore, studies were carried out to purify titanium dioxide obtained as a result of pyrohydrolysis from accompanying impurities. To ensure the most complete removal of impurities from the composition of the product, the influence of the conditions for carrying out hydrochloric acid treatment was studied.

At the first stage, *the effect of hydrochloric acid concentration* was studied. Studies were carried out with hydrochloric acid solutions with concentrations of 5, 7.5, 10, and 15 wt% while maintaining the following constant conditions: temperature 30 °C, ratio S:L = 1:8, and duration 30 min. The results of studies of this series of experiments are presented in Figure 6.

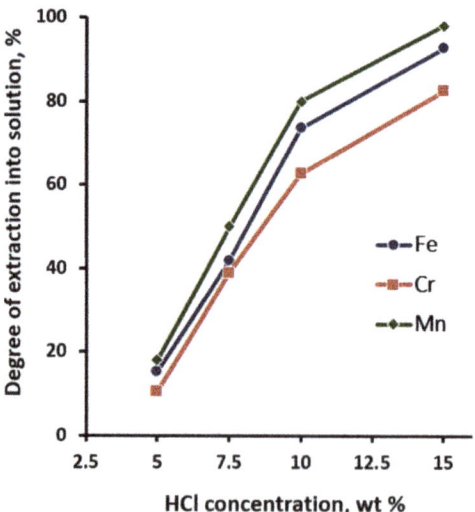

Figure 6. Dependence of the purification degree of titanium dioxide from iron, chromium, and manganese on the concentration of hydrochloric acid in solution.

Figure 6 shows that an increase in the concentration of hydrochloric acid in the solution from 5 to 10 wt% resulted in a sharp increase in the degree of leaching of controlled impurity components into the solution of manganese from 18.0% to 80.1%, iron from 15.2% to 73.6%, chromium from 10.5% to 62.7%. A further increase in the concentration of hydrochloric acid to 15% made it possible to completely purify titanium dioxide from manganese and significantly reduce the content of iron and chromium. The optimal concentration of hydrochloric acid for the purification of titanium dioxide from impurities should be considered the range of 12.5–15 wt%.

At the next stage, *the influence of the S:L ratio on the process of impurities leaching* was studied. The effect of the ratio of the pyrohydrolysis product to the solution of 12.5 wt% hydrochloric acid at 30 °C, duration 30 min was studied. Figure 7 shows that with an increase in the ratio of S:L, there was a slight increase in the recovery of the studied impurity components into the solution, and at ratios S:L = 1:6÷8, the degree of purification of iron, manganese, and chromium reached acceptable values and amounted to ~80, ~87, and ~70%, respectively. The optimal ratios of S:L should be considered to be from 1:6 to 1:8.

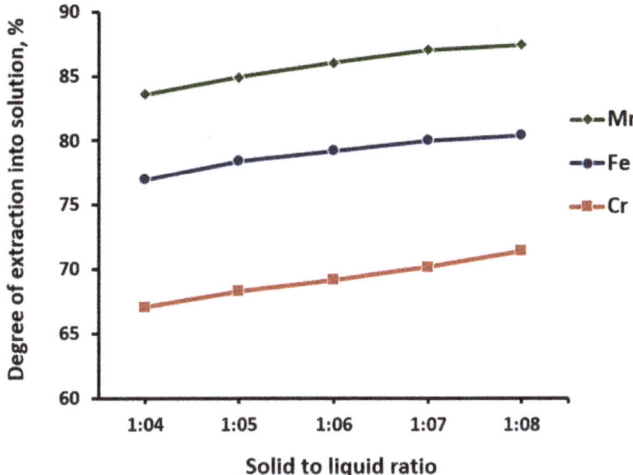

Figure 7. Dependence of the degree of extraction of manganese, iron, and chromium into the solution on the solid to liquid ratio.

The next stage in studying of the effect of the duration of leaching of manganese, iron, and chromium from the pyrohydrolysis product of titanium fluorides was performed within 5–60 min at 30 °C, S:L ratio = 1:8. The results are shown in Figure 8.

Figure 8 shows that the leaching of manganese, iron, and chromium had already reached 96.1, 91.0, and 81.2%, respectively, at the initial stages of the process, and the degree of leaching reached 98.0, 93.1, and 83.1% after 30 min. The increase in the acid leaching duration did not have a significant effect. The optimal duration of the titanium dioxide purification process was determined to be 20–30 min.

Titanium dioxide, purified after hydrochloric acid treatment under selected optimal conditions, was studied using physical and chemical methods of analysis. According to XRD results, titanium dioxide consisted of two modifications—anatase 94.0 and rutile 6.0% (Figure 9).

Titanium dioxide with an anatase structure has a wide range of applications. However, in order to obtain a rutile modification, purified titanium dioxide must be subjected to an appropriate treatment. Titanium dioxide that consists of two modifications has low pigment qualities, according to [2]. High quality is provided by rutile. In this regard, a sample of titanium dioxide with anatase base was calcined at a temperature of 900 °C during 120 min. The calcined product was studied using XRD analysis, the results of which showed that titanium dioxide was obtained with rutile modification (Figure 10).

The composition of titanium dioxide produced under optimal conditions and calcined at 900 °C is shown in Table 4.

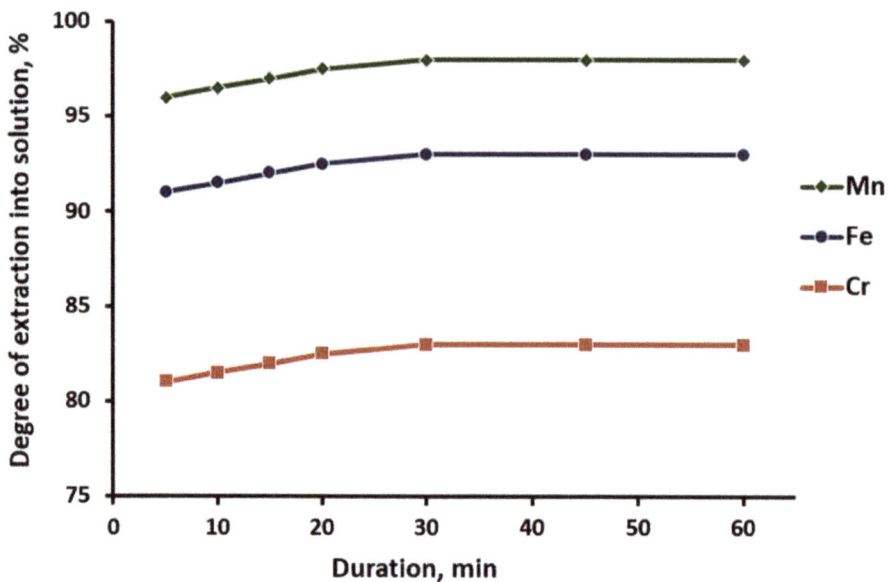

Figure 8. Effect of the duration of the leaching process of manganese, iron, and chromium on hydrochloric acid solution.

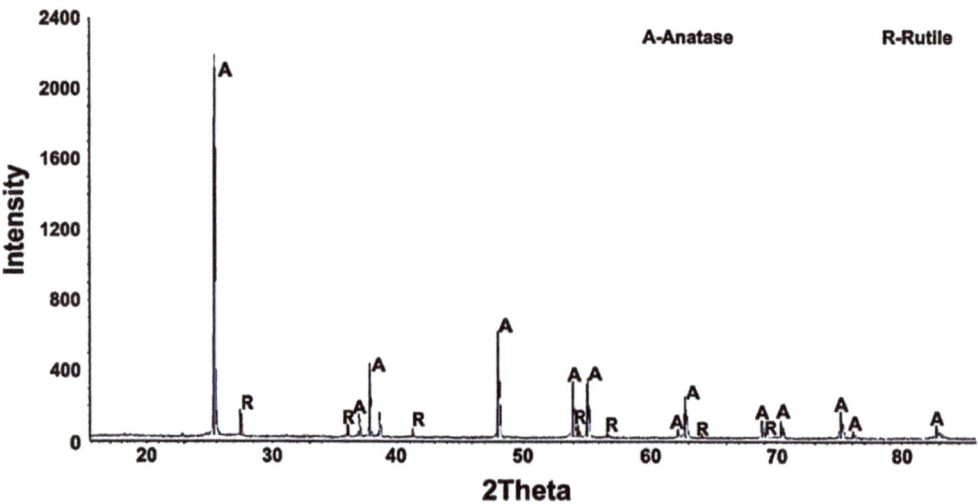

Figure 9. XRD pattern of titanium dioxide after treatment with hydrochloric acid.

Table 4. The content of the main components in pigmentary titanium dioxide, wt%.

TiO_2	SiO_2	Cr_2O_3	MnO_2	Fe_2O_3	F
99.8	0.0005	0.032	0.005	0.039	n/d

Thus, the optimal conditions for purification of the pyrohydrolysis product of titanium fluoride and production of titanium dioxide pigment include a concentration of hydrochloric acid solution of 12.5–15 wt%, a temperature range of 25–30 °C, a solid to liquid ratio = 1:6÷8, and a duration of 20–30 min. Calcination should be performed at 900 °C for 120 min.

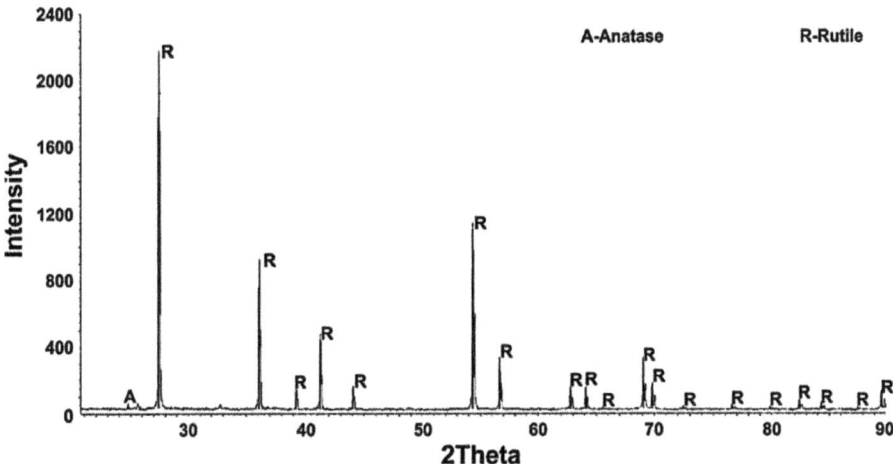

Figure 10. XRD pattern of rutile (900 °C, 120 min calcination time).

4. Conclusions

Fine dusts of ore-thermal smelting of ilmenite concentrate, containing up to 50 wt% TiO_2 in their composition, cannot be returned back to the electric smelting process due to their high silica content (up to 20 wt%) and are a waste of titanium production.

The ammonium fluoride processing method allows for the selective separation of a valuable component from others present in the raw material.

The possibility to produce titanium dioxide from electrosmelting dusts of ilmenite concentrate was studied; while silicon was previously removed from the dusts by the ammonium fluoride method, sublimates of titanium fluorides $(NH_4)_{0.8}TiOF_{2.8}$ and $(NH_4)_2TiF_6$ were obtained. Titanium dioxide with the anatase structure was obtained by pyrohydrolysis of sublimates.

After purification of the product of pyrohydrolysis of titanium fluorides from impurities of iron, chromium, and manganese using hydrochloric acid solutions, pigmentary titanium dioxide of rutile modification with a content of 99.8 wt% TiO_2 was obtained.

Author Contributions: Conceptualization, A.U., Z.K. and N.L.; methodology, N.L. and K.K.; software, A.Y. and A.M.; validation, A.U., Z.K. and N.L.; formal analysis, A.U., Z.K. and N.L.; investigation, A.U., N.L., A.Y., K.K. and A.M.; resources, A.Y., K.K. and A.M.; data curation, Z.K., A.U., N.L. and K.K.; writing—original draft preparation, A.U. and N.L.; writing—review and editing, Z.K. and A.Y.; supervision, Z.K.; project administration, Z.K. and A.U.; funding acquisition, Z.K. and A.U. All authors have read and agreed to the published version of the manuscript.

Funding: This research is funded by the Science Committee of the Ministry of Science and Higher Education of the Republic of Kazakhstan, Grant Project No. AP08855505 and Grant Project No. AP09258788.

Institutional Review Board Statement: Not applicable.

Informed Consent Statement: Not applicable.

Data Availability Statement: The data and results presented in this study are available in the article.

Conflicts of Interest: The authors declare that there are no conflicts of interest regarding the publication of this manuscript.

Abbreviations

IMOB JSC	Institute of Metallurgy and Ore Beneficiation Joint Stock Company
UKTMP JSC	Ust-Kamenogorsk Titanium and Magnesium Plant Joint-Stock Company
XRD	X-ray phase analysis
Solid to liquid ratio (S:L)	the ratio of the weight of the solid phase (in grams) to the volume of the liquid phase (in mL)

References

1. World Titanium Market: Trends and Prospects. Available online: http://www.ereport.ru/articles/commod/titanium.htm (accessed on 12 May 2022).
2. Qiongsha, L.; Phil, B.; Hanyue, Z. Titanium sponge production technology in China. In Proceedings of the 13th World Conference on Titanium, San Diego, CA, USA, 16–20 August 2015. [CrossRef]
3. Feng, G.; Zuoren, N.; Danpin, Y.; Boxue, S.; Yu, L.; Xianzheng, G.; Zhihong, W. Environmental impacts analysis of titanium sponge production using Kroll process in China. *J. Clean. Prod.* **2018**, *174*, 771–779. [CrossRef]
4. Chervony, I.F.; Listopad, D.A.; Ivashchenko, V.I.; Sorokina, L.V. On the physical and chemical laws of the formation of titanium sponge. Collection of research papers Donetsk National Technical University. *Metallurgy* **2008**, *10*, 37–46.
5. O'Regan, B.; Grätzel, M. A low-cost, high-efficiency solar cell based on dye-sensitized colloidal TiO_2 films. *Nature* **1991**, *353*, 737–740. [CrossRef]
6. Janus, M.; Choina, J.; Morawski, A.W. Azo dyes decomposition on new nitrogen-modified anatase TiO_2 with high adsorptivity. *J. Hazard. Mater.* **2009**, *166*, 1–5. [CrossRef]
7. Heidenau, F.; Mittelmeier, W.; Detsch, R.; Haenle, M.; Stenzel, F.; Ziegler, G.; Gollwitzer, H. A novel antibacterial titania coating: Metal ion toxicity and in vitro surface colonization. *J. Mater. Sci. Mater. Med.* **2005**, *16*, 883–888. [CrossRef] [PubMed]
8. MiarAlipour, S.; Friedmann, D.; Scott, J.; Amal, R. TiO_2/porous adsorbents: Recent advances and novel applications. *J. Hazard. Mater.* **2017**, *341*, 404–423. [CrossRef] [PubMed]
9. Wang, R.; Hashimoto, K.; Fujishima, A.; Chikuni, M.; Kojima, E.; Kitamura, A.; Shimohigoshi, M.; Watanabe, T. Light-induced amphiphilic surfaces. *Nature* **1997**, *388*, 431–432. [CrossRef]
10. Fujishima, A.; Honda, K. Electrochemical Photolysis of Water at a Semiconductor Electrode. *Nature* **1972**, *238*, 37–38. [CrossRef] [PubMed]
11. Ngo, T.Q.; Posadas, A.; Seo, H.; Hoang, S.; McDaniel, M.D.; Utess, D.; Triyoso, D.H.; Mullins, C.B.; Demkov, A.A.; Ekerdt, J.G. Atomic layer deposition of photoactive $CoO/SrTiO_3$ and CoO/TiO_2 on Si (001) for visible light driven photoelectrochemical water oxidation. *J. Appl. Phys.* **2013**, *114*, 084901. [CrossRef]
12. Frank, S.N.; Bard, A.J. Heterogeneous photocatalytic oxidation of cyanide ion in aqueous solutions at titanium dioxide powder. *J. Am. Chem. Soc.* **1977**, *99*, 303–304. [CrossRef]
13. Korina, E.; Stoilova, O.; Manolova, N.; Rashkov, I. Polymer fibers with magnetic core decorated with titanium dioxide prospective for photocatalytic water treatment. *J. Environ. Chem. Eng.* **2018**, *6*, 2075–2084. [CrossRef]
14. Sraw, A.; Kaur, T.; Pandey, Y.; Sobti, A.; Wanchoo, R.K.; Toor, A.P. Fixed bed recirculation type photocatalytic reactor with TiO_2 immobilized clay beads for the degradation of pesticide polluted water. *J. Environ. Chem. Eng.* **2018**, *6*, 7035–7043. [CrossRef]
15. Livraghi, S.; Elghniji, K.; Czoska, A.M.; Paganini, M.C.; Gia-mello, E.; Ksibi, M. Nitrogen-doped and nitrogen-fluorine-codoped titanium dioxide. Nature and concentration of photoactive species and yheir role in determining the photocatalytic activity under visible light. *J. Photochem. Photobiol. A Chem.* **2009**, *205*, 93–97. [CrossRef]
16. Wu, G.; Wen, J.; Nigro, S.; Chen, A. One-step synthesis of N- and F-codoped mesoporous TiO_2 photocatalyst with high visible light activity. *Nanotechnology* **2010**, *21*, 085701. [CrossRef]
17. Weintraub, G. Process of Obtaining Titanic Oxid. U.S. Patent 1014793A, 16 January 1912.
18. Blumenfeld, J. Titanium Compound. U.S. Patent 1504669A, 12 August 1924.
19. Weizmann, C.; Blumenfeld, J. Improvements Relating to the Treatment of Solutions for the Separation of Suspended Matter. UK Patent 228814A, 3 February 1925.
20. Werner, M. Production of Titanium Dioxide. U.S. Patent 1758528A, 13 May 1930.
21. Belenkiy, E.F.; Riskin, I.V. *Chemistry and Technology of Pigments*; Goshimizdat: Leningrad, Russia, 1960; p. 756.
22. Jelks, B. *Titanium: Its Occurrence, Chemistry and Technology*, 2nd ed.; Ronald Press: New York, NY, USA, 1966; p. 691.
23. Svendsen, S.S. Manufacture of Titanium Compounds. U.S. Patent 2042434, 6 June 1931.
24. Svendsen, S.S. Treatment of Titanium-Bearing Materials. U.S. Patent 2042435, 27 September 1934.
25. Medkov, M.A.; Krysenko, G.F.; Epov, D.G. Ammonium hydrodifluoride is a promising reagent for the complex processing of raw materials. *Bull. Far East. Branch Russ. Acad. Sci.* **2011**, *5*, 60–65.
26. Rakov, E.G. *Ammonium Fluorides*; Itogi Nauki i Tekhniki; Ser. Inorganic Chemistry; All-Russian Institute of Scientific and Technical Information: Moscow, Russia, 1988; Volume 15, 154p.
27. Ultarakova, A.A.; Yessengaziyev, A.M.; Kuldeyev, E.I.; Kassymzhanov, K.K.; Uldakhanov, O.K. Processing of titanium production sludge with the extraction of titanium dioxide. *Metalurgija* **2021**, *3–4*, 411–414.

28. Yessengaziyev, A.M.; Ultarakova, A.A.; Burns, P.C. Fluoroammonium method for processing of cake from leaching of titanium-magnesium production sludge. *Complex Use Miner. Resour.* **2022**, *320*, 67–74. [CrossRef]
29. Dmitriev, A.N.; Smorokov, A.A.; Kantaev, A.S. Fluoroammonium method of titanium slag processing. Proceedings of higher educational institutions. *Ferr. Metall.* **2021**, *3*, 178–183. [CrossRef]
30. Andreev, A.A.; Dyachenko, A.N. Method for Processing Titanium-Containing Raw Materials. Patent RF 2365647, 27 August 2009.
31. Andreev, A.A. *Development of Fluoride Technology for the Production of Pigment Titanium Dioxide from Ilmenite*; Abstract of the Dissertation of the Candidate of Technical Sciences; Tomsk Polytechnic University (TPU): Tomsk, Russia, 2008; p. 22.
32. Gordienko, P.S. Method for Producing Titanium Dioxide Using an Aqueous Solution of Fluoride. RF Patent RU 2007132063/15, 20 June 2010.
33. Gerasimova, L.G.; Nikolaeva, A.I.; Sklokin, L.I.; Shestakov, S.V.; Polyakov, E.G.; Zots, N.V. Method for Processing Fluorotitanium-Containing Solutions after Opening Loparite and other Titanium-Containing Concentrates to Obtain Titanium Dioxide. RF Patent RU 2175989 C1, 20 November 2001.
34. Alekseiko, L.N.; Goncharuk, V.K.; Maslennikova, I.G. Method for Producing Titanium Dioxide. RF Patent RU 2539582 C1, 20 January 2015.
35. Ultarakova, A.A.; Karshigina, Z.B.; Lokhova, N.G.; Yessengaziyev, A.M.; Kassymzhanov, K.K.; Tolegenova, S.S. Extraction of amorphous silica from waste dust of electrowinning of ilmenite concentrate. *Metalurgija* **2022**, *61*, 377–380.
36. Ultarakova, A.A.; Karshyga, Z.B.; Lokhova, N.G.; Naimanbaev, M.A.; Yessengaziyev, A.M.; Burns, P. Methods of silica removal from pyrometallurgical processing wastes of ilmenite concentrate. *Kompleksnoe Ispol'zovanie Mineral'nogo syr'â = Complex Use Miner. Resour.* **2022**, *322*, 79–88. [CrossRef]
37. Karshyga, Z.; Ultarakova, A.; Lokhova, N.; Yessengaziyev, A.; Kassymzhanov, K. Processing of Titanium-Magnesium Production Waste. *J. Ecol. Eng.* **2022**, *23*, 215–225. [CrossRef]
38. Laptash, N.M.; Maslennikova, I.G. Fluoride Processing of Titanium-Containing Minerals. *Adv. Mater. Phys. Chem.* **2012**, *2*, 21–24. [CrossRef]
39. Karshyga, Z.B.; Ultarakova, A.A.; Lokhova, N.G.; Yessengaziyev, A.M.; Kuldeyev, E.I.; Kassymzhanov, K.K. Study of fluoroammonium processing of reduction smelting dusts from ilmenite concentrate. *Metalurgija* **2023**, *62*, 145–148.
40. Rakov, E.G.; Teslenko, V.V. *Pyrohydrolysis of Inorganic Fluorides*; Energoatomizdat: Moscow, Russia, 1987; 152p.
41. Watanabe, A.; Nishimura, Y.; Watanabe, N. Obtaining Method of Titanium Oxide from Compound Containing Titanium and Fluorine. Patent JPS57183325A, 11 November 1982.
42. Chen, D.M.; Jiang, Z.Y.; Geng, J.Q.; Zhu, J.H.; Yang, D. A facile method to synthesize nitrogen and fluorine co-doped TiO2 nanoparticles by pyrolysis of $(NH_4)_2TiF_6$. *J. Nanopart. Res.* **2009**, *11*, 303–313. [CrossRef]
43. *Interstate Standard GOST 9808-84*; Pigmentary Titanium Dioxide. Specifications. Decree of the State Standard of the USSR of 19 December 1984, No 4693. Interstate Standard: Moscow, Russia, 1984.

Article

Hot Deformation Behavior of TA1 Prepared by Electron Beam Cold Hearth Melting with a Single Pass

Zhibo Zhang [1,2,†], Weiwei Huang [1,†], Weidong Zhao [1,2], Xiaoyuan Sun [2], Haohang Ji [1], Shubiao Yin [1], Jin Chen [1,*] and Lei Gao [1,*]

1. Faculty of Metallurgical and Energy Engineering, Kunming University of Science and Technology, Kunming 650093, China
2. Science and Technology Innovation Department of Kunming Iron & Steel Co., Ltd., Kunming 650302, China
* Correspondence: jinchen@kust.edu.cn (J.C.); leigao@kust.edu.cn (L.G.)
† These authors contributed equally to this work.

Abstract: The Gleeble-3800 thermal simulator was used for hot compression simulation to understand the hot deformation performance of TA1 prepared by the single-pass electron beam cold hearth (EB) process. The deformation degree is 50% on a thermal simulator when the temperature range is 700–900 °C, with a strain rate of 0.01–10^{-1} s. According to the thermal deformation data, the true stress-strain curve of TA1 was studied. Meanwhile, the constitutive model and processing map were established through the experimental data. These results indicate that the deformation temperature negatively affects strain rate and flow stress. The heat deformation activation energy of EB produced TA1 sample was lower than that of VAR produced TA1 sample in the studied range. The best processing areas of EB-produced TA1 were strain rates of 0.05–0.01 s^{-1}, within 700–770 °C; or strain rates of 0.01–0.15 s^{-1}; 840–900 °C. The results of this paper enrich the fundamental knowledge of the thermal deformation behavior of TA1 prepared by EB furnaces.

Keywords: electron beam cold hearth melting; hot deformation; TA1; constitutive model; processing maps

Citation: Zhang, Z.; Huang, W.; Zhao, W.; Sun, X.; Ji, H.; Yin, S.; Chen, J.; Gao, L. Hot Deformation Behavior of TA1 Prepared by Electron Beam Cold Hearth Melting with a Single Pass. *Materials* **2023**, *16*, 369. https://doi.org/10.3390/ma16010369

Academic Editor: Javier Gil

Received: 19 October 2022
Revised: 21 December 2022
Accepted: 25 December 2022
Published: 30 December 2022

Copyright: © 2022 by the authors. Licensee MDPI, Basel, Switzerland. This article is an open access article distributed under the terms and conditions of the Creative Commons Attribution (CC BY) license (https:// creativecommons.org/licenses/by/ 4.0/).

1. Introduction

Titanium is an important metal applied in modern industry and has characteristics including high strength, lightweight, strong corrosion resistance, and good temperature endurance [1,2]. The TA1 plate is generally obtained by rolling titanium slab as the most used commercial titanium material. However, the shape and surface quality control of the TA1 titanium slab is problematic, resulting in the high cost of the TA1 plate [3]. To reduce the production cost, the electron beams cold hearth melting technique (EB) was introduced to produce TA1 slab, with which high-quality TA1 products can be formed with single pass melting and casting. Compared with the traditional vacuum consumable electrode arc (VAR) furnace melting technology which requires several passes, the titanium slab melted by a single pass EB furnace is more energy saving. The prepared EB product can be used for direct rolling without forging and perforation molding [4,5]. Meanwhile, the EB technique was known to have a strong purification ability, especially for high-density inclusion (HDI), which can hardly be removed by the VAR technique [6,7]. Thus, the EB technique can reduce titanium waste, reducing the production cost of titanium plates [8].

However, in the actual production process, TA1 titanium slabs using EB furnaces usually adopt the same hot rolling process scheme as the TA1 titanium slab melted by the VAR furnace. Since the thermal deformation behaviors of TA1 slabs produced using EB and VAR are different, surface defects, including peeling and scratching, appear during the hot rolling process of TA1 slabs melted by EB furnace, which increases the edge cutting process of titanium coil and production cost.

Many scholars have studied the thermal deformation behavior of TA1 slabs melted by a VAR furnace. Ma et al. [9] established a processing map for the TA1 plate melted by VAR. The results showed that TA1 slab was suitable for rolling under the conditions of a strain rate at 5 s^{-1}, a temperature at 700–750 °C, and the deformation of each pass is greater than or equal to 25%. Li et al. [10] conducted thermal compression simulation experiments on the intermediate billet of hot-rolled TA1 titanium melted by VAR, and the processing map was formed. The best conditions for smelting TA1 in the VAR furnace included a strain rate and temperature range of 1–20 s^{-1} and 750–850 °C. However, few studies are devoted to the thermal deformation behavior of TA1 slabs melted by the EB furnace. Huang et al. [11] researched the thermal deformation performance of TA1 slabs prepared by residual titanium in an EB furnace. The influences of thermal deformation on the microstructure were discussed, whilst the impacts of thermal deformation on rheological stress were not considered. Thus, it is necessary to study the influence of rheological stress during the hot deformation of TA1 to find out the hot deformation characteristics of EB furnace casting billet and effectively predict and formulate the thermoplastic processing technology of TA1.

Therefore, this manuscript researches the thermal deformation behavior of TA1 prepared by EB furnaces, focusing on the impacts of strain rate and deformation temperature on the flow stress. Meanwhile, the constitutive model and processing map are founded according to the stress-strain data, which provides a reference for the improvement of the EB furnace melting technique and the subsequent TA1 hot rolling process.

2. Materials and Methods

The test material was an industrial TA1 slab produced by an EB furnace in a factory in Kunming, China. The chemical composition is indicated in the quality assurance certificate of the enterprise, and has been verified according to the GB/T4698 standard method [12]. The chemical composition is displayed in Table 1 [13]. The Gleeble sample, sized of Φ 8 × 12 mm, is cut from a single pass EB melted slab using wire cutting. To ensure the uniformity of the samples, 10cm of the slab was removed from the end before cutting the sample. After sampling, Gleeble-3800 was used for the thermal simulation compression test samples. The initial microstructure of the material before the Gleeble test is shown in Figure 1. It can be seen from the figure that the original structure of the single pass EB furnace direct melting billet is equiaxed α phase. Still, the grain structure is coarse, and the average grain size is about 3–5 cm, reaching the centimeter level.

Table 1. Chemical composition of the sample (Wt%).

Fe	O	N	C	H	Ti
≤0.20	≤0.18	≤0.03	≤0.08	≤0.015	Bal

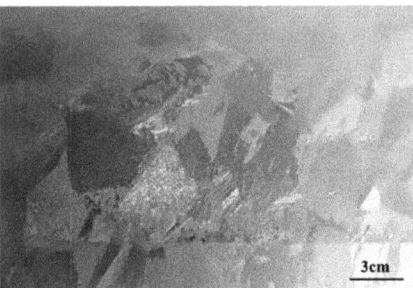

Figure 1. Original structure of slab directly melted and cast in EB furnace.

The heat treatment temperatures of the sample were selected at 700 °C, 750 °C, 800 °C, 850 °C, and 900 °C, according to the thermal simulation compression deformation temperature, combined with TA1 phase transformation temperature and empirical rolling temperature range. The technical details about mechanical testing are: The temperature is measured by a thermocouple fixed in the middle of the sample. The temperature was recorded by the data acquisition system of the thermal simulator. The heating rate is 20 °C/s. Since the technology of hot compression deformation test using Gleeble testing machine is relatively mature, using one sample for each specific temperature and strain rate can meet the research requirements. But it also depends on the ability of the tester to control the machine. When the sample deformation is not uniform, that is, the obtained sample has irregular "drum shape" or even non- "drum shape", the collected data will also appear obvious anomalies. At this point, another sample will be made to correct the corresponding process.

The corresponding strain rates were 0.01 s^{-1}, 0.1 s^{-1}, 1.0 s^{-1}, 5.0 s^{-1}, and 10 s^{-1}, separately. 50% was the set value of compression deformation. The settled temperature is maintained for 5 min, and water-cooling was carried out immediately after compression treatment.

3. Results

3.1. True Stress-Strain Curve of TA1 Sample Produced by Single Pass EB

The friction effect in the process of thermal deformation is reduced by placing graphite sheets at both ends of the sample, the most important variables are the friction force and the geometry of the sample [14,15]. To reduce the influence of friction, a piece of 0.5 mm tantalum foil was placed at the ends of the cylindrical sample during the compression test. The flow stress is the stress that must be applied to cause a material to deform at a constant strain rate in its plastic range. Individual data with obvious deviation were removed when processing the flow stress curve. Figure 2 displays the true stress-strain curve of the TA1 sample at various deformation temperatures and strain rates. With the rise of strain rate, the flow stress increases at a certain deformation temperature from Figure 2. For example, at 0.01 s^{-1} strain rate and 700 °C, the maximum flow stress is about 58 MPa. While the maximum flow stress is about 160 MPa at a 10 s^{-1} strain rate. At high strain rates, the strain hardening trend is obvious, and the degree of dynamic softening is low, so the flow stress is enormous. Figure 3 shows the typical metallographic structure at 750–900 °C and 1 s^{-1} strain rate. When deformed below 850 °C, α-phase equiaxed polyhedral grain structure is obtained, which is mainly different in grain size and uniformity. The non-uniformity of grains obtained after deformation at 750 °C is obvious. It can be seen that some grains have not grown up after recrystallization. Then, with the increase of temperature, the grains grow up gradually and distribute uniformly in the region. However, when the temperature reaches 900 °C for deformation, as the deformation temperature is already in the β-phase zone, the sample structure has a flaky nature, the grains are irregular, and the sawtooth structure of the β-phase grain boundary appears. So, when the strain rate is low, the flow stress is low, which is because the sample is affected by dynamic recovery and dynamic recrystallization softening [16]. Under the condition of constant strain rate, the flow stress declines with the rise of deformation temperature. For example, when the temperature is 700 °C, the maximum flow stress is about 160 MPa at the strain rate of 10 s^{-1}, while 30 MPa is the maximum flow stress at 900 °C. Therefore, when the deformation temperature grows gradually, the softening impact of TA1 is more obvious, and the trend of flow stress reduction is more significant.

At high-temperature compression deformation, the relationship between peak stress and strain rate as a function of temperature is presented in Figure 4a,b. With a constant temperature, the peak stress grows with the rise of the strain rate as shown in Figure 4a; With a constant strain rate, the peak stress reduces with the increase of temperature, as shown in Figure 4b. When the temperature is 700–850 °C, the sensitivity of peak stress to strain rate tends to be consistent. However, when the temperature reaches 900 °C, the sensitivity of peak stress decreases obviously, and the strain rate changes tend to be gentle. This can be attributed to the fact that with the increase in temperature, the average atomic kinetic energy of the material increases, and the thermal activation increases, which reduces the critical shear stress of grain boundary slip and weakens the resistance to slip and dislocation [17].

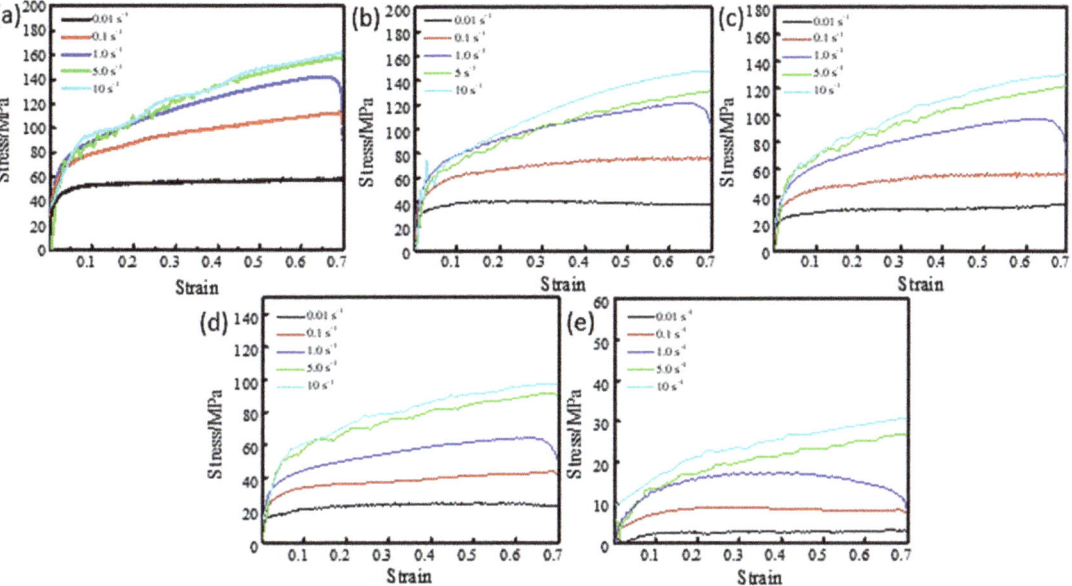

Figure 2. The curves of true stress and true strain at different temperatures: (a) 700 °C; (b) 750 °C; (c) 800 °C; (d) 850 °C; (e) 900 °C.

During the hot deformation of TA1, the hardening impact of strain hardening, dynamic recovery, and dynamic recrystallization softening mechanism will cause the variation of flow stress. The stress-strain curve has two stages: unsteady deformation and steady deformation. At the initial stage of deformation, namely unsteady deformation, the flow stress growths quickly within a very small range of strain increases; The strain endures to growth while the flow stress grows gradually in the steady-state deformation process. The flow stress rises slowly when the deformation temperature is at a low level, and the strain rate is high. The flow stress grows slowly to the peak value, then declines slowly, and finally tends to be stable under the condition of a low strain rate and high deformation temperature. Plastic deformation occurs in the stage of unsteady deformation. The interaction of epistatic faults in the slip system causes the dislocation density to increase speedily, resulting in dislocation entanglement and stacking, and finally forms cellular dislocation structure, resulting in higher stress for material deformation [18]. Hardening is dominant, and the flow stress grows rapidly in the stage of unsteady deformation.

As the strain increases, the material accumulates more energy in the steady-state deformation stage. Cellular dislocations form substructures, and dislocations of opposite signs are canceled out. The softening influence of dynamic recovery rises continuously, and the growing trend of flow stress becomes gentle. The material will have enough time to accumulate more energy at a large deformation temperature and a small strain rate [19]. Dynamic recrystallization will arise during the deformation after the material accumulates enough energy. At this time, the softening effect is more pronounced. The flow stress declines slowly after getting the peak value. Finally, the hardening effect affected by the rise of strain is balanced with the softening effect of dynamic recovery and dynamic recrystallization, and the flow stress stays unwavering [20]. However, with a low deformation temperature and a significant strain rate, there exists not enough time to accumulate energy for the material. Simultaneously, the temperature cannot reach the critical temperature required for dynamic recrystallization. Therefore, the softening effect of the material is almost not noticeable, which shows that the flow stress continues to increase slowly.

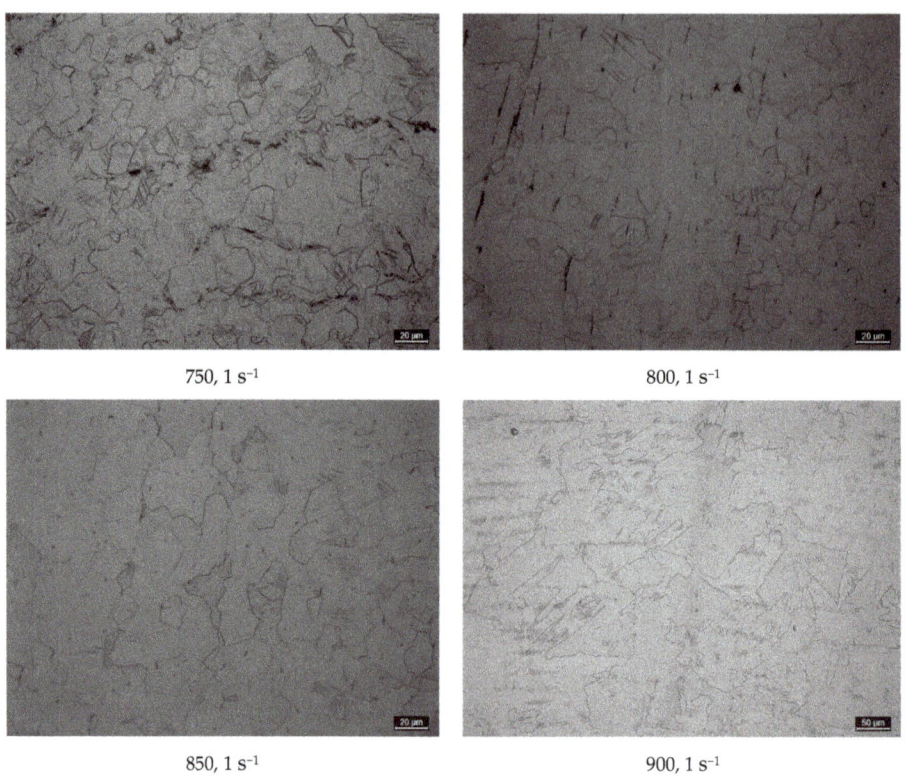

Figure 3. Typical metallographic structure at 750–900 °C and 1 s^{-1} strain rate.

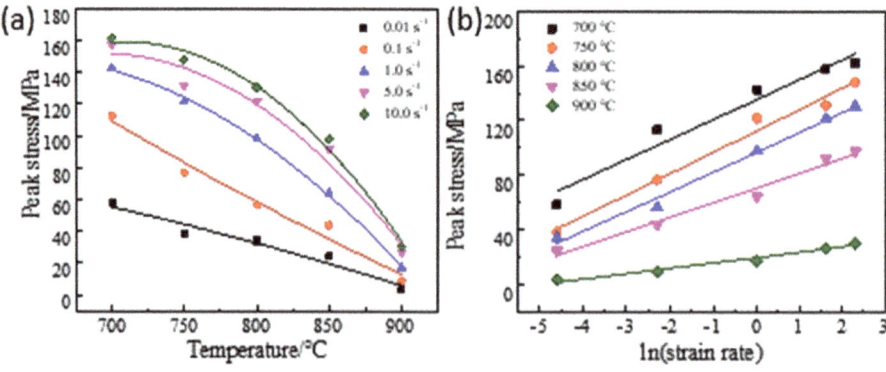

Figure 4. Dependence of peak stress on temperature and strain rate: (**a**) T–σ; (**b**) ln $\dot{\varepsilon}$– σ.

3.2. Constitutive Model of TA1 Sample Produced by Single Pass EB

When studying the mechanical properties of metal materials during plastic deformation, the constitutive model can frequently describe the relationship between strain, stress, temperature, and strain rate. As stated by stress-strain data of TA1 melted and cast in the EB furnace obtained in the experiment, the constitutive model was derived by using the constitutive equation. Arrhenius constitutive equation was exploited to derive, which was summarized and improved from various models, and can accurately express the relationship between flow stress, strain rate, and temperature [21]. The three expressions of the Arrhenius equation are present in Equations (1)–(3). Equation (1) applies to low stress where the strain rate and stress are power functions; Equation (2) applies to the high-stress situation where there is an exponential connection between strain rate and stress; When the connection of strain rate and stress remains a hyperbolic sine function, Equation (3) is applicable.

$$\dot{\varepsilon} = A_1 \sigma^{n_1} exp\left(-\frac{Q}{RT}\right) \quad (1)$$

$$\dot{\varepsilon} = A_2 \exp(\beta\sigma) \exp\left(-\frac{Q}{RT}\right) \quad (2)$$

$$\dot{\varepsilon} = A[\sinh(\alpha\sigma)^n] exp\left(-\frac{Q}{RT}\right) \quad (3)$$

where, $\dot{\varepsilon}$ represents the strain rate, (s^{-1}); σ represents the stress, (MPa); T stands for deformation temperature, (K); Q means the activation energy of thermal deformation, (J/mol); R means the gas constant, (J/mol·K), the value indicates 8.314 [22]; A, A_1, A_2, n, n_1, β, α represent the material constant, and $\alpha = \frac{\beta}{n_1}$.

The natural logarithms were taken on both sides of Equations (1)–(3), and the equations were transformed into Equations (4)–(6). The $\ln \sigma$ was settled as abscissa, $\ln \dot{\varepsilon}$ was settled as the ordinate, the graph was drawn in Figure 5a, where the true strain was 0.69; With σ was settled as abscissa, and $\ln \dot{\varepsilon}$ was settled as the ordinate, the graph with the true strain of 0.69 was drawn in Figure 5b. In Figure 5a, n_1 represented the average slope of the five straight lines (σ-$\ln \dot{\varepsilon}$), and the average slope of the five straight lines in Figure 5b is β. n_1 was reckoned to be 4.89534, β was 0.106564, and the α was 0.021768.

$$\ln \dot{\varepsilon} = n_1 \ln \sigma + \ln A_1 - \frac{Q}{RT} \quad (4)$$

$$\ln \dot{\varepsilon} = \beta\sigma + \ln A_2 - \frac{Q}{RT} \quad (5)$$

$$\ln \dot{\varepsilon} = n\ln[\sinh(\alpha\sigma)] + \ln A - \frac{Q}{RT} \quad (6)$$

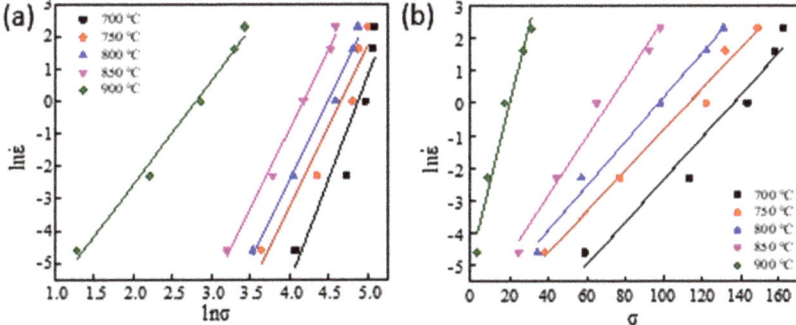

Figure 5. Relationship between (a) $\ln \dot{\varepsilon} - \ln \sigma$, (b) $\sigma - \ln \dot{\varepsilon}$.

The relationship of $ln[\sin h(\alpha\sigma)] - ln\,\dot{\varepsilon}$ is drawn into a curve in Figure 6a. The average n of the straight-line slope of $ln[\sin h(\alpha\sigma)]$ was calculated, which was 2.982678. In a certain range, the activation energy of thermal deformation is determined when the strain rate $\dot{\varepsilon}$ is certain. The partial derivative of $1/T$ on both sides of Equation (6) was calculated, expressed as [23]:

$$Q = Rn\frac{\partial ln[\sinh(\alpha\sigma)]}{\partial(\frac{1}{T})} \qquad (7)$$

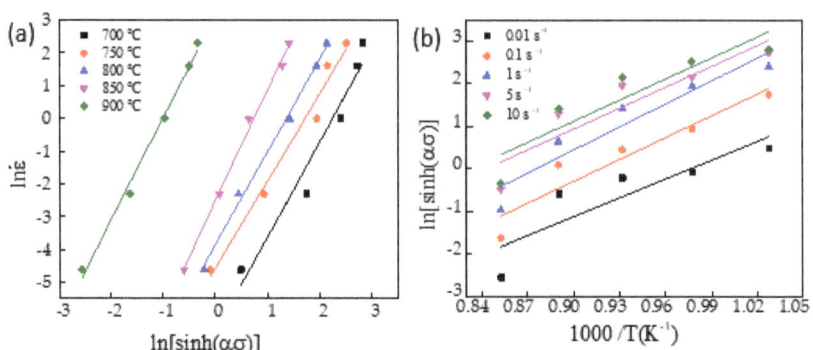

Figure 6. Relationship between (**a**): $ln\dot{\varepsilon} - ln[\sin h(\alpha\sigma)]$, (**b**): $ln[\sin h(\alpha\sigma)] - 1000/T(K^{-1})$.

Figure 6b shows the relation curve of $ln[\sin h(\alpha\sigma)] - 1000/T$ when the true strain was 0.69. Under different strain rates, the average straight-line slope value was 16.55988, which was substituted into Equation (7) to reckon the thermal deformation activation energy (Q), which was 410.655 kJ/mol.

In general, when characterizing the impact of temperature and strain rate on the deformation behavior of materials, the Zener-Holloman parameter of temperature-compensated strain rate factor is often used in Equation (8) [24].

$$\ln Z = \ln A + n\,ln[\sinh(\alpha\sigma)] \qquad (8)$$

According to Equation (3):

$$\dot{\varepsilon}\exp(\frac{Q}{RT}) = A[\sinh(\alpha\sigma)]^n \qquad (9)$$

Figure 7 is the relation curve of $\ln Z - ln[\sin h(\alpha\sigma)]$ when the true strain is 0.69. When $ln[\sin h(\alpha\sigma)]$ was 0, lnA was equal to $\ln Z$, and the value was 43.42325.

The constitutive model can be obtained by replacing the above parameters with the hyperbolic sinusoidal constitutive equation:

$$\sigma = \frac{1}{\alpha(\varepsilon)}\cdot \ln\{(\frac{Z}{e^{\ln A(\varepsilon)}})^{\frac{1}{n(\varepsilon)}} + [(\frac{Z}{e^{\ln A(\varepsilon)}})^{\frac{2}{n(\varepsilon)}} + 1]^{\frac{1}{2}}\} \qquad (10)$$

where $Z = \dot{\varepsilon}\exp(\frac{Q(\varepsilon)}{T\cdot 8.314\ J/(mol\cdot K)})$.

Arrhenius's constitutive model only thinks about the peak stress but ignores the influence of true strain [25]. To obtain the material constants under different strain variables in the constitutive model, the strain variables are taken from 0.05 to 0.70 with an interval of 0.05. The calculation results of α, n, Q, and lnA are shown in Table 2.

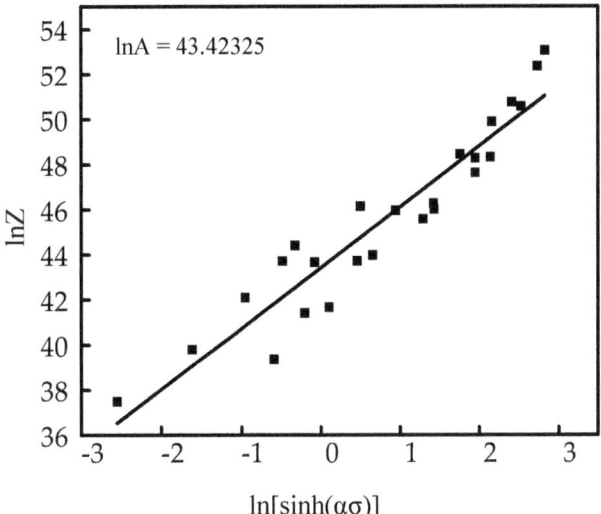

Figure 7. Relationship between $\ln Z$ and $\ln[\sin h(\alpha\sigma)]$.

Table 2. Calculation results of α, n, Q, and $\ln A$, when the true strain is in the range of 0.05–0.70.

True Strain	α	n	Q (kJ/mol)	$\ln A$
0.05	0.0367310	4.537762	628.09658	67.68917
0.1	0.0311358	4.123178	517.24992	55.28302
0.15	0.0287303	4.188236	517.89557	55.30470
0.2	0.0356319	3.242180	483.62878	50.14605
0.25	0.0260507	3.625776	474.42582	50.45797
0.3	0.0260062	3.431418	459.28553	48.67368
0.35	0.0242674	3.438538	448.01972	47.60311
0.4	0.0244092	3.174006	432.84269	45.82066
0.45	0.0237639	3.135992	437.09739	44.38648
0.5	0.0247517	2.947084	422.53110	44.47413
0.55	0.0233285	2.910886	409.14240	43.21425
0.6	0.0216964	3.020520	419.02809	44.54717
0.65	0.0213428	2.996986	418.49972	44.51738
0.7	0.0221200	2.429494	358.97080	38.67918

A polynomial fitting curve of the seventh degree between the strain and the material constants in the constitutive model is shown in Figure 8. Using strain and material constants in the constitutive model α, n, Q, and $\ln A$ in the process of fitting attempt, at least 70% of the calculated points fall on the fitting curve, and the points are evenly distributed on both sides of the curve, guarantee the accuracy of the fitting. Through fitting, it is found that the result of seven times fitting is more accurate. Table 3 displays the parameters in the equation. A polynomial of degree seven can express the relationship between α, n, Q, $\ln A$, and true strain ε, as shown below:

$$\alpha(\varepsilon) = \alpha_0 + \alpha_1\varepsilon + \alpha_2\varepsilon^2 + \alpha_3\varepsilon^3 + \alpha_4\varepsilon^4 + \alpha_5\varepsilon^5 + \alpha_6\varepsilon^6 + \alpha_7\varepsilon^7 \tag{11}$$

$$n(\varepsilon) = n_0 + n_1\varepsilon + n_2\varepsilon^2 + n_3\varepsilon^3 + n_4\varepsilon^4 + n_5\varepsilon^5 + n_6\varepsilon^6 + n_7\varepsilon^7 \tag{12}$$

$$Q(\varepsilon) = Q_0 + Q_1\varepsilon + Q_2\varepsilon^2 + Q_3\varepsilon^3 + Q_4\varepsilon^4 + Q_5\varepsilon^5 + Q_6\varepsilon^6 + Q_7\varepsilon^7 \tag{13}$$

$$\ln A(\varepsilon) = A_0 + A_1\varepsilon + A_2\varepsilon^2 + A_3\varepsilon^3 + A_4\varepsilon^4 + A_5\varepsilon^5 + A_6\varepsilon^6 + A_7\varepsilon^7 \tag{14}$$

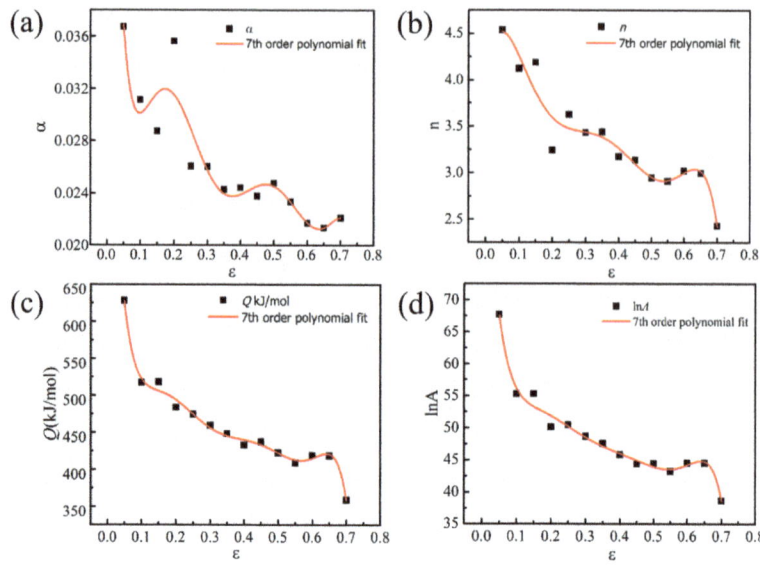

Figure 8. Relationship between material constants and true strain by fitting with seventh-order polynomial: (**a**) α; (**b**) n; (**c**) Q; (**d**) lnA.

Table 3. Data of material constants (α, n, Q, and lnA).

α	n	Q (kJ/mol)	lnA
$\alpha_0 = 0.08047$	$n_0 = 3.9809$	$Q_0 = 1050.80096$	$A_0 = 104.97502$
$\alpha_1 = -1.57336$	$n_1 = 25.52243$	$Q_1 = -14,406.71512$	$A_1 = -1213.97161$
$\alpha_2 = 18.63559$	$n_2 = -388.12653$	$Q_2 = 156,765.29615$	$A_2 = 12,090.3470$
$\alpha_3 = -107.21173$	$n_3 = 2069.28187$	$Q_3 = -890,952.59121$	$A_3 = -64,932.77841$
$\alpha_4 = 328.63345$	$n_4 = -5187.62503$	$Q_4 = 2,814,010.0$	$A_4 = 198,218.43158$
$\alpha_5 = -550.47869$	$n_5 = 6070.79641$	$Q_5 = 4,982,800.0$	$A_5 = -345,319.45288$
$\alpha_6 = 475.65222$	$n_6 = -2494.63078$	$Q_6 = 4,624,390.0$	$A_6 = 319,534.03285$
$\alpha_7 = -165.87799$	$n_7 = -246.27687$	$Q_7 = -1,749,330.0$	$A_7 = -121,622.70438$

The above parameters are substituted into the hyperbolic sinusoidal constitutive equation. The constitutive model of TA1 considering strain shadow is:

$$\sigma = \frac{1}{\alpha(\varepsilon)} \cdot ln\{(\frac{Z}{e^{ln\,A(\varepsilon)}})^{\frac{1}{n(\varepsilon)}} + [(\frac{Z}{e^{ln\,A(\varepsilon)}})^{\frac{2}{n(\varepsilon)}} + 1]^{\frac{1}{2}}\} \quad (15)$$

where $Z = \dot{\varepsilon}exp(\frac{Q(\varepsilon) \cdot 1000 \text{ J/mol}}{T \cdot 8.314 \text{ J/(mol·K)}})$.

3.3. Processing Map of EB Furnace Produced TA1

To reflect the influence of deformation parameters on processing properties, the processing map is drawn to lead the hot deformation process of materials in actual production, where the dynamic materials model (DMM) is generally adopted [26,27]. When the deformation temperature and strain rate remain unchanged, the dynamic relationship between the stress σ and the strain rate of the hot-worked workpiece is as follows:

$$\sigma = K(\dot{\varepsilon})^m \quad (16)$$

The dynamic material model treats the thermal deformation process as an energy dissipation element. During the plastic deformation of the workpiece, P represents the total energy absorbed [28,29]. During plastic deformation, the total energy absorbed can be expressed as [30]:

$$P = \sigma \cdot \dot{\varepsilon} = \int_0^{\dot{\varepsilon}} \sigma d\dot{\varepsilon} + \int_0^{\sigma} \dot{\varepsilon} d\sigma \qquad (17)$$

The energy absorbed by materials is mainly scattered in two methods: (1) The energy dissipation affected by plastic deformation is expressed as power consumption (G), mostly converted to heat; (2) The power dissipation covariance (J) is the energy consumed by microstructure evolution [31,32]. The strain rate sensitivity index(m) refers to the proportion of power dissipation co-quantity and power dissipation quantity [33]. m is calculated as follows:

$$m = \frac{dJ}{dG} = \frac{d(\ln\sigma)}{d(\ln\dot{\varepsilon})} = \frac{\Delta\lg\sigma}{\Delta\lg\dot{\varepsilon}} \qquad (18)$$

During the material forming process, the proportional relationship between the energy (J) consumed by the evolutionary process of the microstructure and the linear dissipation energy is usually expressed by the power dissipation exponent η [34,35]. η is defined as:

$$\eta = \frac{J}{J_{max}} = \frac{2m}{m+1} \qquad (19)$$

The processing instability was established according to the Prasad instability criterion in the dynamic material model, [36], based on Ziegler's extreme value principle of irreversible thermodynamics in equation (20) [37]:

$$\zeta = \frac{\partial \ln(\frac{m}{m+1})}{\partial \ln \dot{\varepsilon}} + m \leq 0 \qquad (20)$$

Along with the established constitutive model, the instability zone and power dissipation index are calculated through Equations (19) and (20). The thermal processing diagram of the dynamic material model is obtained by drawing the principle of superposition of the power dissipation diagram and instability diagram, as shown in Figure 9.

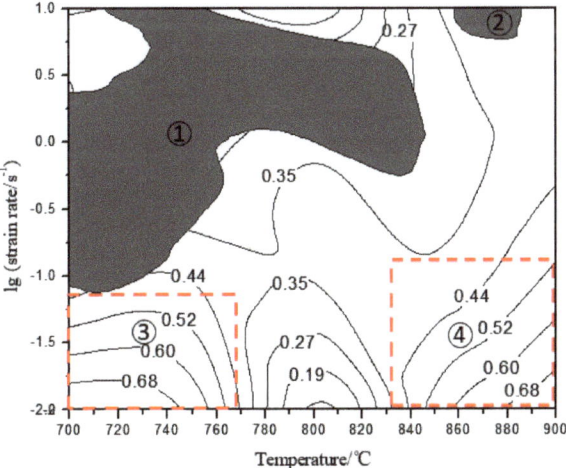

Figure 9. Processing map of TA1. ①, ② are Plastic instability region; ③, ④ are Suitable processing area.

In Figure 9, the gray part of "① and ②" is the plastic instability area. The power dissipation index in this area is less than 0.35, which is not suitable for machining deformation. The dotted line area of "③ and ④" is a suitable processing area. In this area, most of the energy is used for the transformation of the microstructure. The processing map shows that the appropriate hot working areas for industrial pure titanium TA1 are: The range of deformation temperature and strain rates are 700–770 °C and 0.01–0.05 s^{-1}, separately; The range of strain rate and deformation temperature is 0.01–0.15 s^{-1} and 840–900 °C, individually. Plastic flow instability region: the range of strain rate and deformation temperature are 3–10 s^{-1} and 860–880 °C, respectively; The range of strain rate and deformation temperature range are 1–10 s^{-1} and 700–840 °C, individually. To effectively reduce the rolling mill load, in combination with the actual industrial production, the processing area ④ is selected for the hot working deformation test of TA1 titanium billet. The preferred 860 °C open rolling annealing organization is shown in the following Figure 10. The structure of finished titanium material is equiaxed α Titanium, the grain size is uniform, and the grain size is about grade 5. The tensile strength of the transverse tension test is 375 MPa, RP$_{0.2}$ is 275 MPa, elongation after fracture is 44%, and the strength and toughness are good.

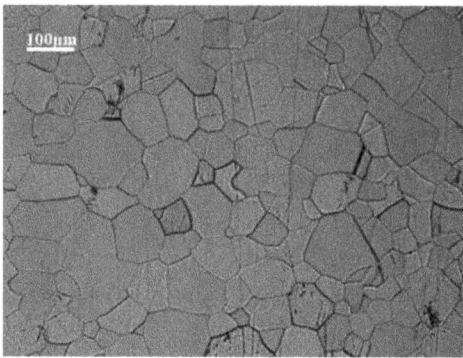

Figure 10. The preferred 860 °C open rolling annealing organization.

4. Discussion

The VAR process is difficult to eliminate high-density and low-density inclusions during the titanium melting process [6]. In the EB furnace smelting process, the cold hearth can be used to separate the three processes of melting, refining, and crystallization (i.e., melting, refining and solidification). The liquid metal first enters the smelting area for melting and preliminary refining. Then it flows into the refining area for complete refining to eliminate the high-density and low-density inclusions that may be mixed in the raw materials. Finally, the purified melt flows into the crystallizer and condenses into slabs. Therefore, the obtained flat slab phase has a more uniform composition and fewer impurity elements than vacuum consumable smelting [7]. In addition, since the EB furnace works under a vacuum with a pressure less than 0.1 Pa, oxygen and nitrogen impurities formation in the smelting process was suppressed. The single-pass EB process has a good dehydrogenation capacity, and the deformation activation energy and the degree of structural segregation of the EB slab are lower than that of the VAR slab. Li Jun et al. [10] studied that the thermal deformation activation energy of the TA1 sample prepared by VAR is 643.3116 kJ/mol. The thermal deformation activation energy of the TA1 sample prepared by the EB furnace was reduced by about 36.2% compared with the thermal deformation activation energy of the TA1 sample prepared by VAR. By comparing the thermal deformation activation energy, we found that the thermal deformation activation energy of the TA1 sample prepared by EB is lower than that of the sample prepared by VAR. The thermal deformation activation energy of TA1 cast in the EB furnace is 410.655 kJ/mol.

Therefore, the TA1 sample prepared by the EB furnace is easier to deform than the TA1 slabs produced by the VAR furnace.

Table 4 compares the processing maps of TA1 slabs produced by the EB furnace and VAR furnace. We found that the machinable range of TA1 slabs produced by VAR is more significant than that produced by EB, whist for the unstable region, the trend is the opposite. The instability area of the processing chart for the slabs produced by EB furnace is relatively large in the region with a large strain rate. Thus, the machinable range of TA1 slab produced by EB is mainly concentrated on a low strain rate, while that of the VAR-produced TA1 slab is relatively concentrated on a high strain rate.

Table 4. Hot deformation of TA1 produced by EB and VAR.

Smelting Method	Suitable Hot Deformation Region
EB	700~760 °C, 0.05~0.01 s^{-1}; 840~900 °C, 0.1~0.01 s^{-1};
VAR [10]	750~850 °C, 1~20 s^{-1};
VAR [38]	strain rate < 10 s^{-1}, T > 725 °C;
VAR [9]	650~750 °C, 5~10 s^{-1};

5. Conclusions

(1) TA1 prepared by electron beam cold hearth melting in just one pass was very sensitive to strain rate and temperature. The strain rate and temperature of TA1 cast in the EB furnace were very significant. Rising the temperature and lessening the strain rate will reduce the flow stress. In the unsteady deformation stage, the flow stress increased rapidly; The flow stress raised slowly or even fell in the steady-state deformation stage.

(2) The constitutive model of TA1 for EB furnace casting considering the effect of strain was Equation (15).

(3) Based on the processing map of the TA1 sample produced by the EB furnace, the suitable processing areas were obtained: The range of strain rate was 0.01–0.05 s^{-1} at the deformation temperature range was 700–770 °C; When the deformation temperature range was 840–900 °C, the strain rate range was 0.01–0.15 s^{-1}.

(4) Compared with VAR produced TA1 sample, EB produced TA1 sample had lower activation energy of thermal deformation and a smaller suitable range for thermal processing.

Author Contributions: Conceptualization, J.C. and W.H.; methodology, W.Z. and Z.Z.; validation, J.C. and L.G.; formal analysis, J.C.; investigation, X.S. and H.J.; resources, L.G. and S.Y.; data curation, J.C. and S.Y.; writing—original draft preparation, W.Z. and Z.Z.; writing—review and editing, L.G. and S.Y.; visualization, X.S. and H.J.; supervision, W.H. and L.G.; project administration, J.C. and S.Y. All authors have read and agreed to the published version of the manuscript.

Funding: This research was funded by National Natural Science Foundation of China (No: 52104351), the Scientific Research Project of the Education Department of Yunnan Province (No: 2020J0652), the Yunnan Fundamental Research Projects (No: 202101AU070088), and the Science and Technology Major Project of Yunnan Province(No. 202202AG050007).

Institutional Review Board Statement: Not applicable.

Informed Consent Statement: Not applicable.

Data Availability Statement: Data presented in this article are available at request from the corresponding author.

Acknowledgments: The authors are grateful for the financial support of all the foundations.

Conflicts of Interest: The authors declare no conflict of interest.

References

1. Boyer, R. Titanium for aerospace: Rationale and applications. *Adv. Perform. Mater.* **1995**, *2*, 349–368. [CrossRef]
2. Vakhrusheva, S.; Hruzin, V. Effect of deformation ratio on texture and properties of Titanium alloy tubes under cold rolling. *Metallofiz. Noveishie Tekhnologii* **2019**, *41*, 1303–1314. [CrossRef]
3. Kotkunde, N.; Deole, A.D.; Gupta, A.K.; Singh, S.K.; Aditya, B. Failure and formability studies in warm deep drawing of Ti-6Al-4V alloy. *Mater. Des.* **2014**, *60*, 540–547. [CrossRef]
4. Gao, L.; Li, X.M.; Huang, H.G.; Sui, Y.D.; Zhang, H.M.; Shi, Z.; Chattopadhyay, K.; Jiang, Y.H.; Zhou, R. Numerical study of aluminum segregation during electron beam cold hearth remelting for large-scale Ti-6wt%Al-4wt%V alloy round ingot. *Int. J. Heat Mass Transf.* **2019**, *139*, 754–772. [CrossRef]
5. Gao, L.; Huang, H.G.; Kratzsch, C.; Zhang, H.M.; Chattopadhyay, K.; Jiang, Y.H.; Zhou, R. Numerical study of aluminum segregation during electron beam cold hearth melting for large-scale Ti-6wt%Al-4wt%V alloy slab ingot. *Int. J. Heat Mass Transf.* **2020**, *147*, 118976. [CrossRef]
6. Gao, L.; Huang, H.G.; Jiang, Y.H.; Chen, G.; Chattopadhyay, K.; Zhou, R. Numerical study on the solid-liquid interface evolution of large-scale titanium alloy ingots during high energy consumption electron beam cold hearth melting. *JOM* **2020**, *72*, 1953. [CrossRef]
7. Gao, L.; Huang, H.G.; Zhang, Y.Q.; Zhang, H.M.; Shi, Z.; Jiang, Y.H.; Zhou, R. Numerical modeling of EBCHM for large-scale TC4 alloy round ingots. *JOM* **2018**, *70*, 2934–2942. [CrossRef]
8. Xu, G.; Tao, J.; Deng, Y.; Zheng, B.; Zhang, Y.; Jiang, Y. Multi-stage hot deformation and dynamic re-crystallization behavior of low-cost Ti–Al–V–Fe alloy via electron beam cold hearth melting. *J. Mater. Res. Technol.* **2022**, *20*, 1186–1203. [CrossRef]
9. Ma, R.X.; Wang, Y.C.; Luo, W. Processing map establishment and hot rolling process optimization of commercial pure titanium. *Titan. Ind. Prog.* **2016**, *33*, 29–32. (In Chinese)
10. Li, J.; Shi, Q.N.; Yu, H.; Ren, W. Thermal deformation behavior and processing diagram of pure titanium TA1. *Iron Steel Vanadium Titan.* **2016**, *37*, 37–43. (In Chinese)
11. Huang, Y.D. Thermal Deformation Behavior and Microstructure of TA1 Prepared by EB Furnace Melting and Casting Residual Titanium. Ph.D. Thesis, Kunming University of Science and Technology, Kunming, China, 2021. (In Chinese).
12. *GB/T4698*; China Nonferrous Metals Industry Association. Methods for Chemical Analysis of Titanium Sponge, Titanium and Titanium Alloys. China Standards Press: Beijing, China, 2020.
13. Huang, W.; Qiu, H.; Zhang, Y.; Zhang, F.; Gao, L.; Omran, M.; Chen, G. Microstructure and phase transformation behavior of Al_2O_3-ZrO_2 under microwave sintering. *Ceram. Int.* **2022**, *49*, 4855–4862. [CrossRef]
14. Wan, Z.; Hu, L.; Sun, Y.; Wang, T.; Li, Z. Hot deformation behavior and processing workability of a Ni-based alloy. *J. Alloys Compd.* **2018**, *769*, 367–375. [CrossRef]
15. Evans, R.W.; Scharning, P.J. Axisymmetric compression test and hot orking properties of alloys. *Mater. Sci. Technol.* **2001**, *17*, 995–1004. [CrossRef]
16. Wang, W.; Gong, P.; Zhang, H.; Shi, Y.; Wang, M.; Zhang, X.; Wang, K. Hot deformation behavior of TC4 Ti-alloy prepared by electron beam cold hearth melting. *Chin. J. Mater. Res.* **2020**, *34*, 665–673. (In Chinese)
17. Liang, H.C.; Yan, X.S.; Zhang, Z.Q.; Bai, C.G.; Li, D.F.; Yang, L. Thermal deformation behavior of Ti-4Al-3V alloy at high temperature. *J. Shenyang Ligong Univ.* **2022**, *41*, 59–65.
18. Ma, T.; Wang, Y.; Yang, C.; Alamusi; Deng, Q.; Liu, Y.; Li, X.; Wei, Q.; Hu, N. Effect of strain rate on microscale formability and microstructural evolution of TA1 foil. *Mater. Sci. Eng. A* **2021**, *817*, 141338. [CrossRef]
19. Zhao, T.; Zhang, B.; Zhao, F.; Zhang, Z.; Dang, X.; Ma, Y.; Cai, J.; Wang, K. Hot deformation behavior of multilayered Ti/Ni composites during isothermal compression. *J. Mater. Res. Technol.* **2022**, *18*, 4903–4917. [CrossRef]
20. Zhanggg, X.B.; Yu, Y.B.; Liu, B.; Zhao, Y.; Ren, J.; Yan, Y.; Cao, R.; Chen, J. Microstructure characteristics and tensile properties of multilayer Al-6061/Ti-TA1 sheets fabricated by accumulative roll bonding. *J. Mater. Process. Technol.* **2020**, *275*, 116378. [CrossRef]
21. Zhou, S.W.; Dong, H.B.; Jiang, Z.W. Flow stress analysis and constitutive equation of TB17 titanium alloy under hot compression. *J. Plast. Eng.* **2018**, *25*, 218–223. (In Chinese)
22. Sun, J.; Chen, Y.; Liu, F.; Yang, E.; Wang, S.; Fu, H.; Qi, Z.; Huang, S.; Yang, J.; Liu, H.; et al. Calibration of arrhenius constitutive equation for B_4C_p/6063Al composites in high temperatures. *Materials* **2022**, *15*, 6438. [CrossRef]
23. Xu, Y.; Yang, X.J.; He, Y.; Du, D.N. Constitutive equation for hot deformation behavior of TC_4 titanium alloy. *Rare Met. Mater. Eng.* **2017**, *46*, 1320–1326.
24. Narayana, P.L.; Li, C.L.; Hong, J.K.; Choi, S.W.; Park, C.H.; Kim, S.W.; Kim, S.E.; Reddy, N.S.; Yeom, J.-T. Characterization of hot deformation behavior and processing maps of Ti–19Al–22Mo alloy. *Met. Mater. Int.* **2019**, *25*, 1063–1071. [CrossRef]
25. Liu, Q.; Wang, Z.; Yang, H.; Ning, Y. Hot deformation behavior and processing maps of Ti-6554 alloy for aviation key structural parts. *Metals* **2020**, *10*, 828. [CrossRef]
26. Xu, M.; Jia, W.J.; Zhang, Z.; Xie, J. Hot compression deformation behavior and processing map of TA15 alloy. *Rare Met. Mater. Eng.* **2017**, *46*, 2708–2713.
27. Yang, Z.; Yu, W.; Lang, S.; Wei, J.; Wang, G.; Ding, P. Hot Deformation behavior and processing maps of a new Ti-6Al-2Nb-2Zr-0.4B Titanium alloy. *Materials* **2021**, *14*, 2456. [CrossRef]
28. Gu, B.; Chekhonin, P.; Xin, S.; Liu, G.; Ma, C.; Zhou, L.; Skrotzki, W. Effect of temperature and strain rate on the deformation behavior of Ti5321 during hot-compression. *J. Alloys Compd.* **2021**, *876*, 159938. [CrossRef]

29. Sun, B.Q.; Yan, X.D.; Yang, S.L.; Yao, Y. Study on hot deformation characteristics and machining diagram of Al-Li-Cu-Mg-Zn-Ag alloy. *Rare Met.* **2017**, *41*, 949–954. (In Chinese)
30. Quan, G.-Z.; Wen, H.-R.; Pan, J.; Zou, Z.-Y. Construction of processing maps based on expanded data by BP-ANN and identification of optimal deforming parameters for Ti-6Al-4V alloy. *Int. J. Precis. Eng. Manuf.* **2016**, *17*, 171–180. [CrossRef]
31. Lin, X.; Huang, H.; Yuan, X.; Wang, Y.; Zheng, B.; Zuo, X.; Zhou, G. Study on hot deformation behavior and processing map of a Ti-47.5Al-2.5V-1.0Cr-0.2Zr alloy with a fully lamellar microstructure. *J. Alloys Compd.* **2022**, *901*, 163648. [CrossRef]
32. Li, C.; Huang, L.; Zhao, M.; Guo, S.; Su, Y.; Li, J. Characterization of hot workability of Ti-6Cr-5Mo-5V-4Al alloy based on hot processing map and microstructure evolution. *J. Alloys Compd.* **2022**, *905*, 164161. [CrossRef]
33. Yang, Y.; Xu, D.; Cao, S.; Wu, S.; Zhu, Z.; Wang, H.; Li, L.; Xin, S.; Qu, L.; Huang, A. Effect of strain rate and temperature on the deformation behavior in a Ti-23.1Nb-2.0Zr-1.00 titanium alloy. *J. Mater. Sci. Technol.* **2021**, *73*, 52–60. [CrossRef]
34. Liu, J.L.; Zeng, W.D.; Shu, Y.; Xie, Y.; Yang, J. Hot working parameters optimization of TC4-DT titanium alloy based on processing maps considering true strain. *Rare Met. Mater. Eng.* **2016**, *45*, 1647–1653.
35. Ji, X.M.; Xiang, C.; Hu, Y.N. Microstructure evolution and parameters optimization during hot deformation of TA12 titanium alloy. *J. Funct. Mater.* **2015**, *46*, 08081–08085. (In Chinese)
36. Prasad, Y.; Seshacharyulu, T. Modelling of hot deformation for microstructural control. *Metall. Rev.* **1998**, *43*, 243–258. [CrossRef]
37. Liu, J.H.; Liu, J.S.; Xiong, Y.S.; He, W.W.; Zhang, P.; Liu, X.M. Study on hot deformation behavior and machining diagram of TC4-DT titanium alloy. *Rare Met. Mater. Eng.* **2013**, *42*, 1674–1678. (In Chinese)
38. Vinod, B.; Ananthi, V.; Ramanathan, S. Hot deformation behavior of Titanium alloy-reinforced solid/agro waste materials: Constitutive analysis and processing maps. *J. Bio-Tribo-Corros.* **2021**, *7*, 119. [CrossRef]

Disclaimer/Publisher's Note: The statements, opinions and data contained in all publications are solely those of the individual author(s) and contributor(s) and not of MDPI and/or the editor(s). MDPI and/or the editor(s) disclaim responsibility for any injury to people or property resulting from any ideas, methods, instructions or products referred to in the content.

Article

Numerical and Experimental Study on Carbon Segregation in Square Billet Continuous Casting with M-EMS

Pengchao Li [1,2], Guifang Zhang [1,3,*], Peng Yan [1,*], Peipei Zhang [1], Nan Tian [1] and Zhenhua Feng [1]

1. Faculty of Metallurgical and Energy Engineering, Kunming University of Science and Technology, Kunming 650093, China; 13696392999@163.com (P.L.); zpei861@163.com (P.Z.); tian1852558@163.com (N.T.); fengzhenhua666@126.com (Z.F.)
2. Linyi Iron and Steel Investment Group Special Steel Co., Ltd., Linyi 276000, China
3. Key Laboratory of Clean Metallurgy for Complex Iron Resources in Colleges and Universities of Yunnan Province, Kunming University of Science and Technology, Kunming 650093, China
* Correspondence: guifangzhang65@163.com (G.Z.); yanp_km@163.com (P.Y.)

Abstract: Electromagnetic stirring (M-EMS) has been extensively applied in continuous casting production to reduce the quality defects of casting billets. To investigate the effect of continuous casting electromagnetic stirring on billet segregation, a 3D multi-physics coupling model was established to simulate the internal heat, momentum, and solute transfer behavior, to identify the effect of M-EMS on the carbon segregation of a continuous casting square billet of 200 mm × 200 mm. The results show that M-EMS can move the high-temperature zone upward, which is favorable for the rapid solidification of the billet, and can promote the rotational flow of the molten steel in the horizontal direction. When the electromagnetic stirring current is varied in the range of 0–500 A, the degree of carbon segregation first decreases and then increases, with the best control of segregation at 300 A. In the frequency range of 3–5 Hz, the degree of carbon segregation degree increases with frequency. Meanwhile, the simulation and experimental results show that 3 Hz + 300 A is the best electromagnetic stirring parameter for improving the carbon segregation of casting billets with a size of 200 mm × 200 mm. So, a reasonable choice of the M-EMS parameters is crucial for the quality of the billet.

Keywords: M-EMS; continuous casting billet; multi-physical field model; carbon segregation

Citation: Li, P.; Zhang, G.; Yan, P.; Zhang, P.; Tian, N.; Feng, Z. Numerical and Experimental Study on Carbon Segregation in Square Billet Continuous Casting with M-EMS. *Materials* **2023**, *16*, 5531. https://doi.org/10.3390/ma16165531

Academic Editor: Elena Pereloma

Received: 19 June 2023
Revised: 31 July 2023
Accepted: 7 August 2023
Published: 9 August 2023

Copyright: © 2023 by the authors. Licensee MDPI, Basel, Switzerland. This article is an open access article distributed under the terms and conditions of the Creative Commons Attribution (CC BY) license (https://creativecommons.org/licenses/by/4.0/).

1. Introduction

Continuous casting is an essential link in the production of high-quality steel and special steel. The segregation behavior of continuous casting billets has an essential impact on the internal quality and performance uniformity of subsequent steel products [1,2]. The continuous casting process of the billet is dominated by highly complex and interrelated phenomena, including molten steel flow, heat transfer, solute transport, solidification nucleation, and crystal growth and transport [3,4]. For the typical segregation behavior in the continuous casting process, the reported causes include the flow, solute transfer, and relative motion of the solid and liquid phases. Electromagnetic stirring (EMS) is a process that uses electromagnetic force to stir processed molten metal. It generates an alternating magnetic field through a copper coil and an electromagnetic force through the interaction of an induced current and magnetic field in a liquid metal [5]. It can be used to control the flow, heat transfer, and solidification of molten steel by electromagnetic forces, thus enabling optimization of the casting process [6]. Electromagnetic stirring in mold (M-EMS) technology is a vital application form of EMS, which is an effective means to optimize the flow of molten steel in the continuous casting process. It is widely used in the continuous casting production of high-quality and unique steel. The slab's solidification structure, center segregation, and inclusion distribution can be improved by controlling the flow and heat transfer of molten steel. However, M-EMS also has some adverse consequences, such as negative segregation [7]. This segregation, also known as a white band, will affect the hardenability, surface hardness, and mechanical properties of steel, and it is often challenging to eliminate in the subsequent process of continuous casting [8,9]. There is no unified explanation for the formation mechanism of negative segregation bands [10], and the widely accepted theory is mainly the solute washing mechanism. It is believed that the electromagnetic force makes the molten steel flow, flushing the solidification front and removing the solute-enriched molten steel [11,12]. In this case, a negative segregation band is formed. Kor et al. [13] reported that when the billet enters and leaves the electromagnetic stirring zone, the billet's sudden slow and fast solidification rate is the main reason for negative segregation bands. Ayata et al. [14] found that the subcutaneous negative segregation of the continuous casting billet gradually deteriorated with the increase in M-EMS intensity. It is considered that the electromagnetic stirring drives the flow of molten steel and the impact on the nozzle. These phenomena bring the high-concentration solute discharged from the solidification front to the central liquid phase area, causing the formation of subcutaneous negative segregation. Wu et al. [15] reported that the carbon segregation index reduced from 1.412 to 1.201 with electromagnetic torque of M-EMS from 230 to 400 cN·cm. Tao et al. [10] found that a suitable current of EMS could reduce the carbon segregation in industrial tests, and the segregation became severe with the progressive increase in current. The segregation behavior in the continuous casting process involves multiple processes, such as molten steel flow, solute migration, heat transfer, and solidification [16]. For molten steel's opacity and high-temperature characteristics, the liquid phase flow state and solute transport behavior of molten steel would not be directly understood by traditional casting industrial tests. Therefore, numerical simulation methods have been adopted by many researchers to carry out relevant studies [17]. Jiang et al. [9] reported that M-EMS with a higher stirring current would not continue to increase the equiaxed crystal ratio of the continuous casting billet, but more serious negative segregation appears. The effects of equiaxed grain settlement and liquid hot melt flow lead to negative segregation bands and positive segregation near the center of the billet. Okazawa et al. [18] adopted a numerical simulation method to quantitatively describe the effect of EMS on the molten steel flow structure by using the homogeneity index and found that the different installation positions and logarithms of agitators would significantly impact the flow structure in the mold

Sun et al. [19] established a segmented three-dimensional electromagnetic–thermal coupling solute transport model to better understand the macroscopic segregation formation in the billet during continuous casting. The effect of M-EMS operating parameters on the segregation was studied by conducting industrial tests. It was found that the simulated W-shaped segregation distribution along the casting thickness was in good agreement with the measured distribution. Zhang et al. [20] used a multi-physical field numerical model to study the macroscopic transfer behavior in the billet. Then, it was found that with the increase in M-EMS current density, the high-temperature region would move upward, and the negative segregation under the skin would finally become more severe due to the intense flushing of the initial solidified shell caused by the flow. Yu et al. [21] numerically simulated the magnetic field, temperature field, and inclusion trajectory of M-EMS under different parameters for round billet continuous casting. Based on the numerical simulation results, industrial experiments were carried out to investigate the solidification structure obtained under different M-EMS parameters and the macro-segregation behavior of high-carbon steel during continuous casting. Fang et al. [22] established a multi-physical field numerical model to investigate the flow, temperature field, and solute concentration field in bloom (380 mm × 280 mm) casting under the action of M-EMS. It was found that M-EMS made the distribution of temperature, solute, and solidified shell more uniform. However, M-EMS has not been found to improve the central segregation of this bloom.

In order to explore the causes and improvement measures of segregation in continuous casting, many researchers have developed multi-physical field models to reveal the macroscopic transport phenomena of the continuous casting electromagnetic stirring process.

However, due to the differences in steel grades and equipment in the continuous casting production processes, a unified conclusion has not yet been formed to fully understand the heat transfer, flow structure, and solute distribution of molten steel under the action of M-EMS. Therefore, in this paper, the electromagnetic–thermal-flow multi-physical field coupling model is established to simulate the temperature field, flow field, and distribution characteristics of carbon in the molten steel of a 200 mm × 200 mm billet, and to explore the influence of M-EMS on the segregation behavior of the billet under different parameters, to provide a technical reference for continuous casting production.

2. Model Description

2.1. Assumptions

The flow and heat transfer in the numerical model of EMS correlates with the magnetic field. Thus, the Navier–Stokes, Maxwell, heat transfer, and solute transport equations should be coupled solutions to the quantitative analysis of the transport phenomena in the mold. To simplify the complexity and enhance computational efficiency, the assumptions are as follows:

1. The steel in the calculation domain is considered as incompressible Newtonian viscous fluid;
2. The computing domain is considered vertical and the influence of mold oscillation and taper is ignored;
3. The electromagnetic field in the effective zone of M-EMS does not affect the molten steel flow;
4. While solving the solute diffusion, the interaction of solute elements can be ignored;
5. The heat transfer phenomenon between the molten steel and the top slag of the meniscus is ignored.

2.2. Governing Equations

(1) Fluid Flow Model

The continuity and momentum equations for the melt flow pattern under turbulent conditions can be described as follows:

$$\nabla \cdot U = 0 \tag{1}$$

$$\frac{\partial}{\partial t}(\rho U) + \nabla \cdot (\rho U U) = \nabla \cdot ((\mu_l + \mu_t) \nabla \cdot U) - \nabla P + S_p + F_m \tag{2}$$

where U denotes flow velocity (m/s), μ_l denotes laminar viscosity (kg/m^{-1}·s^{-1}), μ_t denotes turbulent viscosity (kg/m^{-1}·s^{-1}), ρ denotes density (kg/m^3), P denotes static pressure (Pa), S_p denotes momentum sink, and F_m denotes electromagnetic force (N/m^3). The standard $\kappa - \varepsilon$ model was used to describe the turbulent flow of molten steel as follows:

$$\frac{\partial}{\partial t}(\rho \kappa) + \nabla \cdot (\rho U \kappa) = \nabla \cdot \left[\left(\mu_l + \frac{U_t}{\sigma_k}\right) \nabla \kappa\right] + G_k + G_b - \rho \varepsilon + S_k \tag{3}$$

$$\frac{\partial(\rho \varepsilon)}{\partial t} + \nabla \cdot (\rho U \varepsilon) = \nabla \cdot \left[\left(\mu_l + \frac{U_t}{\sigma_\varepsilon}\right) \nabla \varepsilon\right] + C_{1\varepsilon} \frac{\varepsilon}{\kappa} (G_\kappa + C_{3\varepsilon} G_b) - C_{2\varepsilon} \rho \frac{\varepsilon^2}{\kappa} + S_\varepsilon \tag{4}$$

G_k denotes the turbulence kinetic energy referring to mean velocity gradients, G_b denotes the turbulence kinetic energy referring to buoyancy, and μt is the turbulent viscosity. $C_{1\varepsilon}$, $C_{2\varepsilon}$, $C_{3\varepsilon}$, and C_μ are constant and σ_k and σ_ε are the turbulent Prandtl numbers for κ and ε, respectively. $C_{1\varepsilon}$, $C_{2\varepsilon}$, C_μ, σ_k, and σ_ε are recommended to be 1.44, 1.92, 0.09, 1.0, and 1.3, respectively [23]. When considering the effects of the porous material, S_k and S_ε denote the source terms.

(2) Heat Transfer Model

The following equation can describe the energy conservation during the continuous casting process:

$$\frac{\partial}{\partial t}(\rho H) + \nabla \cdot (\rho U H) = \nabla \cdot \left(k_{eff} \nabla T\right) \tag{5}$$

where k_{eff} denotes the effective thermal conductivity (W·m^{-1}·K^{-1}), and T and H denote the temperature (K) and the total enthalpy, respectively.

(3) Solute Transport Model

The solute distribution is affected by Fickian diffusion, convection, and phase transport. The solute conservation equation is used to describe the solute transport phenomenon as follows:

$$\frac{\partial(\rho C_i)}{\partial t} + \nabla \cdot (\rho U C_i) = \nabla \cdot (\rho f_s D_{s,i} \nabla C_{s,i}) + \nabla \cdot \left(f_l \left(\rho D_{l,i} + \frac{\mu_t}{Sc_t}\right) \nabla C_{l,i}\right) \\ - \nabla \cdot (\rho f_s (U - U_s)(C_{l,i} - C_{s,i})) \tag{6}$$

where $C_{l,C}$ and $D_{s,C}$ (m^2/s) denote the solute diffusion coefficients of solute elements in liquid and solid phases [24].

(4) Electromagnetic Field Model

Maxwell's equations can describe the electromagnetic field during continuous casting with M-EMS:

$$\nabla \times H = J + \frac{\partial D}{\partial t} = J_s + J_e + \frac{\partial D}{\partial t} \tag{7}$$

$$\nabla \times E = -\frac{\partial B}{\partial t} \tag{8}$$

$$\nabla \cdot B = 0 \tag{9}$$

$$\nabla \cdot D = \rho \tag{10}$$

where H (A·m^{-1}) denotes magnetic field strength, D (C·m^{-2}) denotes electric flux density, J (A·m^{-2}) denotes electric current density, E (V·m^{-1}) denotes electric field strength, B (T) denotes magnetic flux density, and ρ (C·m^{-3}) denotes electric charge density.

Then, the time-averaged electromagnetic force can be calculated by:

$$F_m = 0.5 Re(J \times B) \tag{11}$$

where F_m is the time-averaged electromagnetic force and Re represents the genuine part of a complex number.

2.3. Model Parameters and Boundary Conditions

The parameters of the continuous casting and the physical properties of molten steel are given in Table 1. The finite element model consists of an electromagnetic agitator, a toroidal core, and 12 coil packages with a 30 degree symmetry distribution. In order to ensure the accuracy of the calculations, the area near the mold walls and the outlet of the submerged water is encrypted with grids. The total number of meshes in the model is about 500,000, and the model mesh partition and device photos are shown in Figure 1.

Table 1. Continuous casting parameters and physical properties of molten steel [22,25,26].

Parameter	Value	Parameter	Value
Nozzle inner diameter/mm	40	Equilibrium partition coefficient	0.34
Nozzle outer diameter/mm	60	Magnetic permeability/H·s^{-1}	1.257×10^{-6}
Depth/mm	120	Electric conductivity/S·m^{-1}	7.14×10^{5}
Length of mold/mm	900	Latent heat of fusion/J·kg^{-1}	271,000
Casting speed/m·min^{-1}	1.3	Solidus temperature/K	1762
Frequency/Hz	3, 4, 5	Liquidus temperature/K	1783
Current/A	100, 300, 500	Diffusion coefficient of carbon in liquid phase/cm^2·s^{-1}	$0.0052\exp(\frac{-11700}{8.314 \cdot T})$
Inlet temperature/K	1858	Surface emissivity	0.8
Density of molten steel/kg/m^3	$-0.823 \times T + 7100$	Constant pressure heat capacity of molten steel/J·(kg·K)$^{-1}$	$0.0071 \times T^2 - 15.255 \times T + 8959$
Dynamic viscosity of molten steel/Pa·s	0.0065	Thermal conductivity of molten steel/W·(m·K)$^{-1}$	$10^{-5} \times T^2 - 0.033 \times T + 50.265$

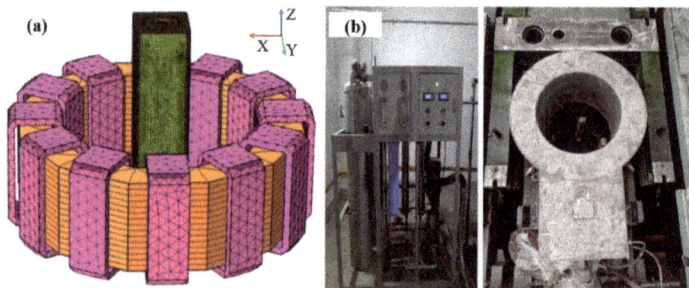

Figure 1. (a) Schematic diagram of model mesh division; (b) equipment photos.

3. Experimental Implementation

A schematic of the billet continuous casting and picture of the M-EMS are shown in Figure 1. The M-EMS with a Klem winding was installed at 0.5 m below the meniscus when the 200 mm × 200 mm carbon steel was produced. A φ10 mm drill bit was used to sample the cross−section of the continuous casting billet, and 9 points were taken on each continuous casting section (the length of the billet is L = 200 mm; the width is W = 200 mm; the thickness is H = 10 mm). Point 5 is the central point, point 2 is W/4, point 3 is W/2, point 4 is W3/4, point 6 is L/4, point 7 is L/2, point 8 is L3/4, and point 1 and point 9 are 10 mm under the skin of the edge of the casting billet, respectively. The sampling diagram is shown in Figure 2, in which the billet sample for analysis was drilled with a tungsten steel drill bit. Carbon content analysis of the billet section in the transverse and longitudinal directions was conducted to evaluate the carbon segregation of the billet. Carbon concentration at the billet subsurface of each point was analyzed with a carbon–sulfur analyzer (EMIA Pro, Horiba Inc., Kyoto, Japan). For this research, 30Cr13 steel was used, and the main chemical components are listed in Table 2.

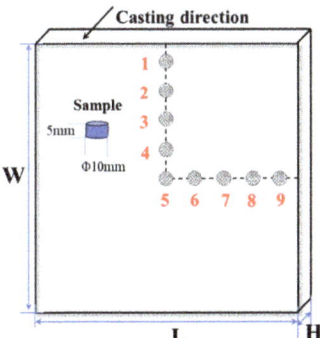

Figure 2. Sampling diagram of the influence of M−EMS on carbon segregation in casting billet.

Table 2. The main chemical components of 30Cr13 steel (wt.%).

C	Mn	P	S	Si	Cr	Ni
0.725	0.3~0.5	≤0.035	≤0.008	0.25–0.50	12.10–12.60	≤0.3

In order to investigate the variation of segregation of the carbon steel, the segregation index was used to characterize the macroscopic segregation of the billet. The carbon segregation index of the billet is calculated according to Equation (12) [27]:

$$a = \frac{C_i}{C_x} \qquad (12)$$

where a is the element segregation index; C_i is the element concentration at the test position i; and C_x is the average element concentration at all detection locations on the specimen section.

4. Results and Discussion

4.1. Model Validation

To verify the reliability of the multi−physical field model describing continuous casting with M−EMS, the internal magnetic distribution results of the billet at the parameters of 240 A current and 4 Hz frequency are presented in Figure 3b. In addition, they are compared with the results reported in the literature under the same conditions [28], as presented in Figure 3a. Then, it can be identified that the minimum flux density predicted by the numerical model in this research occurs in the center of the billet, the magnetic flux density is evenly distributed on the cross−section of the billet, the electromagnetic force rotates along the cross−section, the maximum magnetic flux density is located at the billet corner, and the minimum magnetic flux density is located at the center of the billet. The magnetic field profiles calculated with the model in this paper are consistent with the structure and trends of the magnetic field distribution in the comparative literature. The maximum induced magnetic field intensity calculated by the simulation in this paper is 0.002973 T, while the maximum induced magnetic field intensity in the comparison literature is 0.003021 T. The relative error between the two values for maximum induced magnetic field intensity is only 1.59%, which indicates that the model established in this study can be used for magnetic field prediction. It can be used to assess the influence of EMS on the billet continuous casting process.

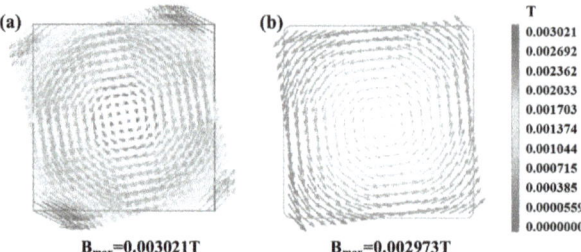

Figure 3. Distribution of magnetic induction intensity inside continuous casting billet under electromagnetic stirring: (**a**) Results of literature reports [28]; (**b**) calculation results of this study.

4.2. Melt Heat Transfer

The temperature field distribution of the vertical center section of the mold under different M−EMS currents is shown in Figure 4. In the beginning, without the action of M−EMS, the molten steel is directed diagonally into the inner cavity of the mold from the ingoing nozzle, and the molten steel diverges around the lower part. Then, the upward reflux of molten steel is reduced and the temperature is lower, leading to a significant temperature drop in the lower part of the meniscus and around the inlet, where the temperature is lower than in other areas. At the same time, the temperature range of the molten steel below the inlet does not favor the solidification of the casting billet. As the action of M−EMS at the current increases from 100 A to 500 A, the temperature of the molten steel around the inlet increases appreciably. At a current of 100 A, the region of influence of the induced magnetic field is limited, and the stirring effect acts only on the molten steel below the ingoing nozzle. When superimposed with inertia, the upward flow of molten steel in this region is weakened, resulting in heat loss of the molten steel below the meniscus, so the temperature of molten steel below the meniscus is lower than that without M−EMS. As shown in Figure 4, the temperature of the molten steel at a current of 300 A is most uniform in the same horizontal plane near the outlet of the mold, and the superheating of the molten steel can be effectively reduced or even eliminated.

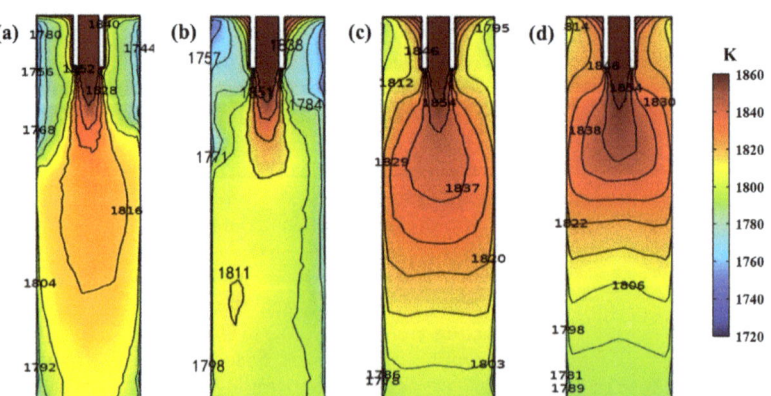

Figure 4. Effect of current on the temperature field of the vertical central section in the continuous casting mold: (**a**) without M−EMS; (**b**) 100 A; (**c**) 300 A; (**d**) 500 A.

Figure 5 shows the temperature field distribution of the vertical central section of the mold at different frequencies, where the current is set to 300 A. As the frequency increases from 3 Hz to 5 Hz, the temperature of the molten steel around the inlet of the mold first decreases and then increases, with values of 1837 K, 1829 K, and 1846 K, respectively. Then, a significant upward shift of the high−temperature zone can be seen, which favors a shortening of the solidification time of the casting billet. At the same time, the high−temperature zone of the molten steel is gradually confined to the upper part of the mold, and more of the heat of the molten steel is likely to be carried away by cooling water in the continuous casting mold. By comparing Figure 5a–c, the isothermal region edge of 1854 K is closer to the submerged entry nozzle with increasing frequency from 3 Hz to 5 Hz. Around the meniscus, the temperature of the molten steel increases with the increase in M−EMS frequency, with values of 1795 K, 1804 K, and 1813 K, respectively. At this moment, the maximum temperature difference is 18 K, and the temperature of the molten steel rises significantly with increasing frequency. When the frequency exceeds 3 Hz, the temperature in the center of the casting billet near the mold outlet is the lowest in the same horizontal plane, which may not be favorable for the solidification of the casting billet. Then, the excessive temperature gradient impairs the formation of an equiaxed crystal structure, which is unfavorable for uniform solute distribution [29].

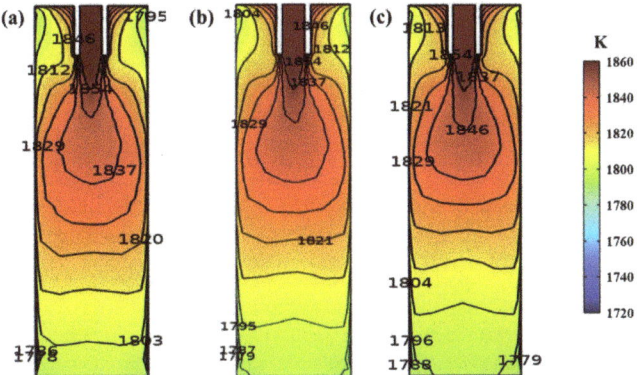

Figure 5. Effect of frequency on the temperature field of the vertical central section in the continuous casting mold: (**a**) 3 Hz; (**b**) 4 Hz; (**c**) 5 Hz.

4.3. Melt Flow Field

Figure 6 shows the distribution of the flow field in the mold under different current and frequency parameters. When the molten steel enters the inner cavity of the mold from the inlet, the molten steel diffuses without M−EMS, and the velocity gradually decreases, forming a narrow vortex at the lower left and right sides of the mold inlet. The flow field at the horizontal section of the center of the electromagnetic stirring device shows that the velocity distribution of the steel flow in the billet center is divided into three parts: the velocity is highest in the center of the billet, followed by the area near the surface of the billet, and the area with the lowest velocity is located between the center and the surface of the billet. When the M−EMS current is 100 A, the vortices on the left and right sides below the inlet are shifted upward, and a minor vortex is gradually formed below the bending moon surface, which is conducive to complete mixing and an improvement in the quality of the casting billet. At the same time, the horizontal profile of the flow field at the center of the electromagnetic stirring 0.5 m below the Mencius scale indicates that the induced magnetic field has a stirring effect on the molten steel. Due to the limited stirring effect of the electromagnetic force generated by the slight current on the molten steel, the molten steel forms a clockwise vortex on the horizontal surface, and the area with a higher flow speed is still located in the billet. As the current gradually increases to 300–500 A, the vortices in the lower part of the inlet gradually move upward and fuse with the vortices in the lower part of the meniscus, eventually forming giant vortices. According to the distribution characteristics of the magnetic field, the electromagnetic force acting on the molten steel near the mold walls is more significant. At the same time, the flow rate of the molten steel near the walls of the mold also increases and is higher than that of the molten steel at the center of the billet and at the four corners of the mold.

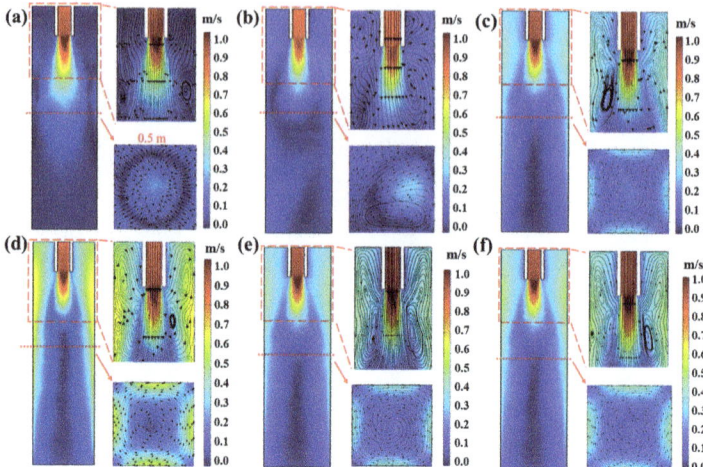

Figure 6. Flow field distribution in the mold under different M−EMS parameters: (**a**) without M−EMS; (**b**) 3 Hz + 100 A; (**c**) 3 Hz + 300 A; (**d**) 3 Hz + 500 A; (**e**) 4 Hz + 300 A; (**f**) 5 Hz + 300 A.

Compared with Figure 6a, Figure 6c,e,f show that a large vortex flow is generated near both sides of the mold inlet with a frequency of 3–5 Hz, and the influence area of the vortex flow can reach near the meniscus. Then, the vortex flow can accelerate the entire flow of molten steel. At the same time, the flow field in the horizontal section of the center of the stirrer 0.5 m below the curved moon indicates that the magnetic field generated by M−EMS has a horizontal rotational stirring effect on the steel. The electromagnetic forces acting on the steel near the mold walls are more significant than at the center of the billet, and the steel flow rate near the mold walls is higher than at the center of the billet and at the corners of the mold. This distribution of electromagnetic force contributes to a higher stirring speed in the area near the billet walls, which is consistent with the results reported in the literature [22], with the different maximum velocities for the different stirring currents. Then, it will facilitate the full mixing flow of the molten steel and promote the formation of equiaxed grain, which is beneficial to the homogenization of temperature and solute components in the molten steel [30].

Figure 7 shows the curve of the tangential velocity of the molten steel in the central cross−section of the electromagnetic agitator at 0.5 m below the meniscus for different current intensities and frequencies. The tangential velocity of molten steel in the flowing area increases with the increase in the distance from the center of the billet, which is beneficial to promoting the rotational flow of the steel in the horizontal plane [31,32]. This is because the electromagnetic force generated by the electromagnetic agitator on molten steel increases with the increase in the distance from the center of the billet. Moreover, the velocity gradually decreases before approaching the casting walls due to the wall effect. Then, the tangential velocity of the molten steel near the walls gradually decreases until it reaches zero. Meanwhile, with the increase in current intensity, the maximum tangential velocity of molten steel is 0.19, 0.31, and 0.38 m/s at 100 A, 300 A, and 500 A, respectively. Because the electromagnetic force near the mold walls area is large and the electromagnetic force away from the mold walls area is small, the flow trend of the molten steel cannot be changed by the simple adjustment of the M−EMS current or frequency parameters; this can affect only the flow speed of the liquid steel. Thus, as the M−EMS frequency is increased from 3 Hz to 5 Hz, the maximum tangential velocities of the molten steel are 0.31, 0.39, and 0.44 m/s, respectively.

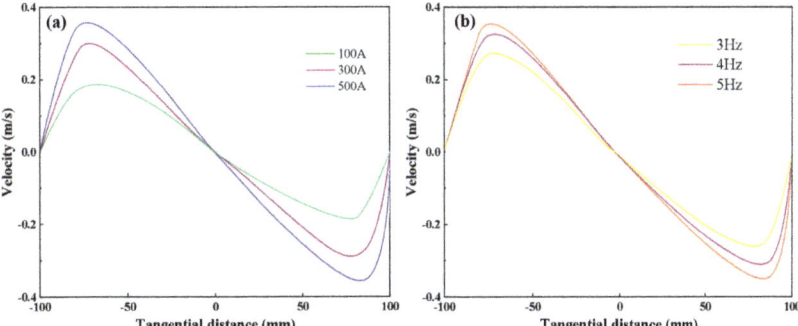

Figure 7. Influence of M−EMS on tangential velocity of molten steel in central cross−section: (**a**) Current at 100, 300, 500 A of 3 Hz; (**b**) frequency at 3, 4, 5 Hz of 300 A.

4.4. Distribution of Solute Carbon

Figure 8 shows the solute distribution cloud of the mold outlet section under different M−EMS parameters, where blue represents low solute concentration, and red represents high solute concentration. As the M−EMS current is continuously increased from 100 A to 300 A, the subcutaneous negative segregation of carbon is improved, and the solute mass percentage increases from 0.6918% to 0.6928%. With the continuous increase in M−EMS current from 300 A to 500 A, the negative segregation degree of the billet increases accordingly, and the solute mass percentage decreases from 0.6928% to 0.6903%. This is because the molten steel's temperature drops rapidly near the mold walls, and the solute transfer decreases. Meanwhile, the rotational velocity continues to increase with the continuous accumulation of electromagnetic force on the molten steel. As a result, the solute at the interface between the high− and the low−temperature regions is unable to transfer to the low−concentration region, forming a negative segregation zone under the skin of the billet.

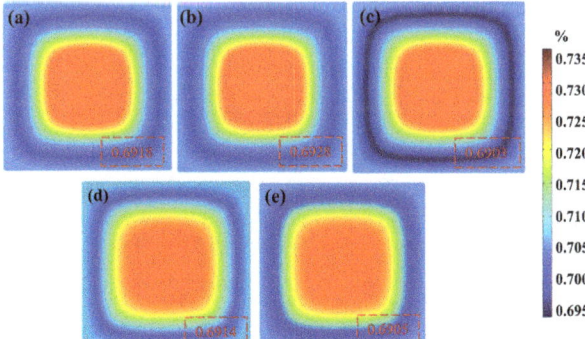

Figure 8. Solute distribution at mold outlet section under different M−EMS parameters: (**a**) 3 Hz + 100 A; (**b**) 3 Hz + 300 A; (**c**) 3 Hz + 500 A; (**d**) 4 Hz + 300 A; (**e**) 5 Hz + 300 A.

Compared with Figure 8b,d,e, the solute carbon concentration in the subcutaneous area of the casting billet decreases from 0.6928% to 0.6905% with the increase in frequency from 3 Hz to 5 Hz. This may be caused by the electromagnetic stirrer's force on the molten steel being limited by the depth of penetration of the electromagnetic field. In the case of a fixed electromagnetic stirring current, the depth of electromagnetic force on molten steel decreases with increasing frequency. Therefore, the effect of electromagnetic forces on the molten steel is minor at a higher stirring frequencies. At a frequency of 3 Hz, the electromagnetic force significantly affects the horizontal rotating flow driven by the electromagnetic force. Then, M−EMS can effectively stir the molten steel and thoroughly mix the flow to reduce the negative segregation.

Figure 9 shows solute carbon distribution in the horizontal center of the mold outlet under different M−EMS parameters. As the steel enters the inner cavity of the mold from the entrance, the solute diffuses from the center to the surrounding walls, and the center of the mold always maintains a high carbon concentration due to the continuous inflow of molten steel. As the distance from the center of the slab decreases, the solute carbon concentration initially decreases significantly and then increases rapidly. Negative segregation degree is largest for a current of 500 A, followed by a current of 100 A. The negative segregation degree is the lowest with a current of 300 A. When the electromagnetic stirring frequency is specific, the stirring power is small at a current of 100 A, and the molten steel can not be thoroughly stirred. When the current is 500 A, the power is overly large, and the electromagnetic force causes an over−stirring of the molten steel. The same trend variation was also reported in Tao et al.'s experimental study [10]. Since the horizontal rotation of the molten steel driven by M−EMS is significant, the scouring effect on the solid–liquid phase interface of the casting billet is increased. As a result, the solute is carried away from the solid–liquid interface, leading to negative segregation below the casting billet. The effect of the electromagnetic force driving the horizontal rotation of the molten steel is too significant, which will increase the scouring effect on the solid–liquid phase interface of the casting billet [33,34]. Therefore, the solute between the solid–liquid interface is taken away, resulting in negative segregation under the casting billet. In addition, the washed carbon accumulates inside the billet as the solidification process progresses, resulting in the high carbon level in the center of the billet shown in Figure 9. Based on the numerical simulations of the carbon distribution shown in Figures 8 and 9, the appropriate M−EMS parameters are a current of 300 A and a frequency of 3 Hz, which can improve the subcutaneous carbon negative segregation of the casting billet.

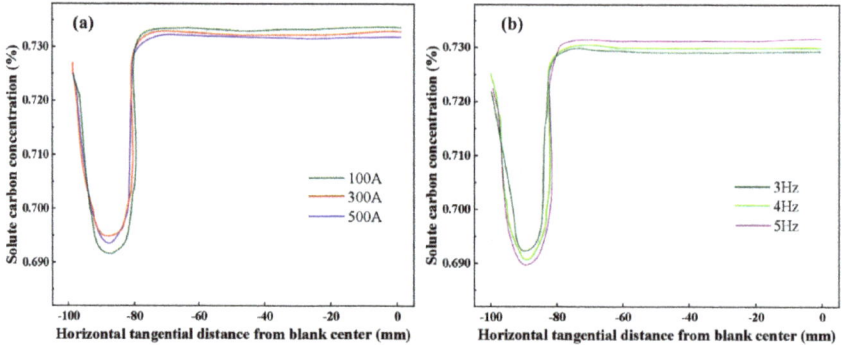

Figure 9. Solute carbon distribution at the horizontal center of mold outlet: (**a**) Current at 100, 300, 500 A of 3 Hz; (**b**) frequency at 3, 4, 5 Hz of 300 A.

4.5. Results for Solute Carbon inside Billet

The conclusions of the previous numerical simulation studies are an essential basis for optimizing casting practice, and it is necessary to optimize continuous casting production using numerically optimal parameters and to verify the effect of reducing carbon sequestration. Due to the focus of the present study on the carbon segregation variation with the action of electromagnetic stirring in the mold, four furnace casting billet samples with or without M−EMS were used for comparative study under a casting speed of 1.3 m/min and superheat degree of 30 °C. According to the optimal parameters of the numerical simulation, the billet cross−sections with 10 mm thickness were collected under steady casting conditions with or without M−EMS. The low−magnification photographs of the collected billets are shown in Figure 10a. A slight negative segregation (bright white) band can be seen in the billet without M−EMS, as shown in Figure 10a (1#). Then, the negative segregation band disappears in a section of the billet obtained under the M−EMS parameter of 3 Hz + 300 A, as shown in Figure 10a (2#,3#,4#). As seen in Figure 10a of the carbon segregation index analysis, the billet shows negative segregation at the edges, positive segregation at 1/2 radius, negative segregation from 1/2 radius to the center, and positive segregation at the center. In order to evaluate the effect of the M−EMS parameters recommended by the numerical model, the carbon segregation index was calculated using Equation (12), with the sampling scheme as shown in Figure 2. According to the carbon concentration of different positions of the billet section shown in Figure 10b, the segregation index of the billet section is calculated to be 0.92~1.10 at the M−EMS parameter of 3 Hz + 300 A. Since the formation of carbon segregation in continuous casting is a highly complex process, and this study focuses only on the effects of M−EMS, there are still some deviations between the results of numerical simulations and the production practice. The extreme difference between the maximum value of the segregation index minus the minimum value is 0.030%, which implies that the segregation of the billet is better optimized under the combination of the mixing parameters. This indicates that 300 A + 3 Hz is an appropriate M−EMS parameter to ameliorate carbon segregation in a billet with a size of 200 mm × 200 mm.

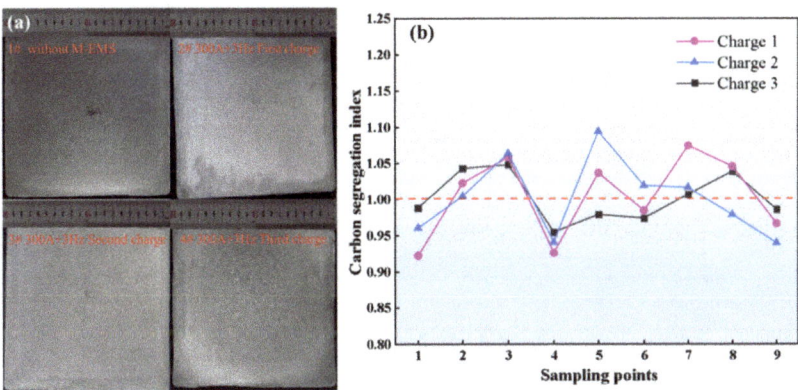

Figure 10. (a) Low magnification photograph of the billets obtained at a speed of 1.3 m/min and superheat of 30 °C: 1#—without M−EMS, 2#/3#/4#—three charge with M−EMS at 3 Hz + 300 A; (b) analysis results of carbon segregation index of the cross−section of the casting billet.

5. Conclusions

(1) With the increase in M−EMS frequency from 3 Hz to 5 Hz, the high−temperature zone is shifted upward, which is conducive to the rapid solidification of the cast billet. At the same time, as the M−EMS current increases from 100 A to 500 A, the maximum horizontal tangential velocity of the molten steel increases from 0.19 m/s to 0.38 m/s, which promotes the rotational flow of the steel in the horizontal plane.

(2) When M−EMS is applied, the temperature of the molten steel near the mold walls decreases and the flow velocity near the mold walls increases. Hence, the wash−out effect on the initial solid created by the fluid flow and the variation of the solute transport rate produces a trend in the concentration of solute carbon first of a significant decrease and then of a rapid increase, with decreases in distance from the billet center.

(3) When the M−EMS current is 300 A at a frequency of 3 Hz, producing a suitable mixing intensity and suitable current penetration depth effect, the negative segregation control of the carbon solute is controlled optimally. The carbon segregation index of the billet section is controlled at 0.92~1.1.

Author Contributions: Conceptualization, P.L. and G.Z.; experimental setup design, P.L., G.Z. and P.Y.; validation, P.L., N.T. and Z.F.; analysis, G.Z., P.Y. and P.Z.; writing, P.L., G.Z., P.Y. and P.Z.; project administration, P.L. and G.Z.; funding acquisition, G.Z. All authors have read and agreed to the published version of the manuscript.

Funding: The work was supported by the National Natural Science Foundation of China (NSFC), grant number "52074140".

Institutional Review Board Statement: Not applicable.

Informed Consent Statement: Not applicable.

Data Availability Statement: The data presented in this study are available on request from the corresponding author.

Conflicts of Interest: The authors declare no conflict of interest.

References

1. Zhang, G.J.; Zhang, X.D.; Zhang, J.F.; Liu, X.H. Effect of CC Process on Macro Carbon Segregation in Steel Bloom. *Contin. Cast.* **2010**, *3*, 43–46.
2. Wang, Y.D.; Zhang, L.F.; Zhang, H.J. Simulation of the macrosegregation in the gear steel billet continuous casting process. *Chin. J. Eng.* **2021**, *43*, 561–568.
3. Zhang, Z.; Wu, M.; Zhang, H.; Hahn, S.; Wimmer, F.; Ludwig, A.; Kharicha, A. Modeling of the as−cast structure and macrosegregation in the continuous casting of a steel billet: Effect of M−EMS. *J. Mater. Process. Technol.* **2021**, *301*, 117434. [CrossRef]
4. Chen, D.F.; Wei, Q.C. The Solute Transport and Segregation during Solidifying Process in Continuous Casting. *J. Rare Earths* **2000**, *18*, 206–208.
5. Tzavaras, A.; Brody, H. Electromagnetic Stirring and Continuous Casting−Achievements, Problems, and Goals. *JOM* **1984**, *36*, 31–37. [CrossRef]
6. Beitelman, L. Effect of mold EMS design on billet casting productivity and product quality. *Can. Metall. Q.* **1999**, *38*, 301–309. [CrossRef]
7. Zhou, D.F.; Fu, J.; Wang, P.; Liu, Q. Effect of CC Process Parameters on Carbon Segregation of Bearing Steel. *IronSteel* **1999**, *34*, 22–26.
8. Sun, H.B.; Zhang, J. Study on the macro segregation behavior for the bloom continuous casting: Model development and validation. *Metall. Mater. Trans. B* **2013**, *45*, 1133. [CrossRef]
9. Jiang, D.; Zhu, M.Y. Solidification Structure and Macro segregation of Billet Continuous Casting Process with Dual Electromagnetic Stirrings in Mold and Final Stage of Solidification: A Numerical Study. *Metall. Mater. Trans. B* **2016**, *47*, 3446–3458. [CrossRef]
10. Tao, J.J.; Zhu, H.G.; Wu, X.L. Effect of Carbon Segregation and Residual B on Hardenability of Bearing Steel 20CrMnTiH1. *Spec. Steel* **2007**, *28*, 58–59.
11. Sasaki, K.; Sugitani, Y.; Kobayashi, S. Effect of fluid flow on the formation of the negative segative segregation zone in steel ingots. *Tetsu Hagane* **1979**, *65*, 60–69. [CrossRef]
12. Asai, S.; Nishio, N.; Muchi, I. Theoretical analysis and model experiments on electromagnetically driven flow in continuous casting. *Trans. Iron Steel Inst. Jpn.* **1982**, *22*, 126. [CrossRef]
13. Guo, D.; Hou, Z.; Peng, Z.; Liu, Q.; Chang, Y.; Cao, J. Influence of Superheat on Macrosegregation in Continuously Cast Steel Billet from Statistical Maximum Viewpoint. *ISIJ Int.* **2021**, *3*, 515. [CrossRef]
14. Ayata, K.; Mori, T.; Fujimoto, T.; Ohnishi, T.; Wakasugi, I. Improvement of macrosegregation in continuously cast bloom and billet by electromagnetic stirring. *Trans. Iron Steel Inst. Jpn.* **1984**, *24*, 931–939. [CrossRef]
15. Wu, H.; Wei, N.; Bao, Y.; Wang, G.; Xiao, C.; Liu, J. Effect of M−EMS on the Solidification Structure of a Steel Billet. *Int. J. Miner. Metall. Mater.* **2011**, *18*, 159–164. [CrossRef]
16. An, H.; Bao, Y.; Wang, M.; Zhao, L. Effects of electromagnetic stirring on fluid flow and temperature distribution in billet continuous casting mould and solidification structure of 55SiCr. *Metall. Res. Technol.* **2018**, *115*, 103–115. [CrossRef]
17. Jiang, D.B. Numerical Investigation of Macrosegregation Formation and Influence Mechanisms of External Field During Continuous Casting Process. Ph.D. Thesis, Northeastern University, Shenyang, China, 2018.
18. Okazawa, K.; Toh, T.; Fukuda, J. Fluid Flow in a Continuous Casting Mold Driven by Linear induction Motors. *ISIJ Int.* **2001**, *41*, 851–858. [CrossRef]
19. Sun, H.B.; Zheng, L.; Li, L.J.; Nie, B.H. A Mathematical Model on Macro−Segregation Formation for Popular Bloom Continuous Casting Process. *Mater. Sci. Forum* **2019**, *944*, 770–777. [CrossRef]
20. Zhang, W.; Luo, S.; Chen, Y.; Wang, W.; Zhu, M. Numerical simulation of fluid flow, heat transfer, species transfer, and solidification in billet continuous casting mold with M−EMS. *Metals* **2019**, *9*, 66. [CrossRef]
21. Yu, H.Q.; Zhu, M.Y. Influence of electromagnetic stirring on transport phenomena in round billet continuous casting mould and macrostructure of high carbon steel billet. *Ironmak. Steelmak.* **2012**, *39*, 574–584. [CrossRef]

22. Fang, Q.; Ni, H.W.; Wang, B.; Zhang, H.; Ye, F. Effects of EMS induced flow on solidification and solute transport in bloom mold. *Metals* **2017**, *7*, 72. [CrossRef]
23. Launder, B.E.; Spalding, D.B. The numerical computation of turbulent flows. *Comput. Methods Appl. Mech. Eng.* **1990**, *3*, 269–289. [CrossRef]
24. Wang, B.; Xie, Z.; Jia, G.L.; Lin, G.Q.; Ji, Z.P. Determination of Electromagnetic stirring Parameters at solidified End and Its Effect on Center Segregation. *Iron Steel* **2007**, *3*, 18–21.
25. Chen, H.B.; Long, M.J.; Chen, D.F.; Liu, T.; Duan, H.M. Numerical study on the characteristics of solute distribution and the formation of centerline segregation in continuous casting (CC) slab. *Int. J. Heat Mass Transf.* **2018**, *126*, 843–853. [CrossRef]
26. Wang, Y.D. Study on the Effect of Electromagnetic Stirring on the Macrosegregation of Steel Continuous Casting Blooms. Ph.D. Thesis, University of Science and Technology Beijing, Beijing, China, 2021.
27. Guo, S.Z.; Yang, W.Y.; Chen, G.A.; Sun, Z.Q. Effect of Nb on deformation enhanced ferrite transformation in low carbon steel. *J. Univ. Sci. Technol. Beijing* **2007**, *29*, 582–590.
28. Li, Y.G.; Wang, H.B.; Bai, X.S.; Sun, Y.H.; Zhang, M.H.; Jia, J.P. Numerical Simulation of Electromagnetic Stirring in Billet Mold. *Iron Steel Vanadium Titan.* **2020**, *41*, 108–114.
29. Bridge, M.R.; Rogers, G.D. Structural effects and band segregation formation during the electromagnetic stirring of strand−cast steel. *Metall. Trans. B* **1984**, *15*, 581–589. [CrossRef]
30. Chen, M.; Zhou, D.W. Application of EMS on Continuous Caster. *Wide Heavy Plate* **2009**, *15*, 7–11.
31. Wang, Y.T.; Yang, Z.G.; Zhang, X.F.; Li, J.F.; Wang, B.; Liu, Q. Effects of electromagnetic stirring on the flow field and level fluctuation in bloom molds. *J. Univ. Sci. Technol. Beijing* **2014**, *36*, 7.
32. Zhang, J.; Wang, E.G.; Deng, A.Y.; Zhang, X.W.; He, J.C. Influence of M−EMS Parameters on Distribution of Magnetic Field in Continuous Casting Bloom. *China Metall.* **2011**, *21*, 15–18.
33. Sun, H.B.; Li, L.J.; Wu, X.X.; Liu, C.B. Effect of subsurface negative segregation induced by M−EMS on componential homogeneity for bloom continuous casting. *Metall. Res. Technol.* **2018**, *115*, 603. [CrossRef]
34. Sun, Z.Q.; Zheng, L.Y.; Luo, B. Effect of Electromagnetic Stirring on Carbon Segregation in Blooms of Medium Carbon Steel. *South. Met.* **2019**, *226*, 8–13.

Disclaimer/Publisher's Note: The statements, opinions and data contained in all publications are solely those of the individual author(s) and contributor(s) and not of MDPI and/or the editor(s). MDPI and/or the editor(s) disclaim responsibility for any injury to people or property resulting from any ideas, methods, instructions or products referred to in the content.

Article

Drying Kinetics of Microwave-Assisted Drying of Leaching Residues from Hydrometallurgy of Zinc

Chunlan Tian [1], Ju Zhou [1], Chunxiao Ren [1], Mamdouh Omran [2,*], Fan Zhang [1,*] and Ju Tang [1]

[1] Kunming Key Laboratory of Energy Materials Chemistry, Yunnan Minzu University, Kunming 650500, China; tianchunlan@ymu.edu.cn (C.T.); zhouju@ymu.edu.cn (J.Z.); renchunxiao@ymu.edu.cn (C.R.); tangju@ymu.edu.cn (J.T.)

[2] Faculty of Technology, University of Oulu, 90570 Oulu, Finland

* Correspondence: mamdouh.omran@oulu.fi (M.O.); zhangfan@ymu.edu.cn (F.Z.)

Abstract: In the hydrometallurgical process of zinc production, the residue from the leaching stage is an important intermediate product and is treated in a Waelz kiln to recover valuable metals. To ensure optimal results during the Waelz kiln process, it is necessary to pre-treat the residues by drying them first due to their higher water content. This work studies the residue's drying process using microwave technology. The study results indicate that microwave technology better removes the residue's oxygen functional groups and moisture. The dehydration process's effective diffusion coefficient increases as the microwave's heating power, the initial moisture content, and the initial mass increase. The Page model is appropriate for imitating the drying process, and the activation energy of the drying process for the residues is −13.11217 g/W. These results indicate that microwave technology efficiently dries the residues from the leaching stage. Furthermore, this study provides a theoretical basis and experimental data for the industrial application of microwave drying.

Keywords: zinc hydrometallurgy; residues from leaching stage; microwave drying; drying rate; kinetics model

Citation: Tian, C.; Zhou, J.; Ren, C.; Omran, M.; Zhang, F.; Tang, J. Drying Kinetics of Microwave-Assisted Drying of Leaching Residues from Hydrometallurgy of Zinc. *Materials* **2023**, *16*, 5546. https://doi.org/10.3390/ma16165546

Academic Editor: Andres Sotelo

Received: 2 July 2023
Revised: 3 August 2023
Accepted: 5 August 2023
Published: 9 August 2023

Copyright: © 2023 by the authors. Licensee MDPI, Basel, Switzerland. This article is an open access article distributed under the terms and conditions of the Creative Commons Attribution (CC BY) license (https:// creativecommons.org/licenses/by/ 4.0/).

1. Introduction

Zinc is highly utilized in various steel, machinery, and chemical industries. It is the third most produced and consumed non-ferrous metal [1,2]. For now, the conventional zinc recovery process is the "roasting—leaching—electrowinning" process, and approximately 80% of total world production comes from this process [3]. The residue from the leaching stage is an essential intermediate product and contains about 30% Zn and 30% Fe. In the zinc recovery process, the residue is primarily subjected to pyrometallurgical treatment and treated in a Waelz kiln to recover valuable metals like zinc, lead, indium, and germanium [4,5]. Therefore, the high moisture content of the residue will lead to difficulties in raising the kiln temperature and the formation of agglomerates in the pre-heat zone, resulting in a lower metal recovery rate. In addition, The dust removal efficiency of the following section and the service life of the dust collection bags will be adversely affected by the high moisture level in the residue [6]. So, the residue needs to be pre-treated by drying before the Waelz kiln process.

Solar irradiation and hot air are usually used to dry materials [7]. Earth receives over 10,000 times more solar energy than the world needs [8,9]. This makes solar drying a sustainable and cost-effective solution for drying various materials and products. However, it is subject to several limitations, such as deterioration of material quality caused by external natural factors and uncontrollable external parameters of the drying process, such as temperature, water content, drying air flow rate, and low drying efficiency [10,11]. Hot air-drying is commonly used in food processing, agriculture, and manufacturing industries to preserve and dry products [12–14]. It involves the circulation of hot air around the material, which causes the moisture to evaporate and be carried away. Yang et al. [15]

studied drying shiitake mushrooms using hot air. The results indicated that the hot air-drying process enhances the lignification of shiitake mushrooms by increasing lignin, affecting the nutritional and edible properties. Hot air drying can effectively reduce the moisture in the material. However, it is an energy-intensive and time-consuming process.

Microwave is an electromagnetic wave with a frequency between 300 MHz and 300 GHz, a wavelength between 1 mm and 1 m, and a penetrating nature [16]. The materials harvest electromagnetic energy and convert it into thermal energy to remove moisture in the microwave field [17]. Microwave drying is a new method with quick heat and higher thermal efficiency. The principle feature of microwave drying is completely different from other traditional heating methods [18]. The materials are uneven heating via thermal radiation in conventional drying methods. The surface of materials first absorbs the thermal energy, and the internal temperature of the materials increases slowly [19]. However, moisture evaporation mainly occurs on the material's surface, and the direction of the moisture gradient is from inside to outside. The direction of diffusion of moisture in the materials is directly opposite to the temperature gradient, resulting in blocked diffusion, extended drying time, and reduced drying efficiency. In the microwave drying process, the material is heated evenly. The temperature gradient has the same direction as the moisture gradient, resulting in the same direction of heat and mass transfer, forming an internal pressure gradient, prompting the rapid evaporation of internal moisture, thus allowing the moisture to diffuse to the surface and evaporate quickly, significantly reducing drying time and improving drying effectiveness [20]. Carvalho et al. [21] showed that conventional convection drying (50–70 °C) of malt takes 540–840 min, and the microwave reduces the processing time by about 95%. De Faria et al. [22] investigated the impact of microwave-assisted drying on maize seeds with a wet basis moisture content of 20%, using temperatures of 40, 50, and 60 °C, and power ratings of 0, 0.6, and 1.2 W/g. The results showed that at a temperature of 40 °C and a power rating of 0.6 W/g, the drying time was reduced by approximately 5 h. Demiray Seker et al. [23] also found a significant difference in the drying time of microwave-dried onion flakes compared to convection drying. Convection drying was carried out at different temperatures (50, 60 and 70 °C), 570 min at 50 °C, 360 min at 60 °C and 210 min at 70 °C. Microwave drying was performed at three different microwave power levels, 328, 447 and 557 W. Drying was performed for 66 min at 328 W, 45 min at 451 W and 40 min at 557 W. Compared to traditional drying methods, microwave drying offers advantages such as high energy efficiency, low energy consumption, selective heating, fast heating speed, and absence of by-products [24].

In recent years, a great deal of research has been carried out on microwave drying processes for materials in the food industry, metallurgical industry, etc. Kipcak et al. [25] studied the effect on drying kinetics of microwave power level, rehydration properties, and energy consumption of mussels. The experimental results showed that the trial data were fitted with the Weibull model of relations, and the drying effect was ideal when the microwave power was 360 W. Liu et al. [26] studied the effect of microwaves on the depth drying of Zhaotong lignite. The experimental results showed that the Page model could characterize the microwave depth drying behavior of Zhaotong lignite in diverse settings, and the effective diffusion coefficient enhanced with the power level of the microwave output increases with increasing microwave output power level. Huang et al. [27] have studied the desiccation of ammonium polyvanadate using microwaves and discussed the effects of inert mass, microwave output, and initial moisture content on the drying performance of ammonium polyvanadate. The experimental results showed that the average speed of the drying process was favorably correlated with the output power of the microwaves and that the Modified Page model may accurately designate the process of drying with microwaves of ammonium polyvanadate. Ling et al. [28] studied the microwave drying experiments of zirconium oxide. The experimental results showed that the surface diffusion coefficient increased as the microwave output increased. The sample weight decreased, and the Henderson and Pabis model fit the dehydration process well. Zheng et al. [29] studied natural titanium dioxide's microwave heating properties and kinetics. They found that the particle size of the material had a considerable impact on the drying effect, leading to the observation that microwaves can fragment mineral particles.

Based on the previous research, this study proposed a solution to the problem of using the microwave to dry zinc-leaching residue for the traditional zinc recovery process with high energy consumption and low efficiency, which realised the solution to the problem of high energy consumption and low efficiency, and effectively removed the water in the leaching residue. The study examines the influence of various microwave heating powers, initial zinc-leaching residue moisture contents, and initial masses on the microwave drying characteristics of zinc-leaching residue. Four classical drying kinetic models, namely Page, Lewis, Wang and Singh, Quadratic, were utilized to fit the experimental information, and the best-fitting model was selected. The effective diffusion coefficient and activation energy were also calculated. The results of this study provide valuable research evidence and reference for the discipline of acid-leaching residue drying.

2. Experimental Section

2.1. Experimental Substances

The zinc leaching residue selected for this experiment was collected from a zinc smelter in Yunnan Province, China; the dry basis moisture of the slag is in the region of 9.2%, and heated in a tumble dryer at 60 °C for 24 h to maintain a constant composition. Sulphuric acid is used to recover valuable metals in hydrometallurgical zinc production. In this paper, the leaching residue from an acid leaching stage. The acid residue underwent analysis using the method recommended by the national standard of the People's Republic of China (GB/T 10561-2019) [30], and the main element contents are shown in Table 1.

Table 1. Chemical elements of raw materials for acid leach residue (wt%).

Element	Fe_2O_3	ZnO	SiO_2	PbO	CuO	MnO	K_2O
Contain (wt%)	36.448	29.243	14.617	11.262	1.720	0.637	0.479
Element	CdO	Tr_2O_3	TiO_2	C1	SrO	V_2O_5	Ag_2O	ZrO_2
Contain (wt%)	0.414	0.358	0.289	0.240	0.238	0.084	0.055	0.048

X-ray diffraction (XRD) was performed to analyze the material composition of the residue. The outcomes, as shown in Figure 1, indicate that the primary components of the original material are lead sulfate and zinc ferrite. The copper element is below 5%, and no associated phase is found in the XRD analysis.

The size and distribution of the particles of the acid-leaching residue were analyzed using a laser particle sizer. Figure 2 presents the histogram of the size distribution and the accumulated volume distribution curve of the acid-leaching resin. As can be seen from Table 2, the nominal particle size of the acid wash in the region of 0.938 to 15 μm with a volume mean diameter (Mz) of 4.19 μm and a median particle size (D_{50}) of 3.23 μm.

Table 2. Particle size distribution of acid leaching residue.

Particle Size Distribution Ratio	D_{10}	D_{20}	D_{30}	D_{40}	D_{50}	D_{60}	D_{70}	D_{80}	D_{90}	D_{95}	Mz
Particle size/μm	0.938	1.438	1.943	2.541	3.23	4.18	5.44	7.14	10.3	15	4.19

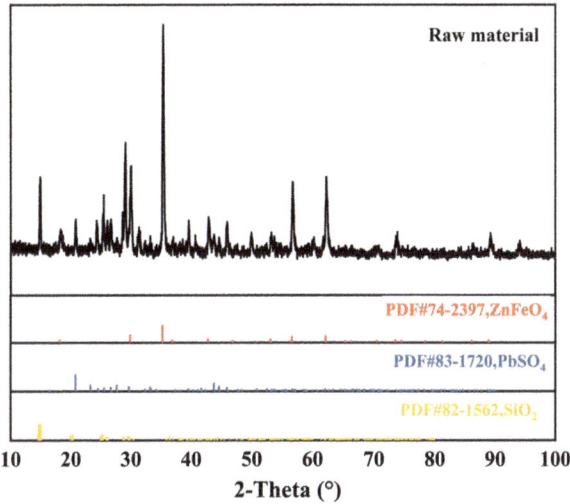

Figure 1. XRD results of acid leaching residue.

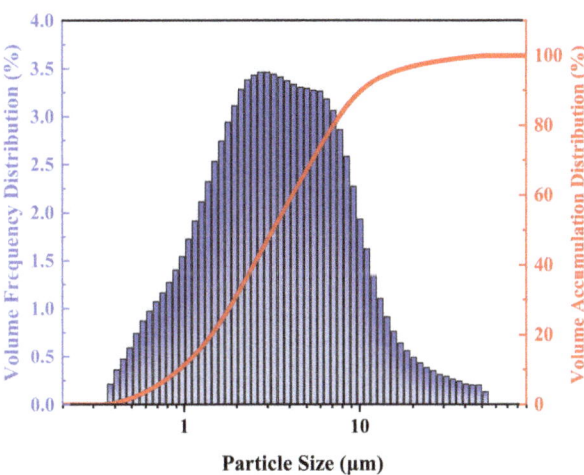

Figure 2. Distribution of particle size of acid leaching residue.

2.2. Experimental Equipment

A diagram of the microwave drying plant is shown in Figure 3. The plant comprises three primary components: the Gas delivery unit, the heating installation, and the information-gathering tool. The gas supply system regulates the gas atmosphere in the reaction chamber during drying. The heating system is a microwave drying oven that heats the sample to dry it, while the data acquisition system comprises a mass sensor and a computer steering gear. The mass sensor detects changes in the material's quality during the microwave heating and drying process, and the computer control system records the data to track changes in quality.

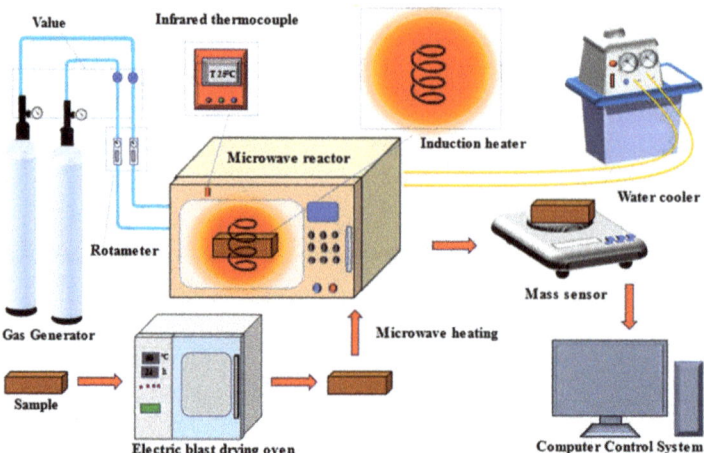

Figure 3. Diagram of microwave drying equipment.

2.3. Experimental Procedures

The zinc-leaching residue from a factory was used as the sample for the experiment. To guarantee the validity of the scores, a specific amount of the residue was mixed with water and placed in a rectangular corundum crucible measuring 9 cm × 6 cm × 1.6 cm. This was done to eliminate any influence that the starting condition of the specimen may have on the microwave drying process during the experiment. The crucible was then placed in a microwave-drying oven for the experiment. This study aimed to investigate the impact of microwave heating power, initial moisture content, and initial mass on the microwave drying of the residue. The experiment involved varying the microwave heating power (160 W, 320 W, 480 W, 640 W, 800 W), initial moisture content (3.25%, 6.55%, 8.15%, 9.75%, 11.35%), and initial mass (10 g, 15 g, 20 g, 25 g, 30 g) of the residue.

Zinc leachates with 8.60% initial moisture and 20 g mass were selected by the controlled variable method and dried at variables in microwave heating power (160 W, 320 W, 480 W, 640 W, 800 W); test samples were made up using the concept of dry basis moisture percent. Zinc leachate Samples with various initial moisture contents (3.25%, 6.55%, 8.15%, 9.75%, 11.35%) were prepared, weighed 20 g, and dried at 480 W microwave heating power; zinc leachate with an initial moisture content of 8.60% and initial masses of 10 g, 15 g, 20 g, 25 g, and 30 g were selected using the controlled variable method. The three single-variable trials were carried out at a heating power of 480 W. The sample masses were recorded at 30 s intervals for all three univariate experiments until the sample masses did not change on three occasions.

3. Methods

3.1. Calculation of the Relevant Parameters

(1) Moisture content

The calculation of the moisture content (M_t) of acid-leaching residue at a given time (t) during the experiment is based on the principle of mass conservation. This is done using the Equation (1) provided.

$$M_t = \left[1 - \frac{m_0 \times (1 - M_0)}{m_t}\right] \times 100\% \tag{1}$$

where: M_t—the water content of the acid leaching residue at time t, %; M_0—initial moisture content of the zinc leaching residue, %; m_0—the initial mass of the acid leaching residue (with water), g; m_t—the mass of the acid leaching residue at time t, g.

(2) Hydration ratios

$$MR = \frac{M_t - M_e}{M_0 - M_e} \times 100\% \qquad (2)$$

where, M_t—the water holding capacity of acid leach residue at time t, %; M_e—equilibrium wet basis moisture content, %; M_0—the initial water content of acid leaching residue, %.

(3) Instantaneous drying rate

The effect of water removal from the acid-leaching residue at time t can be expressed in terms of the instantaneous drying rate, which is calculated according to Equation (3):

$$R = -\frac{M_{(t+\Delta t)} - M_t}{\Delta t} \qquad (3)$$

where R—instantaneous drying rate, g/s; M_t—mass of acid leaching residue from drying to t, g; $M_{(t+\Delta t)}$—mass of acid leaching residue from drying to $(t + \Delta t)$, g.

(4) Average drying rate

$$Ra = -\frac{m - m_0}{t} \qquad (4)$$

where, Ra—average drying rate, g/s; m—mass of acid leaching residue at the end of drying, g; m_0—initial mass of acid leaching residue (with moisture), g; t—time used at the end of drying, s.

3.2. Numerical and Kinetic Models for Thin Layer Drying

The thin-layer drying method involves transferring heat from hot air to a wet material by means of convection. The heated air is then forced to escape through the thin layer of the substance, carrying out the evaporated water vapor [31]. Currently, The primary focus of drying kinetics research is the use of mathematical simulations of thin-layer drying curves to derive thin-layer drying models. Several thin-layer drying models are available, generally categorized as theoretical, semi-theoretical, empirical, and semi-empirical. Empirical models share similarities, given that they rely heavily on laboratory testing conditions and provide limited insight into the drying behavior of the material [32]. Empirical models are developed using experimental data and measurement analysis and can be used to simulate drying processes [33]. Theoretical models consider external conditions and the influence of moisture transport within the material [34]. Semi-theoretical and empirical models which consider only the resistivity of the external resistance to moisture exchange with air can give more accurate results., improve predictions of drying behavior, and reduce assumptions by relying on experimental data [35]. In this work, Four common drying kinetic models, Page, Lewis, Wang and Singh, Quadratic, as shown in Table 3, were selected to fit the experimental data during the microwave drying of zinc leaching residue. To choose the most appropriate drying model, a series of statistical indicators were used, including the fit coefficient (R^2), residual sum of squares (RSS), and F-value values. The expressions for these statistical indicators are shown in Table 4.

Table 3. The trial information on drying kinetics models by fitting.

No.	Model	Model Expression
1	Page	$M_R = exp(-kt^n)$
2	Lewis	$M_R = exp(-kt)$
3	Wang and Singh	$M_R = 1 + at + bt^2$
4	Quadratic	$M_R = a + bt + ct^2$

Table 4. Mathematical model statistical indicators.

Statistical Indicators	Equations
slimming coefficient	$R^2 = \dfrac{\sum(MR_{exp}-\overline{MR}_{exp})(MR_{pre}-\overline{MR}_{pre})}{\sqrt{(MR_{exp}-\overline{MR}_{exp})^2(MR_{pre}-\overline{MR}_{pre})^2}}$
residual sum of squares	$RSS = \sum_{i=1}^{N}(MR_{exp}-MR_{pre})^2$
F-Value	$F\text{-}Value = \dfrac{MS_{treatment}}{MS_{error}}$

4. Results and Discussion

4.1. Effect of Microwave Heating Power on the Drying of Acid-Leaching Residues Using Microwaves

Figure 4 illustrates the moisture content variation curves of the acid-leaching residue with microwave heating time and the variation curves of the average drying rate with the microwave heating power at different levels. The graph shows that below the 160 W, 320 W, 480 W, 640 W, and 800 W microwave heating power, the drying completion time is 960 s, 570 s, 420 s, 390 s, and 390 s, respectively. Drying time gradually decreases as microwave power increases. The acid-leaching residue microwave drying process can be broken down into three stages. The first stage consists of preheating, which involves using microwave heating to evaporate the internal free water in the sample. During this stage, there is a small amount of evaporation, resulting in a minimal change in the moisture content. The second stage is steaming, when the internal bound water evaporates from the sample's surface. The rate of decline in moisture content is gradually accelerated during this stage. The end of evaporation is the third stage, during which only a small amount of bound water is left in the sample. The moisture content shows a slow decline, eventually reaching zero. The study reveals that the removal rate depends on the initial moisture content, mass, and microwave heating power. The higher the microwave heating power, the more microwaves are absorbed by the acid-leaching residue, resulting in an accelerated drying rate [36].

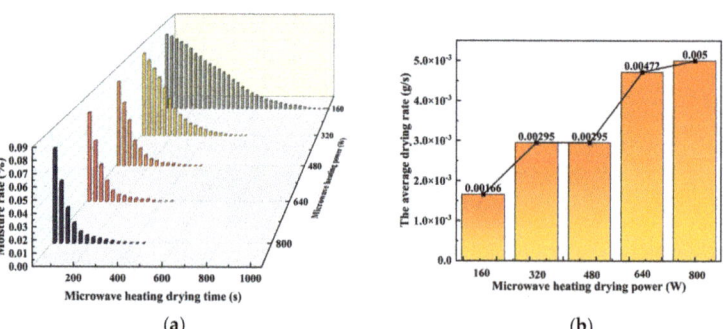

Figure 4. (a) Moisture rate-microwave heating time curves of acid leaching residue of zinc at different microwave heating powers; (b) Average drying rate-microwave heating power curves of acid leaching residue of zinc at different microwave heating powers.

Figure 4b plots the average drying rate for the acid leach residue at different microwave heating powers. At 160 W, 320 W, 480 W, 640 W, and 800 W, the average drying rates were 0.00166 g/s, 0.00295 g/s, 0.00295 g/s, 0.00472 g/s, and 0.005 g/s. Interestingly, the average drying rate was the same at 320 W and 480 W, and it accelerated above 480 W. Insufficient penetration of the microwaves into the acid-leaching residue could be the reason for the lower average drying rate at low microwave heating power [37]. If the microwave heating power exceeds 480 W, the average drying rate of acid-leaching residue increases slowly. This could be attributed to the saturation of microwave absorption by the residue. Further, an increase in microwave heating power causes the acid leach residue to no longer absorb microwaves, resulting in a wastage of microwave energy as the microwaves pass through the residue [38].

4.2. Effect of Initial Moisture Content on the Drying of Acid-Leaching Residues Using Microwaves

Figure 5 displays the moisture content variation curves of the acid leaching residue concerning the change of microwave exposure time and average drying rate variation curves with modification of initial moisture content at different initial moisture contents. The acid-leaching residue's drying completion time was 360 s, 300 s, and 270 s, 300 s, 270 s, as the initial moisture content increased, as shown in Figure 5a. The results in Figure 5b demonstrate that as the initial moisture content of the acid-leaching residue increased from 3.25% to 11.35%, the average drying rate also increased. This can be interpreted as being due to more water molecules in the residue at a higher initial moisture content, which absorbs more microwave energy and reaches boiling point more quickly. According to the study, the higher moisture content in acid-leaching residue leads to more absorption of microwaves and subsequent boiling of water molecules which escape into the environment.

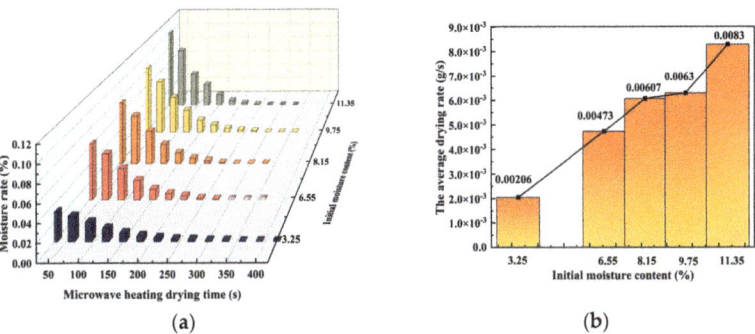

Figure 5. (a) Moisture rate-time curves for different initial moisture contents of the material. (b) Average drying rate-initial moisture content curve for different initial moisture contents of the material.

4.3. Effect of Initial Mass on the Drying of Acid-Leaching Residues Using Microwaves

This study examined the drying behavior of acid-leaching residues with varying initial masses. The residues were dehydrated using a microwave heating power of 480 W, and the corresponding changes in the moisture content and the rate of drying were recorded. The results are shown in Figure 6a, where the average drying rates and completion times varied depending on the initial mass of the residues. The drying rates ranged from 0.00218 g/s to 0.00907 g/s, and the completion times ranged from 300 s to 420 s.

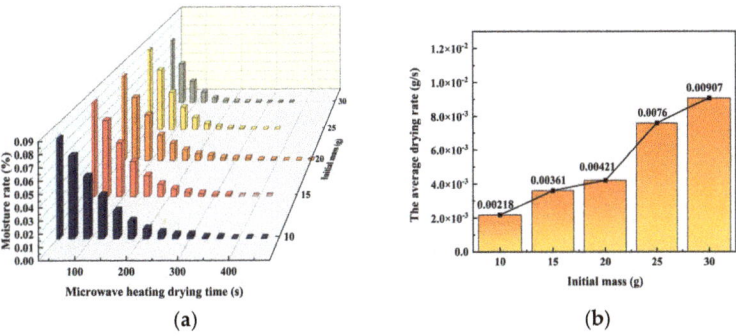

Figure 6. (a) Moisture rate-microwave heating drying time curves for acid leaching residue at different initial masses; (b) Acid leaching residue at different initial mass—average drying rate curve.

The study found that the proportion of drying was the highest, and the completion time was the shortest when the initial mass of the acid-leaching residue was 25 g. Additionally, it can be seen from Figure 6b that the average rate of drying tended to increase in proportion to the increase in the initial mass of the acid leach residue.

4.4. Results and Analysis of the X-ray Diffractometer

The phase composition of the Zn leach residue was essentially the same before and after microwave drying according to XRD tests on the Zn leach residue and after drying, indicating that microwave drying does not affect the phase composition of the material. The XRD pattern of zinc leaching slag before and after drying is shown in Figure 7.

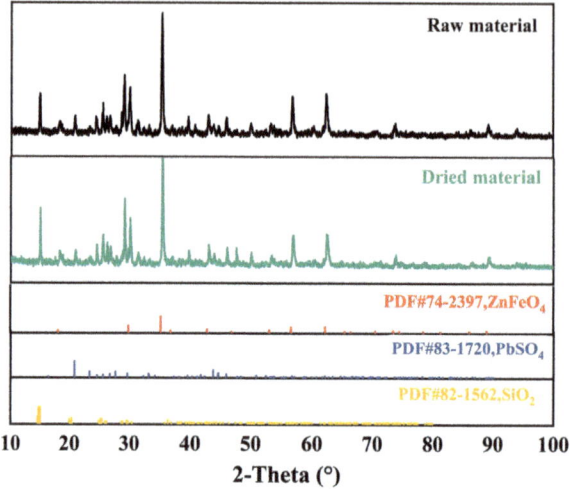

Figure 7. XRD spectra of acid-leaching residue before and after drying.

4.5. Results and Analysis of the Fourier Transform Infrared Spectroscopy

To examine the impact of microwave heating applied to the microscopic functional groups on the surface of samples of zinc leaching residue, FT-IR spectroscopy (NICOLET-IS10, Nicolet, USA) was conducted on both dried and undried samples within the spectral range of 4000~450 cm^{-1} was covered.

Water's absorption peaks and molecular bending-stretching vibrational frequencies appear around 1640 cm^{-1} and 3400 cm^{-1}, respectively [39]. As depicted in Figure 8, the FT-IR spectra of the pre-dried acid leaching residue showed a distinctive absorbance peak at 3414.976 cm^{-1}, which is indicative of stretching vibrations of the O-H bond and at 1645.534 cm^{-1} due to bending vibrations within the H-O-H plane. The FT-IR spectra taken from the dried acid leaching residue showed a characteristic absorcent peak at 3405.816 cm^{-1} arising from the extension vibration of the O-H bond and a characteristic absorcent peak at 1618.052 cm^{-1} arising from the inflexion vibration in the H-O-H plane. In contrast, it was found that the characteristic absorption peak due to the extension vibration of the O-H bond was weaker in the dried acid-leaching residue than in the acid-leaching residue before drying, indicating that the microwave was effective in removing the oxygen-containing functional groups from within the acid leaching residue. This experimental result is also in accordance with the work by Lin et al. [40].

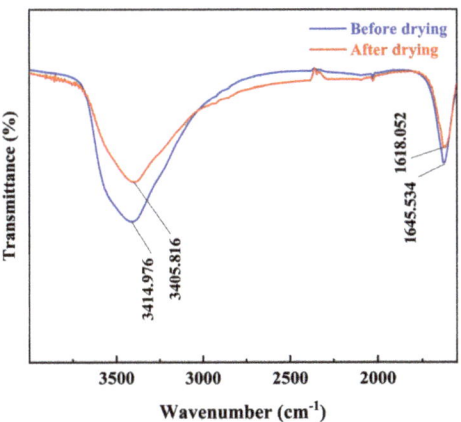

Figure 8. FT-IR spectra of acid-leaching residue before and after drying.

4.6. Drying Kinetics Model Fitting Process

The experimental results of microwave drying of leaching residue obtained for different experimental conditions were fitted using Page, Lewis, Wang and Singh Quadratic models. Tables 5–7 and Figure 9 display the fitted data for the models. Page, Lweis, Wang and Singh's quadratic model was used for microwave heating power of 480 W, an initial moisture content of 3.25%, and an initial mass of 10 g. In addition, the models were used to fit hydration ratio curves and experimental data with varying microwave heating powers, initial moisture content, and initial mass. A normal distribution plot was utilized as a reference for comparison [41]. Von the four models, the Page model was deemed to be the one which best fitted the experimental data.

Table 5. Data on the heating power of 480 W microwave fitted to various models.

Model	Params	R^2	RSS	F-Value
Page	$k = 1.34012 \times 10^{-4}$ $n = 1.87167$	0.99117	0.01233	1139.59521
Lewis	$k = 0.61636$	−0.44709	2.02034	1.38423×10^{-7}
Wang and Singh	$a = -0.00592$ $b = 8.34054 \times 10^{-6}$	0.93963	0.08428	160.80051
Quadratic	$a = 1.06336$ $b = -0.00643$ $c = 9.20115 \times 10^{-6}$	0.94357	0.07878	108.6863

Table 6. Data for different models fitted to acid-leaching residue with a moisture content of 3.25%.

Model	Params	R^2	RSS	F-Value
Page	$k = 8.09485 \times 10^{-5}$ $n = 1.91271$	0.9944	0.00822	1786.59595
Lewis	$k = 0.61636$	−0.67299	2.45557	9.87034×10^{-8}
Wang and Singh	$a = -0.00551$ $b = 7.48395 \times 10^{-6}$	0.95554	0.06526	219.75131
Quadratic	$a = 1.19557$ $b = -0.0073$ $c = 1.09222 \times 10^{-5}$	0.98555	0.02121	375.17481

Table 7. Data for different models fitted to an initial mass of 10 g of acid-leaching residue.

Model	Params	R^2	RSS	F-Value
Page	$k = 7.85963 \times 10^{-5}$ $n = 1.92866$	0.9965	0.00529	2907.07712
Lewis	$k = 0.61636$	−0.56991	2.373341	1.09979×10^{-7}
Wang and Singh	$a = -0.00559$ $b = 7.63484 \times 10^{-6}$	0.95834	0.06299	238.41688
Quadratic	$a = 1.16422$ $b = -0.007$ $c = 1.01621 \times 10^{-5}$	0.98082	0.029	306.8248

The Page model was used to fit the zinc leachate moisture content data at different microwave heating powers (160 W, 320 W, 480 W, 640 W, 800 W), initial moisture contents (3.25%, 6.55%, 8.15%, 9.75%, 11.35%) and initial masses (10 g, 15 g, 20 g, 25 g, 30 g). As shown in Table 8, the results showed that the Page model had higher R^2 values, smaller RSS values, larger F-values, lower Chi-sqr values and better normal distribution than the other three drying fit models. These results indicate that the Page model is the most suitable model to describe the drying kinetics of zinc leachate compared to the other three models.

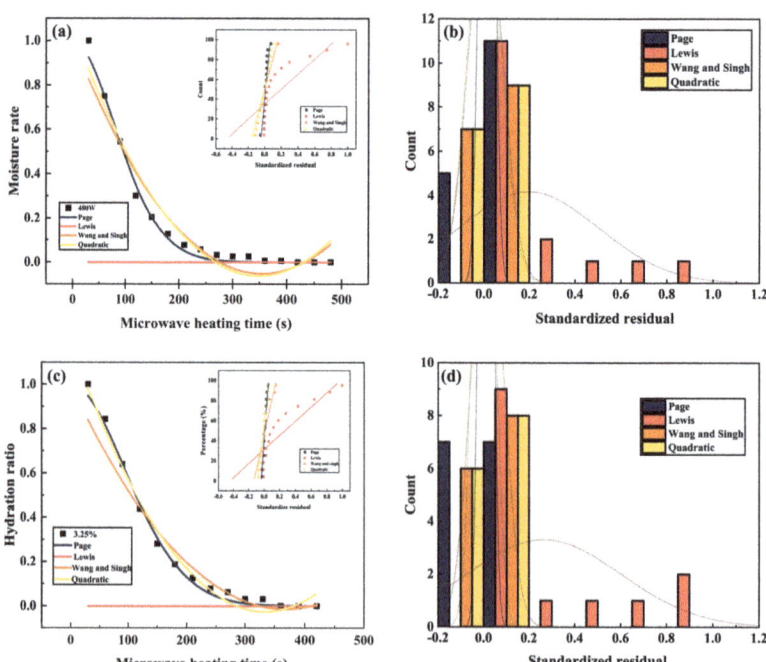

Figure 9. Cont.

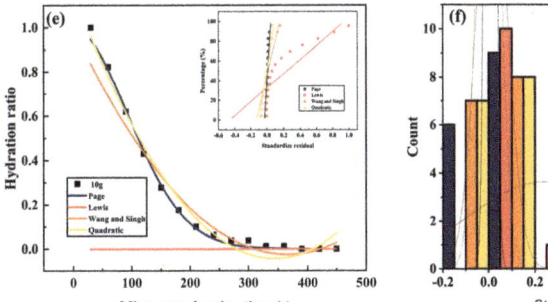

Figure 9. (**a**,**c**,**e**) Fitted different mathematical models to empirical data derived from the microwave drying process of zinc leach residues; (**a**) microwave heating power of 480 W, controlled water content of 8.60%, and mass of zinc leaching residue of 20 g; (**c**) moisture content of 3.25%, controlled microwave heating power of 480 W and mass of zinc leaching residue of 20 g; (**e**) microwave heating power of 480 W, controlled water content of 8.60% and mass of zinc leaching residue of 10 g. (**b**,**d**,**f**) shows the residual histograms for Page, Lewis, Wang and Singh, Quadratic models fitted to the hydration ratio data model: (**b**) residual histograms for Page, Lewis, Wang and Singh, Quadratic models fitted to a microwave heating power of 480 W, a control moisture content of 8.60% and a zinc leaching residue mass of 20 g; (**d**) residual histograms for Page, Lewis, Wang and Singh, Quadratic models fitted to a zinc leaching residue moisture content of 3.25%, a controlled microwave heating power of 480 W and a mass of 20 g. (**f**) residual histograms for Page, Lewis, Wang and Singh, Quadratic models fitted to a zinc leaching residue mass of 10 g, a controlled microwave heating power of 480 W and a moisture content of 8.60%.

Table 8. Experimental factors and coding.

	Condition	Params	R^2	RSS	F-Value
Microwave heating power	160 W	$k = 2.28837 \times 10^{-5}$ $n = 1.75036$	0.99763	0.00891	15,116.38519
	320 W	$k = 5.76718 \times 10^{-5}$ $n = 1.82929$	0.9971	0.00644	5620.87119
	480 W	$k = 1.34012 \times 10^{-4}$ $n = 1.87167$	0.99117	0.01233	1139.59521
	640 W	$k = 1.07697 \times 10^{-4}$ $n = 1.98482$	0.98746	0.0154	715.01768
	800 W	$k = 1.56908 \times 10^{-4}$ $n = 1.91609$	0.98269	0.02051	513.26985
Moisture content	3.25%	$k = 8.09485 \times 10^{-5}$ $n = 1.91271$	0.9944	0.00822	1786.59595
	6.55%	$k = 3.40842 \times 10^{-5}$ $n = 2.15056$	0.99572	0.00574	1874.06956
	8.15%	$k = 3.72183 \times 10^{-5}$ $n = 2.15197$	0.99637	0.00452	2025.6383
	9.75%	$k = 5.95382 \times 10^{-5}$ $n = 2.04029$	0.99566	0.00558	1855.28185
	11.35%	$k = 5.37631 \times 10^{-5}$ $n = 2.10929$	0.99196	0.00957	870.83348

Table 8. *Cont.*

	Condition	Params	R^2	RSS	F-Value
Mass	10 g	$k = 7.85963 \times 10^{-5}$ $n = 1.92866$	0.9965	0.00529	2907.07712
	15 g	$k = 6.82384 \times 10^{-5}$ $n = 1.98976$	0.99511	0.00694	1895.00272
	20 g	$k = 1.34012 \times 10^{-4}$ $n = 1.87167$	0.99117	0.01233	1139.59521
	25 g	$k = 6.04899 \times 10^{-5}$ $n = 2.07446$	0.99423	0.00714	1318.63153
	30 g	$k = 5.17039 \times 10^{-5}$ $n = 2.19605$	0.98961	001151	665.16048

Based on the observations in Figure 10a,c,e, the hydration rate data at different microwave heating powers, moisture contents, and masses can be fitted well with the Page model. Also, as shown in Figure 10b,d,f, the hydration rate data under the same conditions follow a normal distribution. These results also suggest that the Page model can effectively explain the behavior of microwave-dried zinc leachates under different conditions [42].

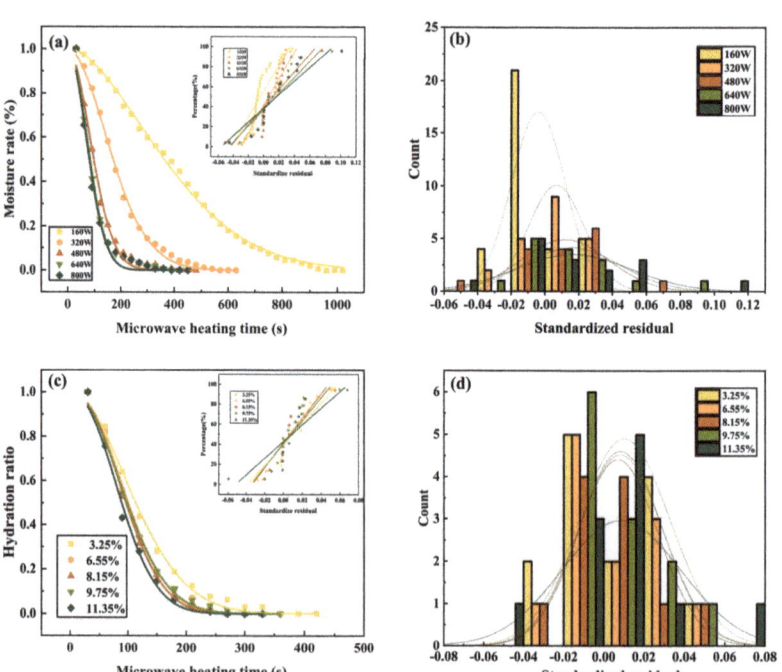

Figure 10. *Cont.*

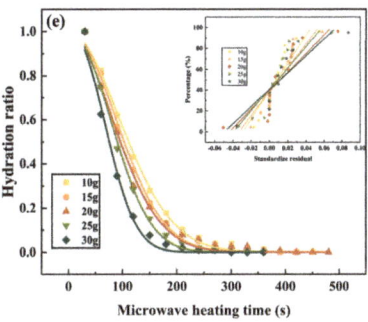
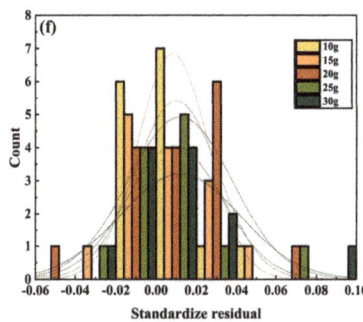

Figure 10. (a,c,e) Adaptation by Page hydration ratio data for variable microwave heating power, initial moisture content and initial mass, with small plots representing normal probability plots: (a) Fitted of Page hydration ratio data for different microwave heating power (160 W, 320 W, 480 W, 640 W, 800 W); (c) Fitted of Page hydration ratio data for different initial moisture content (3.25%, 6.55%, 8.15%, 9.75%, 11.35%); (e) Fitted graphs of Page hydration ratio data for different initial masses (10 g, 15 g, 20 g, 25 g, 30 g). (b,d,f) Histograms of residuals under the Page hydration ratio data model with different microwave heating power, initial moisture content and initial mass: (b) Histograms of residuals under the Page hydration ratio data model with different microwave heating power (160 W, 320 W, 480 W, 640 W, 800 W); (d) Histograms of residuals under the Page hydration ratio data model with different initial moisture content (3.25%, 6.55%, 8.15%, 9.75%, 11.35%); (f) histograms of residuals under the Page hydration ratio data model for different initial masses of 10 g, 15 g, 20 g, 25 g, 30 g).

4.7. Calculating the Diffusion Coefficient and Activation Energy

Fick's second law of diffusion is frequently used to describe the drying process involved in various materials [43]. A longitudinal fit of $\ln MR$ versus t was performed using the obtained microwave drying data, as shown in Figure 11. This experiment assesses the surface diffusion coefficient, obtaining the steepness and intercept from the linear plot formed between $\ln MR$ and t. The diffusion out of water in the zinc leach residue over the drying process is described by Fick's second law of diffusion [44].

$$MR = \frac{6}{\pi^2} \sum_{n=1}^{n=\infty} \frac{1}{n^2} \exp\left[-\frac{4n^2\pi^2 D_e}{d^2}t\right] \quad (5)$$

During a long drying time, $n = 1$, Equation (5) above can be simplified as

$$MR = \frac{6}{\pi^2} \exp\left[-\frac{4\pi^2 D_e}{d^2}t\right] \quad (6)$$

Taking the reciprocals to the logarithm of both sides of Equation (6) simultaneously gives Equation (7):

$$\ln MR = \ln\frac{6}{\pi^2} - \frac{4\pi^2 D_e}{d^2}t \quad (7)$$

From Equation (7), D_e represents the effective diffusion coefficient (m²/s), d refers to the particle size of the material (m), and t denotes the drying time (s).

Suitably, by fitting Equation (7) to relevant empirical data, a slant a of the fitted line can be derived, and the average effective diffusion coefficient may be estimated from Equation (7) using this slant:

$$D_e = -a\frac{d^2}{\pi^2} \quad (8)$$

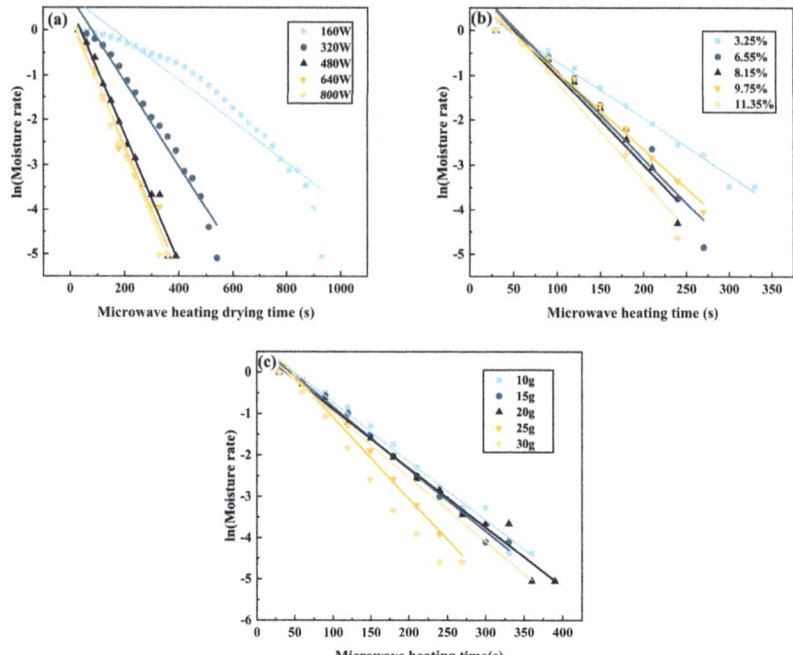

Figure 11. Linear fit of *ln MR* versus *t* for different.(a) Linear fit of *ln MR* versus *t* for various microwave heating powers; (b) Linear fit of *ln MR* versus *t* for initial different moisture contents; (c) Linear fit of *ln MR* versus *t* for several initial masses.

Table 9 shows the effective diffusion coefficients of acid leaching residue for different microwave heating powers, initial moisture content, and initial mass. Figure 12 shows the trend of the effect of different microwave heating power, initial moisture content and initial mass on the effective diffusion coefficient of acid leaching residue. Specifically, adjusting the microwave power to 800 W, the initial moisture content is 11.35%, and the initial mass is 30 g, the diffusion coefficient is the highest. If the microwave heating power is increased, a corresponding increase in the diffusion coefficient of the zinc leach residue is observed. This is because the faster heating rate leads to a more significant diffusion coefficient of the residue [45].

Table 9. Effective diffusion coefficients of acid leaching residue for different microwave heating powers, initial moisture content, and initial mass.

Microwave Heating Powder (W)	D_e (10^{-12} m²/s)	Moisture Content (%)	D_e (10^{-12} m²/s)	Mass (g)	D_e (10^{-12} m²/s)
160	1.89056	3.25	5.18777	10	5.84803
320	3.87545	6.55	8.04206	15	6.1679
480	5.87673	8.15	8.1815	20	5.87673
640	5.81932	9.75	7.06192	25	8.07487
800	6.18021	11.35	8.94838	30	8.63671

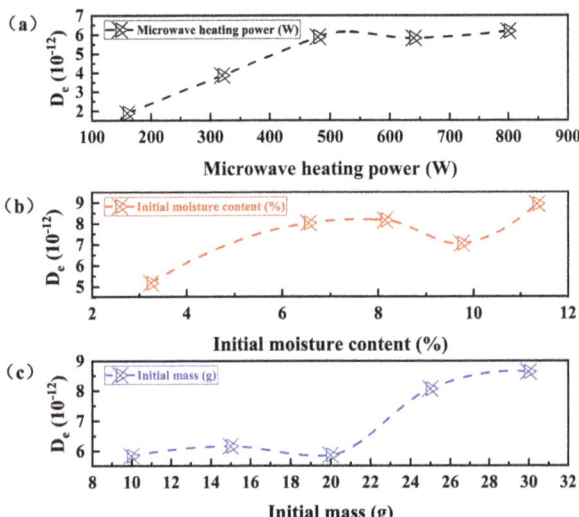

Figure 12. Effect of different microwave heating power, initial moisture content and initial mass on the effective diffusion coefficient of acid leaching residue. (**a**) The effect of different microwave heating power on the effective diffusion coefficient of acid-leaching residue; (**b**) The effect of different initial water content on the effective diffusion coefficient of acid-leaching residue; (**c**) the Influence of different initial mass on the effective diffusion coefficient of acid-leaching residue.

Based on the relationship between hydration ratio and heating time, it was found that the equilibrium time of the hydration ratio is higher for a microwave heating power of 640 W than that of 480 W. The reduction in the diffusion coefficient of zinc leaching residue could be attributed to fitting errors. The diffusion coefficient in zinc leaching residue increases gradually as the moisture content increases, but only up to a certain point. When the moisture content reaches 8.15%, the diffusion coefficient begins to weaken due to the increase in water. This weakening is caused by the water molecules inside the residue becoming weaker. However, when the moisture content is 11.35%, there may be a fitting error that causes an increase in the diffusion coefficient. The efficiency of zinc leaching residue to absorb microwaves and its diffusion coefficient increases gradually as its initial mass increases. This is because a small mass of less than 20 g cannot absorb enough energy, resulting in a low diffusion coefficient [46].

Dadali et al. proposed an alternative method to estimate the activation energy for microwave drying [47]. Activation energy is the energy needed to convert a molecule from its resting state to an active state where chemical reactions can occur. D_e depends on the material mass and the microwave heating power intensity for the Arrhenius-type equation:

$$D_e = D_o \exp\left(\frac{E_a m}{P_m}\right)$$

where D_e is a function of material quality and microwave power level; D_0 is the finger front factor of the Arrhenius equation in m²/s, E_a which is the activation energy (W/g), m which is the mass of the product 20 g and P_m which gives the microwave heating power (W).

The calculation equation can be described as,

$$ln D_e = ln D_0 + \frac{20 E_a}{P_m} \tag{9}$$

As can be seen in Figure 13, the diffusion coefficient increases linearly over the microwave power range extending from 160 to 480 W. As a result, three different microwave heating powers were selected, namely 160 W, 320 W, and 480 W. Using the equivalence formula combining $\ln D_e$ and $20/P_m$, the activation energy was determined to be -13.11217 g/W.

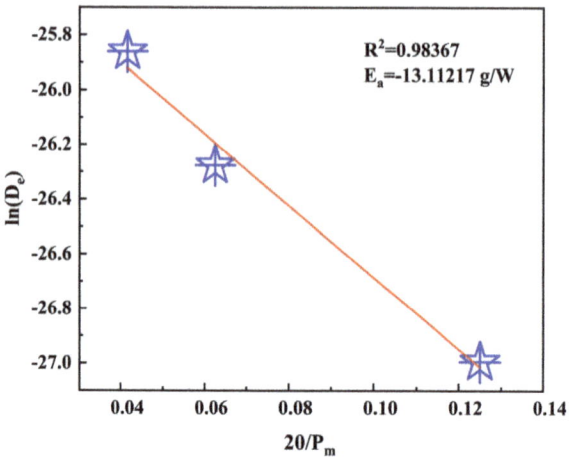

Figure 13. Calculating the activation energy needed for microwave drying.

5. Conclusions

This study examines the importance of microwave heating power, initial moisture content, and initial mass concerning wet zinc leaching residue drying.

The microwave drying of zinc-leaching residue is affected by various factors, including microwave heating power, initial moisture content, and initial mass. As these factors increase, the average drying speed also increases. This can be attributed to the fact that a larger mass of zinc-leaching residue absorbs more microwave energy when other conditions remain constant, resulting in a faster temperature rise and, subsequently, a quicker drying performance. The higher the microwave heating power, the faster the heating rate and the greater the microwave radiation on the leakage residue surface. This results in a step improvement in the microwave drying frequency.

Additionally, a relatively high initial moisture content signifies more water present, resulting in more water molecules absorbing microwaves. This causes additional water molecules to be at the boiling point and released through the exposed surface of the leach residue. The phase of leaching residue did not change before and after drying. Microwave technology effectively removes oxygen-containing functional groups in the leaching residue and reduces its moisture content.

This study used the quadratic model of Page, Lewis, Wang and Singh to fit the zinc leach slag hydration ratio data. The experiments were conducted at a heating power of 480 W, an initial water content of 8.60%, and an initial mass of 20 g, respectively. The results showed that the Page model was the best-fitting model. The Page model was further used to match the data for different masses, microwave heating power, and moisture content to produce R^2, RSS, F-values, and reduced Chi-Sqr values under the Page model. As the microwave heating power, initial moisture content, and initial mass increased, the effective diffusion coefficient tended to grow, and the average drying rate gradually increased.

Fick's second law of diffusion was used to estimate the diffusion coefficient, which showed an overall increase in microwave heating power, initial moisture content, and initial mass. A more significant effective diffusion coefficient indicates a faster reaction rate, indicating that microwave heating can be highly selective in accelerating the reaction rate. The activation energy calculated for microwave-dried leach residues was -13.11217 g/W, based on the relationship between microwave power and activation energy.

Author Contributions: Methodology, M.O.; Investigation, J.Z. and C.R.; Resources, J.T.; Data curation, C.T.; Writing—original draft, C.T.; Writing—review & editing, F.Z.; Funding acquisition, F.Z. All authors have read and agreed to the published version of the manuscript.

Funding: This study was financially supported by the Yunnan Fundamental Research Projects (grant No. 202201AU070044) and Academy of Finland (grant No. 349833).

Institutional Review Board Statement: Not applicable.

Informed Consent Statement: Not applicable.

Data Availability Statement: The data presented in this study are available upon request from the corresponding author. The data are not publicly available due to technical or time limitations.

Conflicts of Interest: The authors declare no conflict of interest.

References

1. Singh, J.; Singh, S.P. Geopolymerization of solid waste of non-ferrous metallurgy—A review. *J. Environ. Manag.* **2019**, *251*, 109571. [CrossRef]
2. Yan, L.Y.; Wang, A.J.; Chen, Q.S.; Li, J.W. Dynamic material flow analysis of zinc resources in China. *Resour. Conserv. Recycl.* **2013**, *75*, 23–31. [CrossRef]
3. Li, J.H.; Chen, Z.F.; Shen, B.P.; Xu, Z.F.; Zhang, Y.F. The extraction of valuable metals and phase transformation and formation mechanism in roasting-water leaching process of laterite with ammonium sulfate. *J. Clean. Prod.* **2017**, *140*, 1148–1155. [CrossRef]
4. Song, S.L.; Sun, W.; Wang, L.; Liu, R.Q.; Han, H.S.; Hu, Y.H.; Yang, Y. Recovery of cobalt and zinc from the leaching solution of zinc smelting slag. *J. Environ. Chem. Eng.* **2019**, *7*, 102777. [CrossRef]
5. Ma, A.Y.; Zheng, X.M.; Zhang, L.B.; Peng, J.H.; Li, Z.; Li, S.; Li, S.W. Clean recycling of zinc from blast furnace dust with ammonium acetate as complexing agents. *Sep. Sci. Technol.* **2018**, *53*, 1327–1341. [CrossRef]
6. Wang, H.J.; Liu, Z.Y.; Liu, Z.H.; Li, Y.H.; Li, S.W.; Zhang, W.H.; Li, Q.H. Leaching of iron concentrate separated from kiln slag in zinc hydrometallurgy with hydrochloric acid and its mechanism. *Trans. Nonferrous Met. Soc. China* **2017**, *27*, 901–907. [CrossRef]
7. VijayaVenkataRaman, S.; Iniyan, S.; Goic, R. A review of solar drying technologies. *Renew. Sustain. Energy Rev.* **2012**, *16*, 2652–2670. [CrossRef]
8. Tiwari, A. A review on solar drying of agricultural produce. *Food Process* **2016**, *7*, 1–12. [CrossRef]
9. Nguyen, M.H.; Price, W.E. Air-drying of banana: Influence of experimental parameters, slab thickness, banana maturity and harvesting season. *J. Food Eng.* **2007**, *79*, 200–207. [CrossRef]
10. El Hage, H.; Herez, A.; Ramadan, M.; Bazzi, H.; Khaled, M. An investigation on solar drying: A review with economic and environmental assessment. *Energy* **2018**, *157*, 815–829. [CrossRef]
11. Rao, T.S.S.B.; Murugan, S. Solar drying of medicinal herbs: A review. *Sol. Energy* **2021**, *223*, 415–436.
12. Lewicki, P.P. Design of hot air drying for better foods. *Trends Food Sci. Technol.* **2006**, *17*, 153–163. [CrossRef]
13. Salim, N.S.M.; Gariepy, Y.; Raghavan, V. Hot Air Drying and Microwave-Assisted Hot Air Drying of Broccoli Stalk Slices (*Brassica oleracea* L. Var. *Italica*). *J. Food Process. Preserv.* **2017**, *41*, e12905. [CrossRef]
14. Ratti, C. Hot air and freeze-drying of high-value foods: A review. *J. Food Eng.* **2001**, *49*, 311–319. [CrossRef]
15. Yang, W.J.; Du, H.J.; Mariga, A.M.; Pei, F.; Ma, N.; Hu, Q.H. Hot air drying process promotes lignification of Lentinus edodes. *Lwt-Food Sci. Technol.* **2017**, *84*, 726–732. [CrossRef]
16. Chaturvedi, P.K. Introduction to Microwaves. In *Microwave, Radar & RF Engineering*; Springer: Singapore, 2018; pp. 1–17.
17. Anwar, J.; Shafique, U.; Waheed-uz-Zaman; Rehman, R.; Salman, M.; Dar, A.; Anzano, J.M.; Ashraf, U.; Ashraf, S. Microwave chemistry: Effect of ions on dielectric heating in microwave ovens. *Arab. J. Chem.* **2015**, *8*, 100–104. [CrossRef]
18. Mishra, R.R.; Sharma, A.K. Microwave–material interaction phenomena: Heating mechanisms, challenges and opportunities in material processing. *Compos. Part A Appl. Sci. Manuf.* **2016**, *81*, 78–97. [CrossRef]
19. Yu, C.-H.; Fu, Q.J.; Tsang, S.C.E. 13—Aerogel materials for insulation in buildings. In *Materials for Energy Efficiency and Thermal Comfort in Buildings*; Elsevier: Amsterdam, The Netherlands, 2010; pp. 319–344.
20. Yang, S.; Yang, Y.Y.; Zhang, J.Y.; Zhang, Z.Y.; Zhang, L.; Lin, X.C. Laser-induced cracks in ice due to temperature gradient and thermal stress. *Opt. Laser Technol.* **2018**, *102*, 115–123. [CrossRef]
21. Carvalho, G.R.; Monteiro, R.L.; Laurindo, J.B.; Augusto, P.E.D. Microwave and microwave-vacuum drying as alternatives to convective drying in barley malt processing. *Innov. Food Sci. Emerg. Technol.* **2021**, *73*, 102770. [CrossRef]
22. de Faria, R.Q.; dos Santos, A.R.P.; Gariepy, Y.; da Silva, E.A.A.; Sartori, M.M.P.; Raghavan, V. Optimization of the process of drying of corn seeds with the use of microwaves. *Dry. Technol.* **2019**, *38*, 676–684. [CrossRef]
23. Demiray, E.; Seker, A.; Tulek, Y. Drying kinetics of onion (*Allium cepa* L.) slices with convective and microwave drying. *Heat Mass Transf.* **2017**, *53*, 1817–1827. [CrossRef]
24. Li, Y.S.; Yang, W.S. Microwave synthesis of zeolite membranes: A review. *J. Membr. Sci.* **2008**, *316*, 3–17. [CrossRef]
25. Kipcak, A.S. Microwave drying kinetics of mussels (*Mytilus edulis*). *Res. Chem. Intermed.* **2017**, *43*, 1429–1445. [CrossRef]

26. Liu, C.H.; Zhang, M.P.; Zhu, X.J.; Wang, Q.; Xiong, H.B.; Chen, M.H.; Liu, M.H. Drying kinetics and upgrading characteristics analysis of Zhaotong lignite with microwave deep drying. *Int. J. Coal Prep. Util.* **2021**, *42*, 3531–3553. [CrossRef]
27. Huang, W.W.; Zhang, Y.Q.; Qiu, H.J.; Huang, J.Z.; Chen, J.; Gao, L.; Omran, M.; Chen, G. Drying characteristics of ammonium polyvanadate under microwave heating based on a thin-layer drying kinetics fitting model. *J. Mater. Res. Technol.* **2022**, *19*, 1497–1509. [CrossRef]
28. Ling, Y.Q.; Li, Q.N.; Zheng, H.W.; Omran, M.; Gao, L.; Chen, J.; Chen, G. Optimisation on the stability of CaO-doped partially stabilised zirconia by microwave heating. *Ceram. Int.* **2021**, *47*, 8067–8074. [CrossRef]
29. Zheng, H.W.; Hao, X.D.; Zhang, S.R.; Omran, M.; Chen, G.; Chen, J.; Gao, L. Modeling of process and analysis of drying characteristics for natural TiO_2 under microwave heating. *Chem. Eng. Process.-Process Intensif.* **2022**, *174*, 108900. [CrossRef]
30. GB/T 10561-2019; Methods for Chemical Analysis of Lead-Zinc Ore [S]. China Standard Publishing House: Beijing, China, 2019.
31. Kucuk, H.; Midilli, A.; Kilic, A.; Dincer, I. A Review on Thin-Layer Drying-Curve Equations. *Dry. Technol.* **2014**, *32*, 757–773. [CrossRef]
32. Jones, D.A.; Lelyveld, T.P.; Mavrofidis, S.D.; Kingman, S.W.; Miles, N.J. Microwave heating applications in environmental engineering—A review. *Resour. Conserv. Recycl.* **2002**, *34*, 75–90. [CrossRef]
33. Fan, J.L.; Chen, B.Q.; Wu, L.F.; Zhang, F.C.; Lu, X.H.; Xiang, Y.Z. Evaluation and development of temperature-based empirical models for estimating daily global solar radiation in humid regions. *Energy* **2018**, *144*, 903–914. [CrossRef]
34. Zhang, Q.Z.; He, J.M.; Song, L.; Hu, K.; Sun, P.S. Theoretical model of water vapor absorption–desorption equilibrium of concrete considering the effect of temperature. *Constr. Build. Mater.* **2023**, *375*, 130968. [CrossRef]
35. Kucuk, H.; Kilic, A.; Midilli, A. Common Applications of Thin Layer Drying Curve Equations and Their Evaluation Criteria. In *Progress in Energy, and the Environment*; Springer: Cham, Switzerland, 2014; pp. 669–680.
36. Meda, V.; Orsat, V.; Raghavan, V. 2—Microwave heating and the dielectric properties of foods. In *The Microwave Processing of Foods*, 3rd ed.; Elsevier: Amsterdam, The Netherlands, 2017; pp. 23–43.
37. Clark, D.E.; Folz, D.C.; West, J.K. Processing materials with microwave energy. *Mater. Sci. Eng. A* **2000**, *287*, 153–158. [CrossRef]
38. Zhang, Y.; Man, R.L.; Ni, W.D.; Wang, H. Selective leaching of base metals from copper smelter slag. *Hydrometallurgy* **2010**, *103*, 25–29.
39. Wu, S.K.; He, M.Y.; Yang, M.; Zhang, B.Y.; Wang, F.; Li, Q.Z. Near-Infrared Spectroscopy Study of Serpentine Minerals and Assignment of the OH Group. *Crystals* **2021**, *11*, 1130. [CrossRef]
40. Hu, L.; Wang, G.H.; Wang, Q.D. Efficient drying and oxygen-containing functional groups characteristics of lignite during microwave irradiation process. *Dry. Technol.* **2018**, *36*, 1086–1097. [CrossRef]
41. Jtte Editorial Office; Chen, J.Q.; Dan, H.C.; Ding, Y.J.; Gao, Y.M.; Guo, M.; Guo, S.C.; Han, B.Y.; Hong, B.; Hou, Y.; et al. New innovations in pavement materials and engineering: A review on pavement engineering research. *J. Traffic Transp. Eng. Engl. Ed.* **2021**, *8*, 815–999.
42. Jha, P.; Meghwal, M.; Prabhakar, P.K. Microwave drying of banana blossoms (*Musa acuminata*): Mathematical modeling and drying energetics. *J. Food Process. Preserv.* **2021**, *45*, e15717. [CrossRef]
43. Koua, B.K.; Koffi, P.M.E.; Gbaha, P. Evolution of shrinkage, real density, porosity, heat and mass transfer coefficients during indirect solar drying of cocoa beans. *J. Saudi Soc. Agric. Sci.* **2019**, *18*, 72–82. [CrossRef]
44. Chen, Y.S.; Wang, J.Z.; Flanagan, D.R. Chapter 9—Fundamental of Diffusion and Dissolution. In *Developing Solid Oral Dosage Forms*, 3rd ed.; Elsevier: Amsterdam, The Netherlands, 2017; pp. 253–270.
45. Nowakowski, B. Reaction rate and diffusion coefficient of reactive Lorentz gas. *Phys. A: Stat. Mech. Its Appl.* **1998**, *255*, 93–119. [CrossRef]
46. Zheng, F.Q.; Chen, F.; Guo, Y.F.; Jiang, T.; Travyanov, A.Y.; Qiu, G.Z. Kinetics of Hydrochloric Acid Leaching of Titanium from Titanium-Bearing Electric Furnace Slag. *JOM* **2016**, *68*, 1476–1484. [CrossRef]
47. Dadali, G.; Demirhan, E.; Ozbek, B. Color change kinetics of spinach undergoing microwave drying. *Dry. Technol.* **2007**, *25*, 1713–1723. [CrossRef]

Disclaimer/Publisher's Note: The statements, opinions and data contained in all publications are solely those of the individual author(s) and contributor(s) and not of MDPI and/or the editor(s). MDPI and/or the editor(s) disclaim responsibility for any injury to people or property resulting from any ideas, methods, instructions or products referred to in the content.

Review

Review of the Preparation and Application of Porous Materials for Typical Coal-Based Solid Waste

Jinsong Du [1,2], Aiyuan Ma [2,*], Xingan Wang [1,*] and Xuemei Zheng [2]

[1] College of Environmental and Chemical Engineering, Dalian University, Dalian 116622, China; 13733025795@163.com
[2] School of Chemistry and Materials Engineering, Liupanshui Normol University, Liupanshu 553004, China; zxm_lpssy19@163.com
* Correspondence: may_kmust11@163.com (A.M.); wangxingan@dlu.edu.cn (X.W.)

Abstract: The discharge and accumulation of coal-based solid waste have caused great harm to the ecological environment recently. Coal-based solid wastes, such as coal gangue and fly ash, are rich in valuable components, such as rare earth elements (REY), silicon dioxide, alkali metal oxides, and transition metal oxides, which can be used to synthesize various functional Si-based porous materials. This article systematically summarizes the physicochemical characteristics and general processing methods of coal gangue and fly ash and reviews the progress in the application of porous materials prepared from these two solid wastes in the fields of energy and environmental protection, including the following: the adsorption treatment of heavy metal ions, ionic dyes, and organic pollutants in wastewater; the adsorption treatment of CO_2, SO_2, NO_x, and volatile organic compounds in waste gas; the energy regeneration of existing resources, such as waste plastics, biomass, H_2, and CO; and the preparation of Li–Si batteries. Combining the composition, structure, and action mechanism of various solid-waste-based porous materials, this article points out their strengths and weaknesses in the above applications. Furthermore, ideas for improvements in the applications, performance improvement methods, and energy consumption reduction processes of typical solid-waste-based porous materials are presented in this article. These works will deepen our understanding of the application of solid-waste-based porous materials in wastewater treatment, waste gas treatment, energy regeneration, and other aspects, as well as providing assistance for the integration of new technologies into solid-waste-based porous material preparation industries, and providing new ideas for reducing and reusing typical Chinese solid waste resources.

Keywords: coal gangue; fly ash; energy and environmental protection; Si-based porous material; resource utilization

Citation: Du, J.; Ma, A.; Wang, X.; Zheng, X. Review of the Preparation and Application of Porous Materials for Typical Coal-Based Solid Waste. *Materials* **2023**, *16*, 5434. https://doi.org/10.3390/ma16155434

Academic Editor: Carlos Leiva

Received: 24 June 2023
Revised: 22 July 2023
Accepted: 25 July 2023
Published: 3 August 2023

Copyright: © 2023 by the authors. Licensee MDPI, Basel, Switzerland. This article is an open access article distributed under the terms and conditions of the Creative Commons Attribution (CC BY) license (https://creativecommons.org/licenses/by/4.0/).

1. Introduction

China is the world's largest producer of coal and electricity. As of 2021, China's power energy structure was still dominated by thermal power generation, with coal-fired power generation as its main component (Figure 1). The large output of coal and the large-scale application of coal-fired power result in a large amount of coal-based solid waste generation in China every year. According to the statistics, the annual output of coal gangue and fly ash in China currently exceeds 700 million tons. The discharge and accumulation of a large amount of coal-based solid waste have caused great harm to the ecological environment [1].

Over the past few decades, multiple countries have dedicated their efforts toward the utilization of coal-based solid waste, applying it in various fields, such as power generation, agriculture, construction materials, and others [2–4]. Recently, to achieve applications of coal-based solid waste with greater added value, numerous researchers have conducted new explorations in the field of functional materials. Several studies indicate that coal-based solid waste contains abundant silicon oxide and aluminum oxide, as well as valuable components, including rare earth elements, alkali metal oxides, and transition metal oxides, which can be used to synthesize a variety of functional porous materials [5–7].

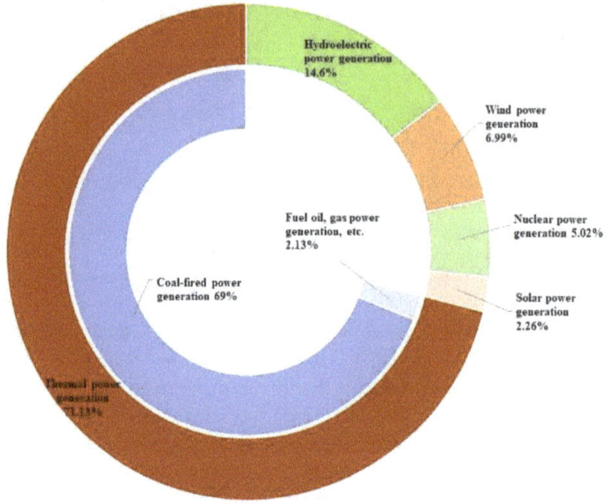

Figure 1. Composition of China's electricity structure in 2021.

Porous materials are materials that contain rich pore structures. They exhibit good photoelectric, propagation, mechanical, adsorption, permeability, and chemical properties. They can be made from various raw materials, such as metals, carbides, borides, nitrides, and silicides. Porous materials based on Si are among the most widely used types of porous materials. Common Si-based porous materials include zeolites, aerogels, and ceramics; these materials are often produced by Si-bearing chemicals or minerals (kaolinite, bentonite, etc.) [8–10].

In recent years, the demand for new materials in the field of energy and environmental protection has increased due to national industrial upgrading and ecological protection needs. Si-based porous materials have a large specific surface area, adjustable structure, strong photoelectric performance, and excellent adsorption and catalytic properties [11–14]. They are essential in the field of energy and environmental protection. As environmental pollution is increasingly severe, many adsorbents are in demand. Si-based porous materials such as functional ceramics, aerogels, zeolites, and geological polymers are widely used to prevent and treat sewage and waste gas pollution due to their strong adsorption capacity [15,16]. As resources such as oil and natural gas are continuously depleted in the energy field, the necessity and urgency of using existing resources such as waste plastics and biomass for energy regeneration are constantly increasing. This indicates the huge application prospects of various Si-based catalysts [17]. In addition, porous silicon anode materials have many advantages in various new energy Li-ion battery electrode materials and have the potential to be used as raw materials for the next generation of Li-ion batteries [18].

However, limited raw material sources restrict Si-based porous materials' production and application. Fortunately, recent studies have found ways to prepare functional Si-based porous materials from Si-rich coal-based solid waste such as coal gangue and fly ash [19,20]. Techniques such as foaming, autoclaving, high-temperature sintering, and hydrothermal synthesis are applied to convert coal-based solid waste into porous materials. The resulting products, including various solid-waste-based zeolites, ceramics, geopolymer, etc., have abundant pore structures, stable properties, and are made from inexpensive and readily available raw materials, which can be applied in practical production [21–24].

This article reviews the advanced processes for converting coal gangue and fly ash into various energy and environmental protection porous materials. It summarizes the strengths and weaknesses of these materials in wastewater treatment, waste gas treatment, energy catalysis regeneration, and battery manufacturing. The application prospects of coal-gangue-based and fly-ash-based porous materials are also discussed at the end of this article. The conclusions will reference the resource utilization of typical coal-based solid waste and contribute positively to achieving the country's 'carbon peaking and carbon neutrality goals'.

2. Physicochemical Characteristics and the Recycling of Coal Gangue and Fly Ash

2.1. Physicochemical Characteristics of Coal Gangue

Coal gangue is a waste material from coal mining and washing in the coal industry. Its main sources include shale, dirt band, and washery rejects. It is a black-gray rock with a lower carbon content and greater hardness than coal. The chemical composition of coal gangue mainly includes SiO_2, Al_2O_3, Fe_2O_3, CaO, etc., as well as some organic components (mainly C) and trace rare metal elements. Table 1 shows the chemical composition of typical coal gangue (LOI is 'loss on ignition') [25]. The main mineral composition of coal gangue varies in different parts of the world. Chinese coal gangue mainly comprises kaolinite, quartz, illite, and a small amount of calcite and iron ore [25,26].

Table 1. Chemical composition of coal gangue. (Unit: %).

Chemical Composition	SiO_2	Al_2O_3	Fe_2O_3	CaO	MgO	Na_2O	K_2O	TiO_2	P_2O_5	LOI
Content	30~65	15~40	2~10	1~4	1~3	1~2	1~2	0.5~4.0	0.05~0.3	20~30

2.2. Physicochemical Characteristics of Fly Ash

Fly ash is a fine ash captured from the flue gas produced by coal combustion. It is the primary solid waste from coal-fired power plants. Depending on the method of boiler combustion, fly ash can be classified as either circulating fluidized bed (CFB) fly ash or pulverized coal (PC) fly ash [27].

A CFB boiler typically uses low-quality coal as fuel and needs to maintain a furnace temperature of 850–950 °C. The resulting fly ash primarily consists of irregularly structured slag-like particles. Due to the low furnace temperature, the interior still contains highly organic components such as sulfur and carbon. In terms of its phase composition, CFB fly ash mainly contains quartz, hematite, magnetite, and small amounts of calcite, periclase, and anorthite [28,29].

PC fly ash is formed by combustion and cooling of high heating value coal powder (particle size less than 100 μm) in a pulverized coal furnace at temperatures above 1400 °C. It is the main type of fly ash in China currently. Because of the high-temperature combustion and rapid cooling process, PC fly ash usually appears as spherical particles of various sizes. In terms of the chemical composition, most of the organic components in coal ash are completely decomposed by combustion at high temperatures. However, there are still traces of residual polycyclic aromatic hydrocarbons (PAHs) remaining after combustion, which have toxic, mutagenic, and carcinogenic properties [30]. Table 2 shows the chemical composition of typical fly ash in China [31]. In terms of phase composition, PC fly ash is mainly composed of spherical amorphous bodies (accounting for 50% to 80% of the total amount), including high-iron amorphous bodies, low-iron amorphous bodies, amorphous SiO_2, and amorphous Al_2O_3. In addition to amorphous bodies, the mineral crystal composition is similar to CFB fly ash but includes some mullite phases formed at high temperatures [32].

Table 2. Chemical composition of fly ash. (Unit: %).

Chemical Composition	SiO_2	Al_2O_3	Fe_2O_3	CaO	K_2O	Na_2O	MgO	SO_3	LOI
Content	40~60	17~35	2~15	1~10	0.5~4	0.5~4	0.5~2	0.1~2	1~26

2.3. General Processing Methods of Coal Gangue and Fly Ash

There are three main methods for the recycling and utilizing of coal gangue and fly ash by preparing Si-based porous materials [33]: (1) extracting valuable elements from coal gangue and fly ash through various physicochemical methods to achieve the reproduction of porous materials; (2) mixing coal gangue and fly ash with other active substances to produce porous materials; (3) directly modifying the solid waste based on its inherent characteristics to produce porous materials. The specific processes are shown in Figure 2.

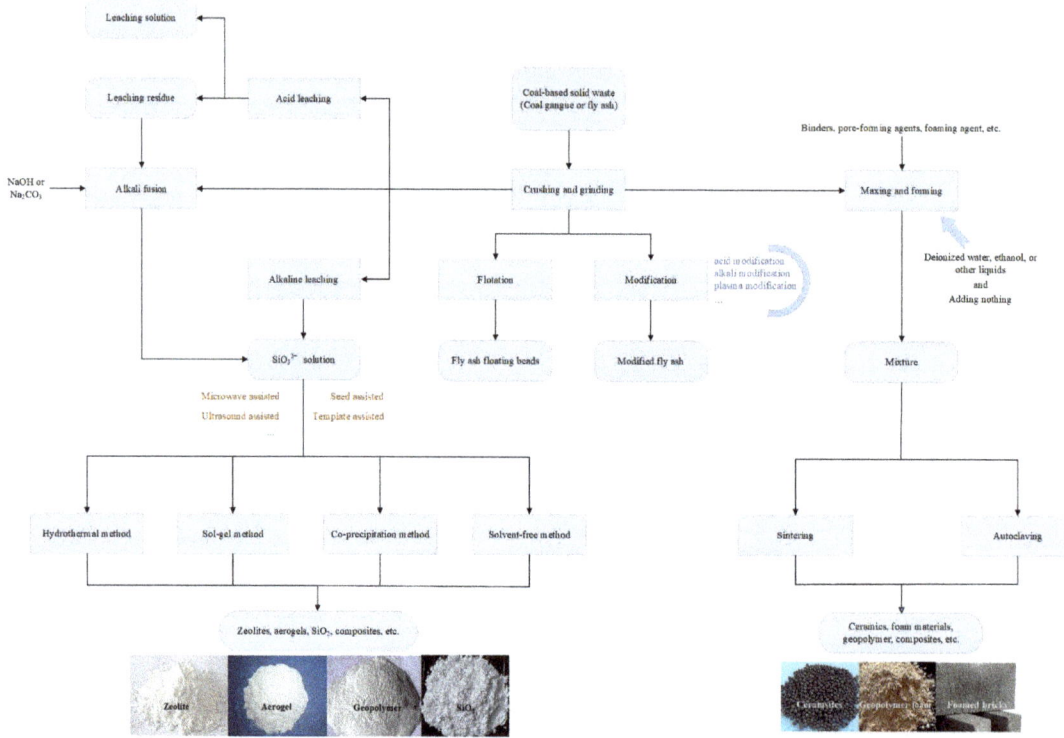

Figure 2. General processing methods of typical coal-based solid wastes (coal gangue and fly ash).

2.3.1. Extraction of Si and Reproduction

To effectively extract the abundant Si contained in various crystal phases of coal gangue and fly ash, it needs to be activated through multiple steps such as crushing, grinding, alkali fusion, and other processes. The silicon source liberated from coal gangue and fly ash can be synthesized into zeolites, aerogels, and other Si-based porous materials through processes such as the hydrothermal method, the sol–gel method, the co-precipitation method, microwave-assisted thermal crystallization, etc. [34]. For example, Zhu et al. [35] designed a three-step scheme of calcination-alkali fusion-acid dissolution to achieve efficient separation of Si from coal gangue and synthesized a low-density nanostructured aerogel. Ozdemir et al. [36] used the method of alkali fusion to extract the silicon source from fly ash and synthesized X-type zeolite through ultrasound-assisted hydrothermal synthesis.

Currently, the most effective process for synthesizing Si-based porous materials is to treat coal-based solid waste with acid leaching to remove impurities, followed by alkali melting and hydrothermal synthesis. This method can produce high-quality products with good performance, and it is one of the most commonly used methods in the field of solid waste resource utilization [23,34,37].

2.3.2. Mixing with Active Substances

Coal gangue and fly ash powder can be mixed with binders, pore-forming agents, foaming agents, and other active substances to form various functional ceramics, foam materials, composites, and other porous materials through techniques such as high-temperature sintering, autoclaving, and foaming. For instance, Wang et al. [38] prepared porous cordierite ceramic through slurry preparation, sponge selection, sponge immersion, drying, and sintering, using the raw materials mixed with 66.2 wt% coal gangue, 22.7 wt% basic magnesium carbonate, and 11.1 wt% bauxite. Using coal gangue powder and $Al(OH)_3$ as the main raw materials, with MoO_3 as an additive, a high-porosity mullite ceramic membrane could be prepared by sintering at 1400 °C [39]. Mixing fly ash with active substances into a slurry and then sintering the slurry in the furnace or autoclaving them in the reactor is also a method to produce porous materials [40,41].

2.3.3. Direct Modification

Commonly, there are a large number of bead-like particles in the fly ash. Due to the gasification of volatile components during the combustion of coal powder, the formed fly ash contains pore structures [42]. Therefore, fly ash itself is a kind of usable porous material. By modification, fly ash can form a pore structure and rougher surface, greatly increasing its specific surface area and effectively enhancing its performance in adsorption, catalysis, and other aspects. The common modification methods include acid modification, alkali modification, and plasma modification [43–45].

Among the various bead-like particles in fly ash, a thin-walled hollow bead can be obtained through fly ash flotation. This is a hollow spherical amorphous body with uniform particle size, corrosion resistance, high temperature resistance, and a large specific surface area. The fly ash floating beads can be used as excellent carriers to process into various porous materials [46].

3. Application of Solid-Waste-Based Porous Materials in the Field of Energy and Environmental Protection

3.1. Application of Solid-Waste-Based Porous Materials in the Field of Wastewater Treatment

The discharge of industrial wastewater has long been a concern for the state. Industries such as electroplating, printing and dyeing, smelting, and pharmaceuticals produce large amounts of wastewater annually. The various types of waste water can severely harm the ecology. To solve the increasingly severe water pollution problem, functional porous materials such as solid-waste-based zeolites, solid-waste-based ceramic particles, solid-waste-based Fenton-like catalysts, solid-waste-based composite photocatalysts, etc., have been developed and applied to treat heavy metal ions, ionic dyes, and organic pollutants in wastewater [47–50]. These materials have physical adsorption capabilities and can efficiently treat specific pollutants through their active components.

3.1.1. Adsorbing Heavy Metal Ions and NH_4^+ through Ion Exchange of Zeolites

Ion exchange is the process of exchanging between ions on an ion exchanger and ions with the same charge in a solution. Zeolite, a traditional porous ion exchanger, has been used for adsorbing various cationic pollutants in wastewater for a long time. Zeolite material consists of SiO_4 and AlO_4 frameworks. Since Al is trivalent, an excess negative charge is needed around it to form an AlO_4 structure, making the entire material negatively charged. The compensating cations outside the silicon–aluminum framework, which are used to maintain electrical neutrality, can exchange with heavy metal ions in wastewater to remove them [51]. Low-cost zeolite materials made from fly ash and coal gangue can effectively adsorb heavy metal ions in wastewater.

Li et al. [47] created an analcite-activated carbon composite from coal gangue and investigated its adsorption mechanism of Pb^{2+} (Figure 3). The research showed that the specific surface area of analcite-activated carbon composite was 20.8202 m^2/g. It could rely on ion exchange to adsorb Pb^{2+} with a maximum adsorption capacity of 125.57 mg/g. Ankrah et al. [52] used fly ash to synthesize mesoporous NaP1 zeolite and achieved the adsorption of multiple metal ions such as Zn^{2+}, Cu^{2+}, and Pb^{2+}. Besides metal ions, NH^{4+} in wastewater can also be adsorbed similarly. By mixing fly ash, red mud, and coffee grounds with water, a slurry was shaped and sintered into ceramsite. Then, making the NaP zeolite crystallized on the surface of the ceramsite by hydrothermal treatment with NaOH and $NaAlO_2$ at 373 K for 24 h, a hierarchically porous NaP zeolite composite was obtained with an NH^{4+} adsorption capacity of 13.1 mg/g at 318 K [53]. Chen et al. [54] used a hydrothermal method to convert fly ash into highly porous converted CFA, with X-zeolite as the main active phase. Through adsorption experiments, it was found that this material's maximum adsorption capacity for NH^{4+} was 138.89 mg/g.

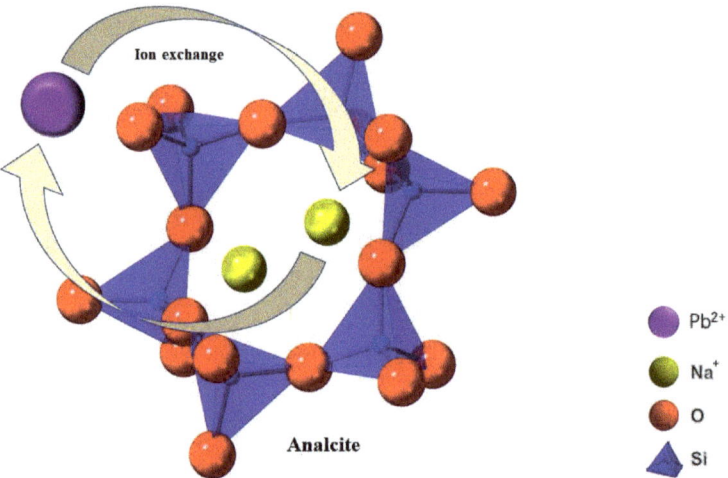

Figure 3. Adsorption mechanism of Pb^{2+} by the analcite-activated carbon composite [47].

Treating heavy metal ions and NH^{4+} based on the ion exchange of solid-waste-based zeolite has advantages such as low cost, large adsorption capacity, simple operation, and good regeneration performance. However, there are also some limitations. Ion exchange must be carried out at an appropriate pH value. A lower pH value will cause H^+ in the wastewater to participate in ion exchange, and a higher pH value will cause metal ions (such as Pb^{2+}, Cu^{2+}, etc.) or NH^{4+} to react with OH^-. Additionally, the difficulty of adsorption depends on the ion exchange affinity of different ions in the wastewater. Generally, the smaller the ion radius and the more positive charge carried by the ion, the greater the ion exchange affinity and the easier the adsorption.

3.1.2. Adsorption of Metal Ions and Ionic Dyes Based on the Interaction of Hydroxyl Groups on the Material Surface

The surface of Si-based porous materials is rich in hydroxyl groups ($\equiv$Si-OH). In an acidic medium, these hydroxyl groups decompose into $\equiv Si^+$, making the material surface easier to adsorb anions. In an alkaline medium, they decompose into $\equiv Si$-O^-, making the material surface easier to adsorb cations. Due to this electrostatic attraction, various Si-based porous materials produced from coal gangue and fly ash can effectively adsorb metal ions and ionic dyes.

Sahoo et al. [55] modified fly ash (MFA) with simple alkali soaking and used it to adsorb heavy metal ions in acidic mine wastewater. The surface area of MFA increased nearly two times through modification. The results showed that the rich functional oxygen-containing groups ($\equiv$Si-OH and $\equiv$Al-OH) on the surface of the MFA played a positive role in metal ion adsorption, and the removal rate of heavy metal ions in the treated wastewater can reach over 90%. Compared to metal ions, ionic dyes have larger volumes and complex compositions of various functional groups, which can easily interact with the hydroxyl groups on the material surface. Sareen et al. [56] created a coral-reef-shaped mesoporous silica material (MS-CFA) with a specific surface area of 350 m^2/g from calcined fly ash by acid leaching, alkali melting, precipitation, and calcination and studied its adsorption mechanism of methylene blue (a cationic dye) (Figure 4). Through adsorption experiments, it was observed that the adsorption rate increased with the pH due to electrostatic attraction and decreased due to the weakening of hydrogen bonds because of the combination of excessive OH$^-$ and amino groups on methylene blue when the pH value reached 9, further demonstrating the mechanism. The results showed that MS-CFA could adsorb up to 20.49 mg/g of methylene blue under optimal conditions. The same conclusion was obtained by Yan et al. in their analysis of the adsorption mechanism of methylene blue and magenta dyes on coal-gangue-based porous ceramic microspheres; that is, the adsorption behavior of ionic dyes based on the surface hydroxyl interaction of Si-based porous materials is caused by electrostatic attraction, hydrogen bonding, and other forces. The maximum adsorption capacity of this ceramic microsphere for methylene blue reached 30.01 mg/g, while for magenta, it reached 24.16 mg/g [48].

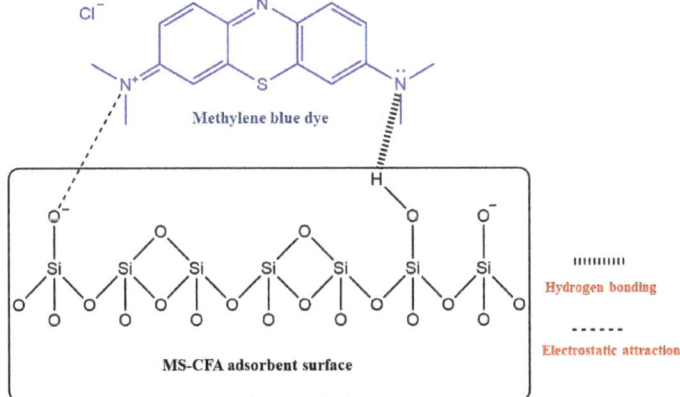

Figure 4. Adsorption mechanism of methylene blue dye by the MS-CFA [56].

Loading substances rich in hydroxyl groups onto porous materials can further enhance their dye adsorption performance. Mohammed et al. [57] dissolved chitosan (CHT) in acetic acid and adjusted the pH value to form CHT microspheres. The CHT microsphere solution and coal fly ash (CFA) were added to the epichlorohydrin (ECH) solution to produce a biocomposite adsorbent of covalently crosslinked chitosan–epichlorohydrin/coal fly ash (CHT-ECH/CFA$_{50}$). The specific surface area of CHT-ECH increased significantly by loading with CFA, and CHT contained a large number of hydroxyl groups, amino groups, and N atoms with lone pair electrons; hence, it had a variety of strong adsorption effects on the Reactive Red 120 dye (RR120) (Figure 5). The maximum adsorption capacity of CHT-ECH/CFA$_{50}$ for RR120 dye can reach 237.7 mg/g.

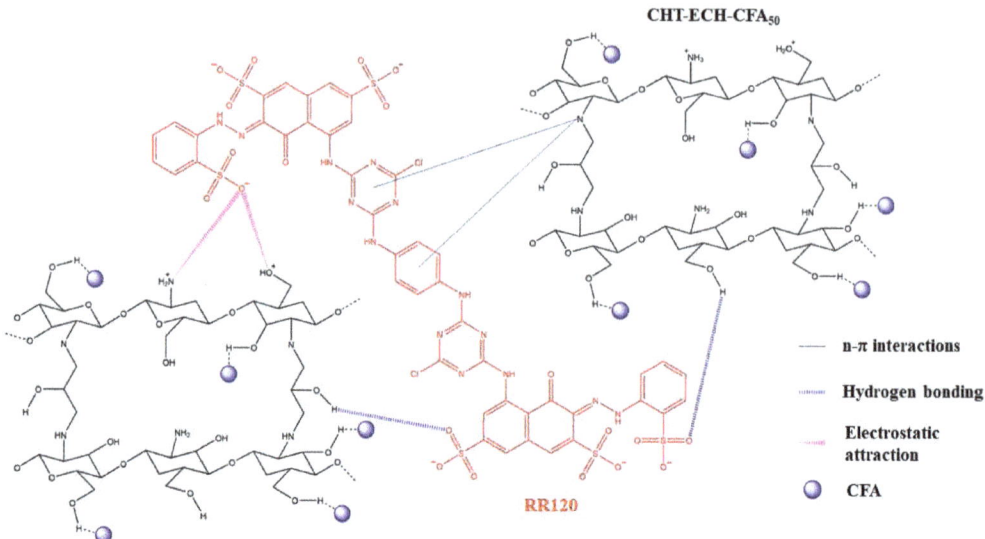

Figure 5. Interaction between CHT-ECH/CFA50 surface and RR120 dye: electrostatic attraction, hydrogen bonding interaction, and n-π interaction [57].

Under conditions of suitable pH, Si-based porous materials made from Si-rich solid wastes such as coal gangue or fly ash can effectively adsorb metal ions and ionic dyes through various adsorption effects generated by surface hydroxyl groups. These materials can be regenerated through chemical elution and other methods. Currently, Si-based materials represented by zeolites can basically achieve the adsorption treatment of heavy metal ion pollutants through the electrostatic attraction and ion exchange effects of surface hydroxyl groups. However, it is difficult to achieve a large amount of adsorption of large molecular ionic dyes just by the limited functional hydroxyl groups on the surface of Si-based porous materials and their small and narrow pore structure. Preparing modified materials by loading substances that are rich in functional active groups, such as chitosan, onto porous materials has become the main development direction of ionic dye adsorbents.

3.1.3. Precipitation and Adsorption of the Pollutants Based on the Alkali Supply Capacity of Materials

Mixing fly ash and coal gangue with other active substances through high-temperature sintering or autoclaving can be made into porous materials that can continuously and stably release OH^- into water, allowing pollutants such as Mn and P, which easily form insoluble substances in an alkaline environment, to be precipitated and adsorbed onto the surface of the material.

Ou et al. [58] mixed fly ash and lime cement to prepare CFA-based ceramsite with an apparent porosity of 49.49%. By studying and analyzing the XRD patterns of the ceramsite before and after the adsorption of Mn^{2+}, the alkali supply mechanism of this type of material was proposed. After the ceramsite adsorbed Mn^{2+}, diffuse peaks appeared in its XRD patterns in the range of $2\theta = 10\sim15°$, indicating the formation of a small amount of amorphous phase. The reason is that the anorthite and gehlenite with lattice imperfections underwent a hydration reaction, as shown in Reactions (1) and (2):

$$CaAl_2Si_2O_8 + 3H_2O = Ca^{2+} + 2OH^- + Al_2[Si_2O_5][OH]_4 \tag{1}$$

$$2Ca_2Al_2SiO_7 + 9H_2O = 4Ca^{2+} + 6OH^- + 2Al[OH]_4^- + Al_2[Si_2O_5][OH]_4 \tag{2}$$

The progress of the reaction promotes the formation of an alkaline environment, causing Mn^{2+} to react with OH^- to form $Mn(OH)_2$ precipitate, which adheres to the surface of the ceramsite. The ability of CFA-based ceramsite to release OH^- is stable and persistent.

Phosphorus removal also has a good effect based on a similar method, and the Ca^{2+} released by the hydration reaction could also be consumed simultaneously during the formation of calcium phosphate salt precipitation. Using cement, clay, calcium oxide, and fly ash as the main raw materials and aluminum powder as the pore-forming agent, after mixing and forming, the TBX porous ceramsite could be obtained by autoclaving in a high-pressure reaction kettle. The experimental results show that under the influence of TBX ceramsite's high alkali supply and calcium release ability, the removal rate of phosphorus can reach more than 98% [41].

Using solid-waste-based porous materials with alkali supply capacity for precipitation adsorption treatment of wastewater has advantages such as low cost, simple operation, and convenient treatment. Furthermore, a large amount of pollutants are concentrated and adsorbed on the surface of the porous material, which is conducive to their recycling and utilization. However, during water treatment, long-term deposition can cause the material surface to be covered by pollutant precipitation, greatly affecting the material's adsorption and alkali supply performance and limiting the application of such materials.

3.1.4. Loading Photocatalytic Materials for the Degradation of Organic Pollutants

Common photocatalytic materials such as ZnO, Fe_2O_3, and TiO_2 can produce a large number of electron holes (h^+) and electrons (e^-) in a short amount of time to assist in the degradation of organic matter under the irradiation of a specific wavelength light. Due to the low quantum activity of the photocatalytic process, these materials are often made into nanoparticles to improve their photocatalytic activity. These particles are also prone to agglomeration during the photocatalytic degradation of organic matter, causing deactivation [59]. To improve the photocatalytic activity and service life, it is a good method to load these nanoparticles onto porous materials to make composite photocatalysts. Various Si-based porous materials made from fly ash and coal gangue are good carriers for photocatalytic materials.

Wang et al. [49] obtained a modified SiO_2-Al_2O_3 aerogel from aluminosilica-sol (produced by alkali melting and acid leaching of fly ash) through gel, aging, solvent exchange, surface modification, washing, and drying. Ti-oxide nanoparticles were loaded onto the aerogel using the sol–gel method to create TiO_2/SiO_2-Al_2O_3 aerogel composites used for the photocatalytic degradation of 4,6-dinitro-2-sec-butylphenol (DNBP). The test results showed that the loading of TiO_2 nanoparticles increased the reaction contact area of TiO_2, improved photocatalytic efficiency, and prolonged the service life of the photocatalyst. Stirring modified coal gangue with rich micropores in $ZnSO_4$ solution for 1 h, after adjusting the pH value of the solution to 9 and boiling refluxing for 3 h, a ZnO/coal gangue composite was prepared for the photocatalytic degradation of methyl orange and methylene blue organic dyes. The degradation of organic dyes was basically achieved within 120 min, and the degradation rate of this catalyst remained above 65% after repeated use for five times [60]. Bai et al. [61] mixed fly ash floating beads with butyl titanate and stirred for a while. After drying and calcining, a TiO_2/fly ash floating bead composite photocatalyst was obtained. Under the conditions of UV light irradiation for 2 h, pH 10.0, and TiO_2 loading amount of 28.57%, the degradation rate of Rhodamine B organic dye was over 80% and had good recycling regeneration.

Several studies have shown that fly ash and coal-gangue-based composite photocatalysts have advantages such as easy availability of raw materials, green and sustainable driving energy, long service life, good regeneration performance, and a wide application range. However, it is difficult to achieve complete degradation of pollutants in wastewater with high COD (chemical oxygen demand) values only by the photocatalytic process. In addition, many photocatalytic materials only exhibit high catalytic degradation performance under specific wavelength light irradiation, further increasing the cost of use. The pH value can also affect the energy level structure of photocatalytic materials and impact the photocatalytic process.

3.1.5. Preparing Heterogeneous Fenton-Like Catalysts for the Degradation of Organic Pollutants

The Fenton reaction is an oxidation process of organic compounds to inorganic states in a mixed solution of Fe^{2+} and hydrogen peroxide. The specific process can be represented by the following reaction [62]:

$$Fe^{2+} + H_2O_2 \rightarrow Fe^{3+} + OH^- + OH\cdot \quad (3)$$

$$H_2O_2 + 2Fe^{3+} \rightarrow 2Fe^{2+} + HO_2\cdot + H^+ \quad (4)$$

$$O_2 + Fe^{2+} \rightarrow Fe^{3+} + O_2^- \quad (5)$$

The strong oxidation of Fenton reagents comes from the hydroxyl radicals (OH·) and peroxide radicals (HO$_2$·) generated in the reaction. Based on the principle of this process, using iron-containing minerals, Fe^{3+}, or transition metals such as Cu, Co, Cd, Ni, etc., to generate strong oxide from H_2O_2 is called a Fenton-like reaction [63]. The conventional Fenton reaction is a homogeneous reaction with strict requirements for conditions such as pH value and H_2O_2 addition amount, and it is difficult to recover the catalyst after the reaction [64]. Connecting active substances such as Fe to the porous materials to prepare heterogeneous Fenton-like catalysts can significantly reduce the loss of active ingredients and make the reaction conditions milder [65]. Fly ash contains rich iron oxides and has a porous structure that is conducive to the adsorption of organic pollutants, making it an excellent resource for preparing heterogeneous Fenton-like catalysts.

Wang et al. [50] studied the catalytic mechanism of H_2SO_4-modified coal fly ash Fenton-like catalysts (MCFA) in degrading Acid Orange 7 (AO7) (Figure 6). It was discovered that in addition to the homogeneous catalysis process, a small amount of Fe^{3+} dissolved in water also carried out a homogeneous catalysis process simultaneously. Experimental results demonstrated that under optimal conditions, the removal rate of AO7 exceeded 95%, and after being reused six times, its removal rate remained around 90%. Coal fly ash was crushed, sieved, and dried by Chen and Du to produce a Fenton-like catalyst for degrading n-butyl xanthate. Under the conditions of reaction temperature of 30 °C, reaction time of 120 min, pH 3.0, H_2O_2 concentration of 1.176 mmol/L, catalyst concentration of 1.0 g/L, and Fe (III) content of 4.14%, the removal rate of n-butyl xanthate exceeded 96.90% [66].

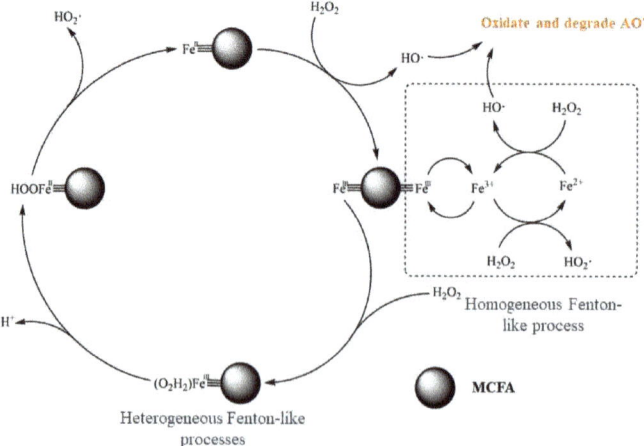

Figure 6. Catalytic mechanism of MCFA in Fenton-like processes [50].

Photo-Fenton technology is a technique that employs photocatalysis to assist the Fenton reaction. Compared to traditional Fenton processes, photo-Fenton technology possesses the advantages of both, and it is more effective in treating organic pollutants. Nadeem et al. [67] added hydrothermally treated coal fly ash (CFA) into a $ZnFe_2O_4$ precursor solution and synthesized $ZnFe_2O_4$-CFA composite Fenton-like catalysts through hydrothermal synthesis. SEM-EDS analysis shows that the surface of composites is very rough and the zinc ferrite is uniformly loaded, making it a photo-Fenton catalyst with good adsorption performance. Figure 7 illustrates the mechanism flowchart of $ZnFe_2O_4$-CFA composite degrading methylene blue dye under UV light. $ZnFe_2O_4$ generates h^+ and e^- under UV light, where h^+ reacts with H_2O to form H^+ and $OH·$ while e^- forms Fe^{2+} with Fe^{3+}. The resulting Fe^{2+} undergoes a Fenton reaction with H_2O_2 to form OH^- and $OH·$. The formed $OH·$ will oxidize and degrade methylene blue due to its strong oxidizing property. In addition, the O_2 produced during the reaction generates superoxide anions ($·O_2^-$) with e^-, which further oxidize into H_2O_2, achieving the recycling of H_2O_2 (Equations (6)–(10)):

$$HO_2· + OH· \rightarrow H_2O + O_2 \tag{6}$$

$$H_2O_2 + 2h^+ \rightarrow O_2 + 2H^+ \tag{7}$$

$$O_2 + e^- \rightarrow ·O_2^- \tag{8}$$

$$H_2O + O_2·^- \rightarrow ·OOH + OH^- \tag{9}$$

$$2·OOH \rightarrow O_2 + H_2O_2 \tag{10}$$

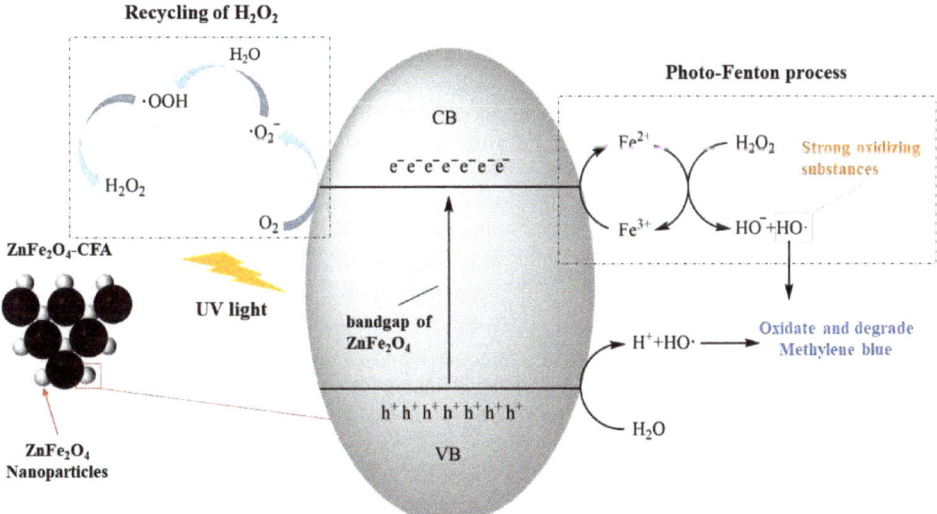

Figure 7. Synergistic mechanism of photocatalysis and Fenton-like process by CFA-$ZnFe_2O_4$ composite [67].

The results showed that the $ZnFe_2O_4$-CFA composite (mCFA:m$ZnFe_2O_4$ = 1:1) had a degradation rate of up to 97% for methylene blue. After being used five times, the degradation rate remained above 85%. $ZnFe_2O_4$-CFA composite is a superior photo-Fenton catalyst with the characteristics of Fe not easily leaching out and renewable reactants.

As a green and efficient water treatment method, Fenton oxidation technology can achieve deep oxidation of wastewater by generating a large amount of $OH·$ and greatly reducing the COD value in the water. The development and appliance of various solid-waste-based heterogeneous Fenton-like catalysts can effectively reduce costs and greatly improve the disadvantages of traditional homogeneous Fenton catalysts, such as limited reaction conditions and poor regeneration. Advanced methods such as light- and electricity-assisted Fenton reactions have further enhanced the degradation efficiency of pollutants. It should be considered that pH value and H_2O_2 concentration are still the main factors affecting the catalytic performance of Fenton-like catalysts due to the limitations of their catalytic mechanism.

The development and appliance of coal-gangue-based and fly-ash-based water treatment materials can effectively reduce raw material costs while properly disposing of a large amount of accumulated coal-based solid waste. Different types of solid-waste-based materials can efficiently adsorb specific pollutants in wastewater based on their unique physicochemical properties and mechanisms of action, and they have unique advantages, such as the green sustainability of composite solid-waste-based photocatalysts and the high-efficiency versatility of solid-waste-based Fenton-like catalysts. Besides this, limited by the action mechanism of different solid-waste-based water treatment materials, external conditions such as pH value, pollutant type, illumination, etc., greatly impact the water treatment performance of the materials. In practical applications, it is necessary to choose suitable materials for treatment according to the specific situation. In addition, solid-waste-based water treatment materials are in contact with water for a long time during the treatment of wastewater. Harmful components such as heavy metal ions in solid waste will slowly escape, causing secondary pollution (especially for structurally unstable materials).

3.2. The Application of Solid-Waste-Based Porous Materials in the Field of Waste Gas Treatment

Common exhaust gases include CO_2, SO_2, NO_x, and volatile organic compounds (VOCs). These gases can directly harm animals and plants, indirectly damage the ecological environment by polluting the atmosphere, or contribute to the greenhouse effect, causing global warming and rising sea levels. The adsorption method is a mature exhaust gas treatment technology with advantages such as low energy consumption, high efficiency, and a simple process. Various coal-gangue-based and fly-ash-based gas adsorbents have been developed for the adsorption treatment of exhaust gases [68–70]. The diameter of gas molecules is generally on the order of 0.1 nm. Strong physical adsorption can be achieved by controlling the formation of a rich and evenly distributed mesoporous structure in solid-waste-based adsorbents [71]. Additionally, by active components contained in solid waste or obtained through loading modification, solid-waste-based materials can further produce various forces on exhaust gases, greatly enhancing the performance of these materials.

3.2.1. The Adsorption of CO_2 Based on Hydroxyl Hydrogen Bonding

Hydroxyl groups on the surface of solid materials can adsorb CO_2. Some researchers have investigated the adsorption mechanism of CO_2 by hydroxyl groups on solid surfaces. The O atom of CO_2 can form hydrogen bonds with hydroxyl groups (as shown in Figure 8, taking kaolinite as an example) [72]. The surface of Si-based gas adsorbents made from coal gangue or fly ash is rich in hydroxyl groups, making these adsorbents ideal materials for CO_2 adsorption treatment.

Yan et al. [68] prepared an alkaline silicate material with a mesoporous structure from fly ash and found that the material's surface was rich in hydroxyl groups through FT-IR analysis. They established an isotherm modeling of CO_2 using adsorption–desorption experiments. They found that it conformed to the Langmuir isotherm model, indicating strong adsorption forces on its surface, and the partial desorption of CO_2 after heating confirmed the existence of hydrogen bonding. Wu et al. [73] hydrothermally synthesized Mg-doped $CuSiO_3$ material by adding a Mg^{2+} solution to a SO_3^{2-} solution made from coal gangue. Although the doping of Mg leads to a decrease in the pore size and specific surface area of the mesoporous $CuSiO_3$ material, it could effectively increase the number of hydroxyl groups on the surface of the $CuSiO_3$ material. The experimental results showed that within a certain range, the higher the doping amount of Mg, the higher the CO_2 adsorption capacity of Mg-doped copper silicate material, and that the maximum adsorption capacity could reach up to 8.38 mg/g.

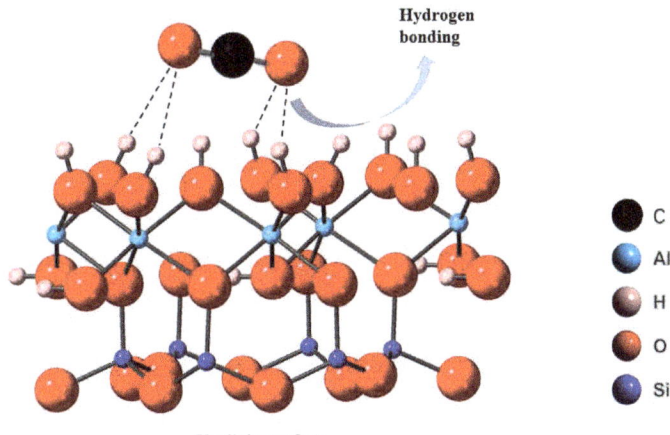

Figure 8. Adsorption model of CO_2 on kaolinite surface [72].

Due to the instability of H-bonds at high temperatures, the adsorbed CO_2 can be desorbed simply by heating. Therefore, porous materials with adsorption capacity based on H-bonding have good regenerability. However, the adsorption efficiency of such materials is extremely low in high-temperature environments, and since H_2O molecules can also form H-bonds with -OH groups, temperature and humidity are both important factors that affect the CO_2 adsorption process.

3.2.2. Introducing Amino Groups by Chemical Modification for CO_2 Adsorption

Amine groups can interact with CO_2 molecules through the production of ion intermediates. The specific process is as follows [74]:

$$CO_2 + 2RNH_2 \rightarrow RNHCOO^- + RNH_3^+ \quad (11)$$

Introducing amine groups into solid-waste-based gas adsorbents through chemical modification methods can effectively improve their CO_2 adsorption performance.

Chandrasekar et al. [69] extracted silicon and aluminum sources from fly ash through alkali fusion. Pluronic P123 ($EO_{20}PO_{70}EO_{20}$) was added as a template agent to synthesize the SBA-15 zeolite with an average pore size of 7.2 nm. Polyethyleneimine (PEI) was introduced into the zeolite through wet impregnation to produce a PEI-loaded SBA-15 zeolite with rich amino groups on its surface. The experiments showed that this material can adsorb up to 110 mg/g of CO_2 at 75 °C. Using SiO_3^{2-} leachate produced from coal gangue as a silicon source and CTAB as a template agent, a $M-SiO_2$ mesoporous silica material with MCM-41 structure was synthesized. An $EDA-M-SiO_2$ porous material was obtained by impregnating in ethylenediamine (EDA) solution and drying, and the material exhibited strong CO_2 adsorption performance, with an adsorption capacity of 83.5 mg/g under optimal conditions [75].

The introduction of amino groups greatly improves the CO_2 adsorption capacity of solid-waste-based gas adsorbents. Since CO_2 interacts with the adsorbent through chemical reactions, these materials have higher stability. However, excessive loading of amine substances can block pore channels and greatly reduce the adsorption performance of the porous material. In addition, H_2O molecules can easily form hydrogen bonds with amino groups, so humidity can affect adsorption.

3.2.3. The Absorption of CO_2 and SO_2 Based on Alkaline Substances of Solid Waste

Alkaline substances such as $CaSiO_3$, K_2SiO_3, CaO, and MgO can effectively absorb acidic gases such as CO_2 and SO_2 through chemical reactions. Coal-based solid wastes such as fly ash and coal gangue contain some of these substances or the raw materials needed to produce them [25,31].

Sanna et al. [76] mixed fly ash (FA) with a certain amount of K_2CO_3 (K) and calcined it to obtain a K-FA porous material with K_2SiO_3 as the main active substance for CO_2 adsorption; 10 wt% Li_2CO_3 was added to form the K-Li eutectic phase, improving the internal diffusion performance of the K-FA material and accelerating the CO_2 adsorption–desorption process. This material mainly absorbs CO_2 through the following reaction:

$$K_2SiO_3 + CO_2 \rightarrow K_2CO_3 + SiO_2 \tag{12}$$

It was measured that at 700 °C, the CO_2 adsorption capacity of the K-FA material (nFA: nK = 1:1, 10 wt% Li_2CO_3) was 104.72 mg/g.

By adding a certain amount of $Ca(OH)_2$ solution to the alkaline-leached silicon solution from fly ash, filtering, and drying the mixture after reacting, an active calcium silicate material (ACS) would be prepared for SO_2 absorption. BET analysis showed that this material contained abundant mesoporous structures. In an adsorption experiment at 50 °C, it was found that the adsorption capacity of this material would be greatly increased when O_2 and water vapor were present in the flue gas. The reason for this is that the ACS will react with SO_2 to promote chemical adsorption when O_2 and H_2O are present. The process is as follows (12)–(17):

Where (g) and (ad) denote the gaseous and adsorbed states.

$$SO_2(g) \rightarrow SO_2(ad) \tag{13}$$

$$SO_2(g) + H_2O \rightarrow H_2SO_3 \tag{14}$$

$$SO_2(ad) + H_2O \rightarrow H_2SO_3 \tag{15}$$

$$H_2SO_3 + CaSiO_3 \rightarrow CaSO_3 \tag{16}$$

$$CaSO_3 + 1/2H_2O + 1/2O_2 \rightarrow CaSO_4 \cdot 1/2H_2O \tag{17}$$

$$CaSO_3 + 1/2O_2 \rightarrow CaSO_4 \tag{18}$$

The maximum adsorption capacity of ACS for SO_2 can reach 38.2 mg/g. After adsorption, the ACS can be reused by regenerating with water vapor at 300 °C for 3 h. After 20 regeneration cycles, its adsorption performance only decreased by 28.8% [77].

In summary, the solid-waste-based porous materials based on alkaline substances to absorb acid waste gas have good adsorption performance and can function in high-temperature and high-humidity environments. Nevertheless, this type of material requires long-term high-temperature regeneration when recycling, consuming more energy.

3.2.4. Denitration Based on NH_3-SCR Technology

NH_3 selective catalytic reduction (NH_3-SCR) technology is a process that selectively reduces NO_x to N_2 within a certain temperature range assisted by catalyst, using NH_3 as a reducing agent. It is one of the main research directions of denitrification technology at present. The main reaction is as follows:

$$4NO + 4NH_3 + O_2 \rightarrow 4N_2 + 6H_2O \tag{19}$$

Common NH_3-SCR catalysts include transition metal catalysts such as Cu-based and Fe-based catalysts [78]. Some studies have found that metal-loaded zeolite catalysts have higher reaction activity and a broader active temperature range for the NH_3-SCR process than metal/metal oxide catalysts [79]. These catalysts also exhibit good hydrothermal stability [80]. Therefore, preparing zeolite carriers from coal-based solid waste and loading metals onto them to synthesize denitration catalysts is an effective method for solid waste's reuse.

Ma et al. [70] extracted silicon and aluminum sources from fly ash and produced SAPO-34 zeolite with a specific surface area of 579 m^2/g and an average pore size of 0.56 nm by template method, and then loaded Cu onto SAPO-34 using the ion exchange method to obtain the Cu-SAPO-34 catalyst. Figure 9 shows the denitration activity of Cu-SAPO-34 with different Cu loading amounts between 100 and 500 °C. The results indicate that the synthesized Cu-SAPO-34 exhibit high NO_x conversion rates, good high-temperature activity, and anti-sintering properties. Similarly, the SSZ-13 zeolite was obtained through one-step hydrothermal synthesis using coal gangue as a raw material, and a certain amount of Cu was loaded to obtain the micropore Cu-SSZ-13 catalyst, which exhibits high denitration activity between 180 and 400 °C, with NO_x conversion rates exceeding 90%, and high hydrothermal stability [81].

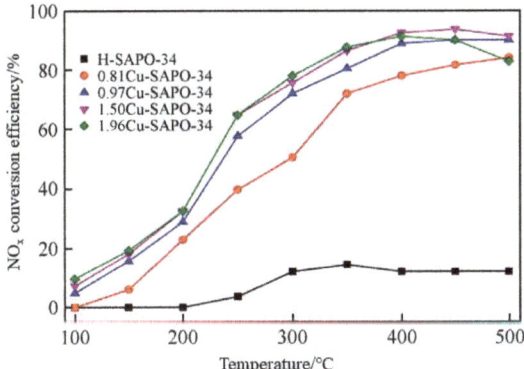

Figure 9. NH_3-SCR activity of Cu-SAPO-34 catalyst with different Cu loading amounts [70].

Solid-waste-based NH_3-SCR zeolite catalysts have numerous acidic sites on their surface that adsorb NH_3 and promote the formation of reaction intermediates, making them possess excellent denitration performance. However, substances (such as H_2O) that cause competitive absorption with NH_3 can influence the catalytic performance of these catalysts. Additionally, SO_2 can react with NH_3 and O_2 to form ammonium salts deposited on the catalyst surface or react with the metal loaded onto the catalyst to form sulfates, causing catalyst deactivation [82].

3.2.5. Absorption of VOCs

Volatile organic compounds (VOCs) are diverse and complex in composition and can directly involve in atmospheric photochemical reactions with NO_x and other pollutants to produce ozone and secondary organic aerosols [83]. Due to significant differences in the physical and chemical properties of different VOCs, VOC adsorbents can only adsorb a few specific volatile organic compounds.

Yuan et al. [84] used high-alumina fly ash as raw material to produce a highly stable silicate material with a specific surface area of 98.47 m^2/g and an average pore size of 31.42 nm. Dynamic adsorption experiments demonstrated that this material had a high toluene adsorption capacity and a low formaldehyde adsorption capacity. Toluene is adsorbed through van der Waals forces, while formaldehyde is adsorbed through hydrogen bonding with hydroxyl groups on the surface of the silicate material.

Using fly ash and coal gangue as raw materials to produce solid-waste-based gas adsorbents has significant economic and environmental benefits. Various solid-waste-based adsorbents have recently been developed to effectively adsorb common waste gases such as CO_2, SO_2, and NO_x [73,77,81]. Some solid-waste-based materials can strongly interact with specific waste gases through their active substances and functional groups, such as the reaction between alkaline substances in solid waste materials and acidic gases and the hydrogen bonding interaction between hydroxyl groups on the surface of solid waste materials and CO_2. Introducing high-activity sites onto solid-waste-based adsorbents through impregnation, ion exchange, and other processes can further increase their adsorption capacity and broaden their range of applications. For instance, they are introducing amino groups onto solid waste adsorbents to increase their CO_2 adsorption capacity, introducing transition metals to obtain denitration capability. However, the adsorption performance of solid-waste-based gas adsorbents is greatly influenced by external factors such as temperature, humidity, gas type, etc., due to the limitations of their adsorption mechanisms, resulting in a significant reduction in the adsorption rate. Additionally, some materials that absorb waste gas through chemical reactions face challenges in regeneration.

3.3. The Application of Solid-Waste-Based Porous Materials in Energy Regeneration

Over the last decade, the consumption of various non-renewable energy sources has been constantly increasing. In addition to finding renewable energy sources as alternatives, existing resources such as waste plastics, waste oil, and biomass can also be regenerated through catalytic cracking, hydrocracking, steam reforming, and other methods [85–87]. To achieve high-quality and efficient regeneration of energy, the use of catalysts is necessary. Materials such as zeolite and silica–alumina gels made from coal-based solid waste are typical solid acid catalysts. Solid alkali components such as CaO and MgO and transition metals such as Fe in solid waste can serve as catalytic active centers. Solid waste with porous structures, such as fly ash, also has the potential to serve as catalytic carriers. Solid-waste-based catalysts have considerable application prospects in the field of energy recovery. Table 3 summarizes the research status of preparing energy regeneration catalyst materials that are prepared from coal gangue and fly ash solid waste in recent years.

Table 3. Research status of coal gangue and fly ash in the field of energy regeneration.

Regenerate Objects	Regeneration Effect	Catalytic Material	Preparation Method	Action Mechanism
Waste polyethylene (PE)	Reaction temperature: 700 °C, ratio of feed to catalyst: 20:1, maximum liquid product yield: 78.2%, with total aromatics yield reaching nearly 22%.	Modified fly ash	The coal fly ash was crushed, sieved, and heat-treated at 800 °C to modify [88].	The fly ash catalyst with high specific surface area and high Si/Al ratio provides a large number of highly selective acidic sites, promoting the cracking of plastics and improving the yield of aromatics [89,90].
Waste linear low-density polyethylene (LLDPE)	Catalyst dosage: 15 wt%, LLDPE's cracking activation energy significantly decreased, and reaction rate and the yield of alicyclic hydrocarbons increased.	Modified fly ash	The coal fly ash was ground to a particle size of less than 150 μm to modify [91].	Solid alkali components such as CaO in fly ash facilitate the formation of carbon anion intermediates, promoting the generation of alicyclic hydrocarbons and accelerating the reaction rate [92]. Transition metal oxide components such as Fe_2O_3 in fly ash also assist the plastic cracking process.

Table 3. Cont.

Regenerate Objects	Regeneration Effect	Catalytic Material	Preparation Method	Action Mechanism
Plastic film residue (PFR)	Reaction temperature: 723 K. Use of catalyst inhibits tar and wax formation, reduces cracking temperature, and achieves a 44% product oil yield, with 70% composed of gasoline hydrocarbons.	HX/CFA zeolite	Coal fly ash (CFA) underwent alkali fusion and hydrothermal synthesis to produce NaX zeolite. After acidification, HX/CFA zeolite was obtained [93].	Acidic sites on synthesized X-type zeolites (solid acid) promote the formation of carbocation intermediates in polymer chains, enabling high-quality and efficient cracking of waste plastics [89,90].
Waste wood pellets	A 10 wt% Ni-loaded catalyst and waste wood pellets were steam reformed to obtain a 54.9% gas yield, with an H_2 production of 7.29 mmol/g.	Ni-ash catalysts	Coal fly ash was impregnated in a $Ni(NO_3)_2 \cdot 6H_2O$ aqueous solution after drying, calcining, reducing it in H_2 atmosphere to prepare the Ni-ash catalyst [94].	The hydrogenation and dehydrogenation functions of metal Ni, as well as the isomerization, cyclization, and hydrogenation cracking functions of acid components (Al_2O_3) in fly ash carrier occur through olefin intermediates during steam reforming. Metals such as Mg and Cu in fly ash also act as catalyst auxiliaries.
Pine sawdust biomass tar	Reaction temperature: 800 °C. The catalyst effectively promoted the decomposition of tar molecules, achieving a conversion rate of 93.5%, and significantly increasing the gas yield.	GC catalysts	The coal gangue (GC) was crushed and sieved, and then calcined at 800 °C for 1 h under N_2 atmosphere to obtain the GC catalyst [95].	The repeated oxidation and reduction of Fe_2O_3 in coal gangue effectively promotes the breaks of C-H bonds and dehydrogenation reactions in catalytic reactions, increasing the yield of H_2 and CO. Alkali metals in coal gangue further facilitate tar decomposition. The formation of Fe^0 during the process enhances the activity and lifespan of the catalyst [96,97].
Soybean oil	Reaction temperature: 65 °C, catalyst concentration: 4%, molar ratio of methanol to oil: 12:1, reaction time: 2 h. The conversion rate of soybean oil methyl ester reached 95.5%.	Zeolite-type sodalite	Coal fly ash was added to an alkali solution and supplemented with an aluminate solution to form a gel. The zeolite-type sodalite was formed through hydrothermal crystallization at 100 °C for 24 h using the gel [98].	Solid alkali components (Si-O-Na groups) in zeolite serve as active sites, promoting transesterification to produce biodiesel.
Jatropha curcas oil	Catalyst (40 wt% CaO loaded) dosage: 0.15 wt%, molar ratio of methanol to oil: 12:1, reaction temperature: 60 °C, and reaction time: 1 h. The maximum biodiesel yield was 94.72%.	CaO/CFA catalysts	Coal fly ash was activated by calcination at 400 °C for 5 h, impregnated with a calcium acetate solution and dried, then calcined at 750 °C for 4 h to obtain CaO/CFA catalyst [99].	The appropriate amount of CaO loading improves the catalyst's pore structure and increases the reaction rate. The introduced CaO (solid alkali) serves as an active site for transesterification, promoting the formation of biodiesel. Si and Al in fly ash act as catalyst carriers to improve catalyst stability.

Table 3. Cont.

Regenerate Objects	Regeneration Effect	Catalytic Material	Preparation Method	Action Mechanism
CO and H_2	Reaction temperature: 523 K, air pressure: 20 bar, and $H_2/CO = 2$. Zeolites carrying more Co particles have higher catalytic activity, resulting in a CO conversion rate of 53.2% and higher selectivity for producing liquid hydrocarbons (C_{5+}).	Co/LTA and Co/FAU catalysts	LTA and FAU zeolites synthesized from high-silicon coal fly ash were used as carriers. The Co/LTA and Co/FAU catalysts were obtained by impregnating with $Co(NO_3)_2$ solutions and calcination in H_2 atmosphere [100].	CO and H_2 are adsorbed on the surface of metal Co, activated through electron effects and interactions with transition metals, and then dissociated, leading to chain growth and termination to produce various hydrocarbons [101]. Fly-ash-based zeolite carriers plays a role in improving pore structure, enhancing catalyst stability, and increasing the activity of active components during the process.
CO and H_2	Reaction temperature: 220 °C and air pressure: 30 bar for FT synthesis. After 130 h of intake, the conversion rates of CO and H_2 were 67% and 73%, respectively. The selectivity for liquid hydrocarbons (C_{5+}) was 66.53%. Among C_{5+} hydrocarbons, the contents of gasoline (C_5–C_{11}), diesel (C_{12}–C_{18}), and high-carbon hydrocarbons (C_{19+}) were 33.19%, 27.64%, and 5.7%, respectively.	Co-Fe/SBA-15 catalysts	Coal fly ash was activated by alkali fusion. A small amount of surfactant was added to hydrothermally synthesize mesoporous SBA-15 zeolite. The Co-Fe/SBA-15 catalyst was prepared by impregnating with $Co(NO_3)_2$ and $Fe(NO_3)_3$ solutions and calcining in H_2 atmosphere [102].	The loading of additional Fe in addition to Co makes the catalyst more advantageous for synthesizing low-carbon hydrocarbons, promoting the generation of gasoline-range products. This catalyst also has a wider CO/H_2 range and higher poison resistance during F-T synthesis.

Overall, coal-gangue- and fly-ash-based catalysts have wide applications in the field of energy regeneration. They can effectively reduce the activation energy required for the regeneration of various resources such as waste plastics, biomass, and gas and improve product yield and quality. The alkali metal oxide and transition metal oxide components in solid waste have high catalytic activity for reactions such as cracking, steam reforming, and dry reforming. The SiO_2 and Al_2O_3 components can not only serve as catalyst carriers to enhance the stability of the catalyst and improve the reaction activity, but they can also synthesize zeolite materials through alkaline fusion and hydrothermal processes to further promote the cracking reaction (solid acid mechanism). Moreover, synthesizing zeolite products is more conducive to the structural adjustment and active metal loading of the catalyst, making solid-waste-based catalysts have higher regeneration efficiency and wider applications. However, the energy regeneration process is often accompanied by a high-temperature and high-carbon environment, which can easily cause coking and carbon deposition on the catalyst, leading to its deactivation. In addition, solid waste raw materials are complex in composition, and it is difficult to explore the specific mechanisms of each component in the catalytic process, making it difficult to control the regeneration process and achieve its large-scale application.

3.4. The Application of Solid-Waste-Based Porous Materials in Battery Manufacturing

As a new energy system, Li-ion batteries have advantages over traditional batteries, such as high energy density, good stability and cyclicality, low self-discharge rate, and voltage hysteresis [103]. Currently, graphite with good conductivity and a low cost is widely used as the anode material for Li-ion batteries. It can combine with Li to form LiC_6

with a theoretical capacity of 372 mAh·g^{-1}. Si-based anode materials are preferred for the next generation of high-capacity Li-ion battery anodes. At room temperature, every 4 Si atoms can react with 15 Li ions to form an alloy phase with a theoretical capacity of 3579 mAh·g^{-1}, and Si-based materials have advantages such as environmental friendliness, abundant sources, and low cost [104,105]. However, the alloying reaction between Si and Li during repeated charging and discharging processes can cause severe volume expansion, resulting in severe battery loss [106]. Studies have shown that the nanostructure design can effectively alleviate the stress changes during the charging and discharging process of Li–Si alloys to maintain the maximum capacity of the battery [107]. At present, there have been several reports on the preparation of porous nano-silicon particles from silicon-rich coal-based solid waste as anode materials for Li-ion batteries.

Liu et al. [108] used Si-rich fly ash as a raw material to produce SiO$_2$ through a process of alkali fusion, deionized water dissolution, acidification precipitation, washing and drying, and calcination. Finally, low-cost Si particles were obtained through a magnesiothermic reaction for use as anode materials. The resulting Si particles were nanoscale in size and interconnected to form a porous structure (Figure 10). The Li-ion battery made using these silicon nanoparticles as anode material provided a capacitance of approximately 3173.1 mAh·g^{-1} and remained stable after 500 cycles at 1 °C. A similar method was used by Xing et al. to synthesize nano-silicon anode materials from fly ash. The prepared Li-ion battery had a capacitance of 1450.3 mAh·g^{-1} at the current density of 4000 mA·g^{-1}, and its reversible capacity remained at 1017.5 mAh·g^{-1} after 100 cycles [109].

Figure 10. (a) SEM; (b) TEM image of Si nanoparticles [108].

Using coal-based solid waste as a low-cost silicon source to prepare porous nano-silicon particles has opened up a new way for the large-scale preparation of Li-ion battery Si-based anode materials and promoted the development of the new energy industry. Additionally, it can effectively reduce the quantity of Si-rich coal-based solid waste and decrease its environmental hazards. However, the volume expansion in Si-based anode materials for Li-ion batteries has not yet been completely solved, resulting in limited cycling performance.

4. Conclusions

In summary, coal-gangue-based and fly-ash-based porous functional materials have broad applications in energy and environmental protection, but there are also certain shortcomings. To further achieve large-scale resource utilization of typical coal-based solid waste, the following improvements can be made:

(1) In the field of wastewater treatment, adopt more efficient pretreatment methods (such as pressurized acid leaching, microwave acid leaching, complex acid leaching, etc.) before preparing water treatment materials to reduce heavy metal content in solid waste and focus research on materials with a more stable structure simultaneously, reducing the possibility of secondary pollution.

(2) In the field of waste gas treatment, pre-treat waste gas with means of temperature control and dehumidification before using solid-waste-based gas adsorption materials,

minimizing the adverse effects of external factors on the gas adsorption performance of the materials. Some gas adsorption materials that are difficult to regenerate can improve their gas diffusion performance by adding modified substances to form eutectic phases, thereby accelerating the regeneration rate.

(3) In the field of energy regeneration, fully explore the mechanism of various effective components in solid-waste-based catalysts in the catalytic process to achieve efficient and controllable energy regeneration. By loading active metals, regulating the silicon–aluminum ratio, and improving the pore structure, this will improve catalyst energy regeneration efficiency and coking resistance.

(4) In the field of battery manufacturing, the volume expansion problem of Si-based anode materials still requires further exploration of new nanostructures and composite systems to be solved, improving the cycling performance of electrode materials.

(5) In the regeneration process of solid waste, apply technical methods (such as microwave-assisted, ultrasonic-assisted, solvent-free methods, etc.) that can effectively reduce energy consumption and increase productivity in the conversion preparation process of solid waste into functional porous materials, reducing the cost of reusing coal-based solid waste and promoting large-scale resource utilization of typical coal-based solid wastes.

Based on current practical applications, porous materials derived from solid waste still exhibit some deficiencies in the fields of energy regeneration and battery manufacturing. As a result, researching and producing various materials for treating waste gas and wastewater is currently the primary method for reusing typical coal-based solid waste. However, their application in energy will be more promising in the future.

China is a typical country with abundant coal reserves but limited oil and natural gas resources. The development of its industries relies heavily on the extensive mining and combustion of coal. However, the production and accumulation of large amounts of coal-based solid waste over the years have severely harmed the ecological environment. In recent years, with industrial upgrading and the need for ecological protection, the significant shortage of Si-based materials in the fields of energy and environmental protection has provided new approaches for reducing and reusing coal-based solid waste. The large-scale application of various types of coal-gangue-based and fly-ash-based energy and environmental protection porous materials will transform coal-based solid waste from valueless waste into the main force in maintaining ecological stability, promoting the upgrading of energy structures, and achieving national green sustainable development.

Author Contributions: J.D.: Writing—Original Draft, Conceptualization, and Methodology, Investigation, Validation; A.M.: Investigation, Validation, Funding Acquisition, Supervision, Writing—Review and Editing; X.W.: Supervision, Writing—Review and Editing; X.Z.: Visualization, Resources. All authors have read and agreed to the published version of the manuscript.

Funding: This work was supported by the Scientific Research and Cultivation Project of Liupanshui Normal University (LPSSY2023KJYBPY06), the Discipline Team of Liupanshui Normal University (LPSSY2023XKTD07), the Guizhou Provincial First-class professional (GZSylzy202103), the Key Cultivation Disciplines of Liupanshui Normal University (LPSSYZDXK202001), and the Jointly Trains Postgraduate Special Scientific Research Project of Liupanshui Normal University (LPSSYLPY202204).

Institutional Review Board Statement: Not applicable.

Informed Consent Statement: Not applicable.

Data Availability Statement: The data presented in this study are available on request from the corresponding author. The data are not publicly available due to technical or time limitations.

Acknowledgments: The authors acknowledge the financial support that helped carry out this work.

Conflicts of Interest: The authors declare no conflict of interest.

References

1. Song, W.; Zhang, J.; Li, M.; Yan, H.; Zhou, N.; Yao, Y.; Guo, Y. Underground Disposal of Coal Gangue Backfill in China. *Appl. Sci.* **2022**, *12*, 12060. [CrossRef]
2. Luo, Y.; Wu, Y.; Ma, S.; Zheng, S.; Zhang, Y.; Chu, P. Utilization of coal fly ash in China: A mini-review on challenges and future directions. *Environ. Sci. Pollut. Res.* **2021**, *28*, 18727–18740. [CrossRef] [PubMed]
3. Liu, H.; Lin, Z. Recycling utilization patterns of coal mining waste in China. *Resour. Conserv. Recycl.* **2010**, *54*, 1331–1340. [CrossRef]
4. Babla, M.; Katwal, U.; Yong, M.; Jahandari, S.; Rahme, M.; Chen, Z.; Tao, Z. Value-added products as soil conditioners for sustainable agriculture. *Resour. Conserv. Recycl.* **2022**, *178*, 106079. [CrossRef]
5. Rybak, A.; Rybak, A. Characteristics of Some Selected Methods of Rare Earth Elements Recovery from Coal Fly Ashes. *Metals* **2021**, *11*, 142. [CrossRef]
6. Querol, X.; Turiel, J.F.; Soler, A.L. The Behaviour of Mineral Matter During Combustion of Spanish Subbituminous and Brown Coals. *Mineral. Mag.* **1994**, *58*, 119–133. [CrossRef]
7. Han, R.; Guo, X.; Guan, J.; Yao, X.; Hao, Y. Activation mechanism of coal gangue and its impact on the properties of geopolymers: A review. *Polymers* **2022**, *14*, 3861. [CrossRef]
8. Hameed, A.; Alharbi, A.; Abdelrahman, E.; Mabrouk, E.; Hegazey, R.; Algethami, F.; Al-Ghamdi, Y.O.; Youssef, H.M. Facile Hydrothermal Fabrication of Analcime and Zeolite X for Efficient Removal of Cd(II) Ions From Aqueous Media and Polluted Water. *J. Inorg. Organomet. Polym. Mater.* **2020**, *30*, 4117–4128. [CrossRef]
9. Yu, Y.; Guo, D.; Fang, J. Synthesis of silica aerogel microspheres by a two-step acid-base sol-gel reaction with emulsification technique. *J. Porous Mater.* **2015**, *22*, 621–628. [CrossRef]
10. Han, L.; Song, J.; Zhang, Q.; Liu, T.; Luo, Z.; Lu, A. Synthesis, Structure and Properties of MgO-Al_2O_3-SiO_2-B_2O_3 Transparent Glass-Ceramics. *Silicon* **2018**, *10*, 2685–2693. [CrossRef]
11. Sun, Z.; Li, M.; Zhou, Y. Recent Progress on Synthesis, Multi-Scale Structure, and Properties of Y-Si-O Oxides. *Int. Mater. Rev.* **2014**, *59*, 357–383. [CrossRef]
12. Sahoo, M.; Kale, P. Restructured porous silicon for solar photovoltaic: A review. *Microporous Mesoporous Mater.* **2019**, *289*, 109619. [CrossRef]
13. Rad, L.R.; Anbia, M. Zeolite-based composites for the adsorption of toxic matters from water: A review. *J. Environ. Chem. Eng.* **2021**, *9*, 106088. [CrossRef]
14. Oenema, J.; Harmel, J.; Vélez, R.; Meijerink, M.; Eijsvogel, W.; Poursaeidesfahani, A.; Vlugt, T.J.; Zečević, J.; de Jong, K.P. Influence of nanoscale intimacy and zeolite micropore size on the performance of bifunctional catalysts for n-heptane hydroisomerization. *ACS Catal.* **2020**, *10*, 14245–14257. [CrossRef]
15. Wang, L.; Zhu, N.; Shaghaleh, H.; Mao, X.; Shao, X.; Wang, Q.; Hamoud, Y.A. The Effect of Functional Ceramsite in a Moving Bed Biofilm Reactor and Its Ammonium Nitrogen Adsorption Mechanism. *Water* **2023**, *15*, 1362. [CrossRef]
16. Xu, J.; Zha, X.; Wu, Y.; Ke, Q.; Yu, W. Fast and highly efficient SO_2 capture by TMG immobilized on hierarchical micro-meso-macroporous AlPO-5/cordierite honeycomb ceramic materials. *Chem. Commun.* **2016**, *52*, 6367–6370. [CrossRef]
17. Yakovenko, R.; Zubkov, I.; Bakun, V.; Papeta, O.; Savostyanov, A. Effects of SiO_2/Al_2O_3 Ratio in ZSM-5 Zeolite on the Activity and Selectivity of a Bifunctional Cobalt Catalyst for Synthesis of Low-Pour-Point Diesel Fuels from CO and H_2. *Petrol. Chem+.* **2022**, *62*, 101–111. [CrossRef]
18. Zhang, Z.; Wang, Y.; Ren, W.; Tan, Q.; Zhong, Z.; Su, F. Low-Cost Synthesis of Porous Silicon via Ferrite-Assisted Chemical Etching and Their Application as Si-Based Anodes for Li-Ion Batteries. *Adv. Electron. Mater.* **2015**, *1*, 1400059. [CrossRef]
19. Li, N.; Han, B. Chinese research into utilisation of coal waste in ceramics, refractories and cements. *Adv. Appl. Ceram.* **2006**, *105*, 64–68. [CrossRef]
20. Asl, S.M.H.; Ghadi, A.; Baei, M.S.; Javadian, H.; Maghsudi, M.; Kazemian, H. Porous catalysts fabricated from coal fly ash as cost-effective alternatives for industrial applications: A review. *Fuel* **2018**, *217*, 320–342. [CrossRef]
21. Li, X.; Liu, L.; Bai, C.; Yang, K.; Zheng, T.; Lu, S.; Li, H.; Qiao, Y.; Colombo, P. Porous alkali-activated material from hypergolic coal gangue by microwave foaming for methylene blue removal. *J. Am. Ceram. Soc.* **2023**, *106*, 1473–1489. [CrossRef]
22. Sun, G.; Zhang, J.; Hao, B.; Li, X.; Yan, M.; Liu, K. Feasible synthesis of coal fly ash based porous composites with multiscale pore structure and its application in Congo red adsorption. *Chemosphere* **2022**, *298*, 134136. [CrossRef] [PubMed]
23. Wang, J.; Fang, L.; Cheng, F.; Duan, X.; Chen, R. Hydrothermal Synthesis of SBA-15 Using Sodium Silicate Derived From Coal Gangue. *J. Nanomater.* **2013**, *2013*, 352157. [CrossRef]
24. Huang, Y.; Hu, N.; Ye, Y.; Fu, F.; Lv, Y.; Jia, J.; Chen, D.; Ou, Z.; Li, J. A novel route for the fabrication of melilite-spinel porous ceramics with ultralow thermal conductivity and sufficient strength. *Ceram. Int.* **2022**, *48*, 37488–37491. [CrossRef]
25. Zhang, C. *New Technology for Comprehensive Utilization of Coal Gangue Resources*; Chemical Industry Press: Beijing, China, 2008; pp. 20–29. (In Chinese)
26. Zhou, C.; Liu, G.; Yan, Z.; Fang, T.; Wang, R. Transformation behavior of mineral composition and trace elements during coal gangue combustion. *Fuel* **2012**, *97*, 644–650. [CrossRef]
27. Pan, J.; Long, X.; Zhang, L.; Shoppert, A.; Valeev, D.; Zhou, C.; Liu, X. The Discrepancy between Coal Ash from Muffle, Circulating Fluidized Bed (CFB), and Pulverized Coal (PC) Furnaces, with a Focus on the Recovery of Iron and Rare Earth Elements. *Materials* **2022**, *15*, 8494. [CrossRef]

28. Seslija, M.; Rosic, A.; Radovic, N.; Vasic, M.; Dogo, M.; Jotic, M. Properties of fly ash and slag from the power plants. *Geol. Croat.* **2016**, *69*, 317–324. [CrossRef]
29. Karayiğit, A.; Oskay, R.; Gayer, R. Mineralogy and geochemistry of feed coals and combustion residues of the Kangal power plant (Sivas, Turkey). *Turk. J. Earth Sci.* **2019**, *28*, 438–456. [CrossRef]
30. Szatyłowicz, E.; Walendziuk, W. Analysis of Polycyclic Aromatic Hydrocarbon Content in Ash from Solid Fuel Combustion in Low-Power Boilers. *Energies* **2021**, *14*, 6801. [CrossRef]
31. Zhou, C.; Li, C.; Li, W.; Sun, J.; Li, Q.; Wu, W.; Liu, G. Distribution and preconcentration of critical elements from coal fly ash by integrated physical separations. *Int. J. Coal Geol.* **2022**, *261*, 104095. [CrossRef]
32. Dai, S.; Zhao, L.; Peng, S.; Chou, C.; Wang, X.; Zhang, Y.; Li, D.; Sun, Y. Abundances and distribution of minerals and elements in high-alumina coal fly ash from the Jungar Power Plant, Inner Mongolia, China. *Int. J. Coal Geol.* **2010**, *81*, 320–332. [CrossRef]
33. Miao, C.; Liang, L.; Zhang, F.; Chen, S.; Shang, K.; Jiang, J.; Zhang, Y.; Ouyang, J. Review of the fabrication and application of porous materials from silicon-rich industrial solid waste. *Int. J. Miner. Met. Mater.* **2022**, *29*, 424–438. [CrossRef]
34. Zhang, X.; Li, C.; Zheng, S.; Di, Y.; Sun, Z. A review of the synthesis and application of zeolites from coal-based solid wastes. *Int. J. Miner. Met. Mater.* **2022**, *29*, 1–21. [CrossRef]
35. Zhu, P.; Zheng, M.; Zhao, S.; Wu, J.; Xu, H. A novel environmental route to ambient pressure dried thermal insulating silica aerogel via recycled coal gangue. *Adv. Mater. Sci. Eng.* **2016**, *2016*, 9831515. [CrossRef]
36. Ozdemir, O.D.; Piskin, S. A Novel Synthesis Method of Zeolite X From Coal Fly Ash: Alkaline Fusion Followed by Ultrasonic-Assisted Synthesis Method. *Waste Biomass Valoriz.* **2019**, *10*, 143–154. [CrossRef]
37. Panitchakarn, P.; Laosiripojana, N.; Viriya-umpikul, N.; Pavasant, P. Synthesis of high-purity Na-A and Na-X zeolite from coal fly ash. *J. Air Waste Manag.* **2014**, *64*, 586–596. [CrossRef]
38. Wang, X.; Xu, H.; Zhang, F.; Li, D.; Wang, A.; Sun, D.; Oh, W.-C. Preparation of porous cordierite ceramic with acid-leached coal gangue. *J. Korean Ceram. Soc.* **2020**, *57*, 447–453. [CrossRef]
39. Liu, M.; Zhu, Z.; Zhang, Z.; Chu, Y.; Yuan, B.; Wei, Z. Development of highly porous mullite whisker ceramic membranes for oil-in-water separation and resource utilization of coal gangue. *Sep. Purif. Technol.* **2020**, *237*, 116483. [CrossRef]
40. Zhang, X.; Mwamulima, T.; Wang, Y.; Song, S.; Gu, Q.; Peng, C. Preparation of Porous Pellets Based on Nano-Zero Valent Iron-Enhanced Fly Ash and Their Application for Crystal Violet Removal. *Res. Environ. Sci.* **2017**, *30*, 1295–1302. (In Chinese)
41. Liu, B.; Zhang, L.; Meng, G.; Zheng, J. Research on Phosphorus Removal from Wastewater by TBX Porous Ceramisite Filter Media. *Acta Sci. Nat. Univ. Pekin.* **2010**, *46*, 389–394. (In Chinese)
42. Mishra, D.P.; Das, S.K. A study of physico-chemical and mineralogical properties of Talcher coal fly ash for stowing in underground coal mines. *Mater. Charact.* **2010**, *61*, 1252–1259. [CrossRef]
43. Xu, K.; Deng, T.; Liu, J.; Peng, W. Study on the phosphate removal from aqueous solution using modified fly ash. *Fuel* **2010**, *89*, 3668–3674. [CrossRef]
44. Mazumder, N.; Rano, R. An efficient solid base catalyst from coal combustion fly ash for green synthesis of dibenzylideneacetone. *J. Ind. Eng. Chem.* **2015**, *29*, 359–365. [CrossRef]
45. Zhang, L.; Wen, X.; Zhang, L.; Sha, X.; Wang, Y.; Chen, J.; Luo, M.; Li, Y. Study on the Preparation of Plasma-Modified Fly Ash Catalyst and Its De-NO_X Mechanism. *Materials* **2018**, *11*, 1047. [CrossRef] [PubMed]
46. Xu, X.; Li, Q.; Cui, H.; Pang, J.; Sun, L.; An, H.; Zhai, J. Adsorption of fluoride from aqueous solution on magnesia-loaded fly ash cenospheres. *Desalination* **2011**, *272*, 233–239. [CrossRef]
47. Li, Q.; Lv, L.; Zhao, X.; Wang, Y.; Wang, Y. Cost-effective microwave-assisted hydrothermal rapid synthesis of analcime-activated carbon composite from coal gangue used for Pb^{2+} adsorption. *Environ. Sci. Pollut. Res.* **2022**, *29*, 77788–77799. [CrossRef]
48. Yan, S.; Pan, Y.; Wang, L.; Liu, J.; Zhang, Z. Synthesis of low-cost porous ceramic microspheres from waste gangue for dye adsorption. *J. Adv. Ceram.* **2018**, *7*, 30–40. [CrossRef]
49. Wang, H.; Qi, H.; Wei, X.; Liu, X.; Jiang, W. Photocatalytic activity of TiO_2 supported SiO_2-Al_2O_3 aerogels prepared from industrial fly ash. *Chin. J. Catal.* **2016**, *37*, 2025–2033. [CrossRef]
50. Wang, N.; Hu, Q.; Hao, L.; Zhao, Q. Degradation of Acid Organic 7 by modified coal fly ash-catalyzed Fenton-like process: Kinetics and mechanism study. *Int. J. Environ. Sci. Technol.* **2019**, *16*, 89–100. [CrossRef]
51. Khosravi, M.; Murthy, V.; Mackinnon, I.D.R. The Exchange Mechanism of Alkaline and Alkaline-Earth Ions in Zeolite N. *Molecules* **2019**, *24*, 3652. [CrossRef]
52. Ankrah, A.; Tokay, B.; Snape, C. Heavy metal removal from aqueous solutions using fly-ash derived zeolite NaP1. *Int. J. Environ. Res.* **2022**, *16*, 1–10. [CrossRef]
53. Zhang, L.; Zhang, S.; Li, R. Synthesis of Hierarchically Porous Na-P Zeotype Composites for Ammonium Removal. *Environ. Eng. Sci.* **2019**, *36*, 1089–1099. [CrossRef]
54. Chen, X.; Song, H.; Guo, Y.; Wang, L.; Cheng, F. Converting waste coal fly ash into effective adsorbent for the removal of ammonia nitrogen in water. *J. Mater. Sci.* **2018**, *53*, 12731–12740. [CrossRef]
55. Sahoo, P.; Tripathy, S.; Panigrahi, M.; Equeenuddin, S. Evaluation of the use of an alkali modified fly ash as a potential adsorbent for the removal of metals from acid mine drainage. *Appl. Water. Sci.* **2013**, *3*, 567–576. [CrossRef]
56. Sareen, S.; Kaur, S.; Mutreja, V.; Sharma, A.; Kansal, S.; Mehta, S. Coral-Reef Shaped Mesoporous Silica Obtained from Coal Fly Ash with High Adsorption Capacity. *Top. Catal.* **2022**, *65*, 1791–1810. [CrossRef]

57. Mohammed, I.A.; Malek, N.N.A.; Jawad, A.H.; Mastuli, M.S.; ALOthman, Z.A. Box-Behnken Design for Optimizing Synthesis and Adsorption Conditions of Covalently Crosslinked Chitosan/Coal Fly Ash Composite for Reactive Red 120 Dye Removal. *J. Polym. Environ.* **2022**, *30*, 3447–3462. [CrossRef]
58. Ou, C.; Dai, S.; Li, S.; Xu, J.; Qin, J. Adsorption performance and mechanism investigation of Mn^{2+} by facile synthesized ceramsites from lime mud and coal fly ash. *Korean J. Chem. Eng.* **2021**, *38*, 505–513. [CrossRef]
59. Zhang, G.; Song, A.; Duan, Y.; Zheng, S. Enhanced photocatalytic activity of TiO_2/zeolite composite for abatement of pollutants. *Microporous Mesoporous Mater.* **2018**, *255*, 61–68. [CrossRef]
60. Xie, J.; Xia, R.; Zhao, S.; Li, M.; Xu, Y.; Du, H. Preparation and Properties of ZnO/Coal Gangue Composite Photocatalysts. *Multipurp. Util. Miner. Resour.* **2019**, *4*, 126–129. (In Chinese)
61. Bai, C.; Fan, X.; Li, G.; Xu, Z.; Sheng, J.; Li, S. Preparation and photocatalytic degradation properties of TiO_2/fly ash cenosphere composite materials. *Environ. Pollut. Control* **2017**, *39*, 735–739. (In Chinese)
62. Song, Q.; Feng, Y.; Liu, G.; Lv, W. Degradation of the flame retardant triphenyl phosphate by ferrous ion-activated hydrogen peroxide and persulfate: Kinetics, pathways, and mechanisms. *Chem. Eng. J.* **2019**, *361*, 929–936. [CrossRef]
63. Wang, J.; Tang, J. Fe-based Fenton-like catalysts for water treatment: Catalytic mechanisms and applications. *J. Mol. Liq.* **2021**, *332*, 115755. [CrossRef]
64. Xing, M.; Xu, W.; Dong, C.; Bai, Y.; Zheng, J.; Zhou, Y.; Zhang, J.; Yin, Y. Metal sulfides as excellent co-catalysts for H_2O_2 decomposition in advanced oxidation processes. *Chem* **2018**, *4*, 1359–1372. [CrossRef]
65. Garza-Campos, B.R.; Guzmán-Mar, J.L.; Reyes, L.H.; Brillas, E.; Hernandez-Ramirez, A.; Ruiz-Ruiz, E.J. Coupling of solar photoelectro-Fenton with a BDD anode and solar heterogeneous photocatalysis for the mineralization of the herbicide atrazine. *Chemosphere* **2014**, *97*, 26–33. [CrossRef] [PubMed]
66. Chen, S.-H.; Du, D.-Y. Degradation of n-butyl xanthate using fly ash as heterogeneous Fenton-like catalyst. *J. Cent. South. Univ.* **2014**, *21*, 1448–1452. [CrossRef]
67. Nadeem, N.; Zahid, M.; Rehan, Z.A.; Hanif, M.A.; Yaseen, M. Improved photocatalytic degradation of dye using coal fly ash-based zinc ferrite ($CFA/ZnFe_2O_4$) composite. *Int. J. Environ. Sci. Technol.* **2022**, *19*, 3045–3060. [CrossRef]
68. Yan, Y.; Gao, Y.; Tang, W.; Li, Q.; Zhang, J. Characterization of high-alumina coal fly ash based silicate material and its adsorption performance to CO_2. *Korean J. Chem. Eng.* **2016**, *33*, 1369–1379. [CrossRef]
69. Chandrasekar, G.; Son, W.J.; Ahn, W.S. Synthesis of mesoporous materials SBA-15 and CMK-3 from fly ash and their application for CO_2 adsorption. *J. Porous Mater.* **2009**, *16*, 545–551. [CrossRef]
70. Ma, Z.; Wang, B.; Lu, G.; Xiao, Y.; Yang, J.; Lu, J.; Li, G.; Zhou, J.; Wang, H.; Zhao, C. Preparation and performance of SAPO-34 based SCR catalyst derived from fly ash. *Chem. Ind. Eng. Prog.* **2020**, *39*, 4051–4060. (In Chinese)
71. Abu Ghalia, M.; Dahman, Y. Development and Evaluation of Zeolites and Metal-Organic Frameworks for Carbon Dioxide Separation and Capture. *Energy Technol.* **2017**, *5*, 356–372. [CrossRef]
72. Wu, D.; Jiang, W.; Liu, X.; Qiu, N.; Xue, Y. Theoretical study about effects of H_2O and Na^+ on adsorption of CO_2 on kaolinite surfaces. *Chem. Res. Chin. Univ.* **2016**, *32*, 118–126. [CrossRef]
73. Wu, Y.; Wu, Z.; Liu, K.; Li, F.; Pang, Y.; Zhang, J.; Si, H. Preparation of nano-sized Mg-doped copper silicate materials using coal gangue as the raw material and its characterization for CO_2 adsorption. *Korean J. Chem. Eng.* **2020**, *37*, 1786–1794. [CrossRef]
74. Zheng, F.; Tran, D.N.; Busche, B.J.; Fryxell, G.E.; Addleman, R.S.; Zemanian, T.S. Ethylenediamine-modified SBA-15 as regenerable CO_2 sorbent. *Ind. Eng. Chem. Res.* **2005**, *44*, 3099–3105. [CrossRef]
75. Du, H.; Ma, L.; Liu, X.; Zhang, F.; Yang, X.; Wu, Y.; Zhang, J. A Novel Mesoporous SiO_2 Material with MCM-41 Structure from Coal Gangue: Preparation, Ethylenediamine Modification, and Adsorption Properties for CO_2 Capture. *Energy Fuels* **2018**, *32*, 5374–5385. [CrossRef]
76. Sanna, A.; Maroto-Valer, M.M. Potassium-based sorbents from fly ash for high-temperature CO_2 capture. *Environ. Sci. Pollut. Res.* **2016**, *23*, 22242–22252. [CrossRef]
77. Wang, F.; Zhang, Y.; Mao, Z. High adsorption activated calcium silicate enabling high-capacity adsorption for sulfur dioxide. *New J. Chem.* **2020**, *44*, 11879–11886. [CrossRef]
78. Liu, S.; He, X. Researches in NH_3-SCR technologies for removal of NO_x from flue gas. *Ind. Catal.* **2021**, *29*, 7–12. (In Chinese)
79. Mao, Y.; Wang, H.; Hu, P. Theoretical investigation of NH3-SCR processes over zeolites: A review. *Int. J. Quantum Chem.* **2015**, *115*, 618–630. [CrossRef]
80. Zhang, W.; Chen, J.; Guo, L.; Zheng, W.; Wang, G.; Zheng, S.; Wu, X. Research progress on NH_3-SCR mechanism of metal-supported zeolite catalysts. *J. Fuel Chem. Technol.* **2021**, *49*, 1294–1315. (In Chinese) [CrossRef]
81. Han, J.; Jin, X.; Song, C.; Bi, Y.; Li, Z. Rapid synthesis and NH_3-SCR activity of SSZ-13 zeolite via coal gangue. *Green Chem.* **2020**, *22*, 219–229. [CrossRef]
82. Isabella, N.; Enrico, T. *Urea-SCR Technology for deNO$_x$ after Treatment of Diesel Exhausts*; Springer: New York, NY, USA, 2014; pp. 507–550.
83. Li, J.; Shi, Y.; Fu, X.; Huang, J.; Zhang, Y.; Deng, S.; Zhang, F. Hierarchical ZSM-5 based on fly ash for the low-temperature purification of odorous volatile organic compound in cooking fumes. *React. Kinet. Catal. Lett.* **2019**, *128*, 289–314. [CrossRef]
84. Yuan, G.; Zhang, J.; Zhang, Y.; Yan, Y.; Ju, X.; Sun, J. Characterization of high-alumina coal fly ash based silicate material and its adsorption performance on volatile organic compound elimination. *Korean J. Chem. Eng.* **2015**, *32*, 436–445. [CrossRef]

85. Lee, H.; Kim, Y.; Lee, I.; Jeon, J.; Jung, S.; Chung, J.; Choi, W.G.; Park, Y.-K. Recent advances in the catalytic hydrodeoxygenation of bio-oil. *Korean J. Chem. Eng.* **2016**, *33*, 3299–3315. [CrossRef]
86. Taufiqurrahmi, N.; Bhatia, S. Catalytic cracking of edible and non-edible oils for the production of biofuels. *Energy Environ. Sci.* **2011**, *4*, 1087–1112. [CrossRef]
87. Qin, T.; Yuan, S. Research progress of catalysts for catalytic steam reforming of high temperature tar: A review—ScienceDirect. *Fuel* **2023**, *331*, 125790. [CrossRef]
88. Gaurh, P.; Pramanik, H. Production of benzene/toluene/ethyl benzene/xylene (BTEX) via multiphase catalytic pyrolysis of hazardous waste polyethylene using low cost fly ash synthesized natural catalyst. *Waste Manag.* **2018**, *77*, 114–130. [CrossRef]
89. Dong, Z.; Chen, W.; Xu, K.; Liu, Y.; Wu, J.; Zhang, F. Understanding the Structure–Activity Relationships in Catalytic Conversion of Polyolefin Plastics by Zeolite-Based Catalysts: A Critical Review. *ACS Catal.* **2022**, *12*, 14882–14901. [CrossRef]
90. Songip, A.R.; Masuda, T.; Kuwahara, H.; Hashimoto, K. Production of high-quality gasoline by catalytic cracking over rare-earth metal exchanged Y-type zeolites of heavy oil from waste plastics. *Energy Fuels* **1994**, *8*, 136–140. [CrossRef]
91. Lai, J.; Meng, Y.; Yan, Y.; Lester, E.; Wu, T.; Pang, C. Catalytic pyrolysis of linear low-density polyethylene using recycled coal ash: Kinetic study and environmental evaluation. *Korean J. Chem. Eng.* **2021**, *38*, 2235–2246. [CrossRef]
92. Inayat, A.; Fasolini, A.; Basile, F.; Fridrichova, D.; Lestinsky, P. Chemical recycling of waste polystyrene by thermo-catalytic pyrolysis: A description for different feedstocks, catalysts and operation modes. *Polym. Degrad. Stab.* **2022**, *201*, 109981. [CrossRef]
93. Cocchi, M.; Angelis, D.D.; Mazzeo, L.; Nardozi, P.; Piemonte, V.; Tuffi, R.; Ciprioti, S.V. Catalytic Pyrolysis of a Residual Plastic Waste Using Zeolites Produced by Coal Fly Ash. *Catalysts* **2020**, *10*, 1113. [CrossRef]
94. Al-Rahbi, A.S.; Williams, P.T. Waste ashes as catalysts for the pyrolysis-catalytic steam reforming of biomass for hydrogen-rich gas production. *J. Mater. Cycles Waste Manag.* **2019**, *21*, 1224–1231. [CrossRef]
95. Du, S.; Dong, Y.; Guo, F.; Tian, B.; Mao, S.; Qian, L.; Xin, C. Preparation of high-activity coal char-based catalysts from high metals containing coal gangue and lignite for catalytic decomposition of biomass tar. *Int. J. Hydrog. Energy* **2021**, *46*, 14138–14147. [CrossRef]
96. Du, S.; Shu, R.; Guo, F.; Mao, S.; Bai, J.; Qian, L.; Xin, C. Porous coal char-based catalyst from coal gangue and lignite with high metal contents in the catalytic cracking of biomass tar. *Energy* **2022**, *249*, 123640. [CrossRef]
97. Du, S.; Mao, S.; Guo, F.; Dong, K.; Shu, R.; Bai, J.; Xu, L.; Li, D. Investigation of the catalytic performance of coal gangue char on biomass pyrolysis in a thermogravimetric analyzer and a fixed bed reactor. *Fuel* **2022**, *328*, 125216. [CrossRef]
98. Manique, M.C.; Lacerda, L.V.; Alves, A.K.; Bergmann, C.P. Biodiesel production using coal fly ash-derived sodalite as a heterogeneous catalyst. *Fuel* **2016**, *190*, 268–273. [CrossRef]
99. Yusuff, A.; Kumar, M.; Obe, B.; Mudashiru, L. Calcium Oxide Supported on Coal Fly Ash (CaO/CFA) as an Efficient Catalyst for Biodiesel Production from *Jatropha curcas* Oil. *Top. Catal.* **2021**, 1–13. [CrossRef]
100. Fotovat, F.; Kazemeini, M.; Kazemian, H. Novel Utilization of Zeolited Fly Ash Hosting Cobalt Nanoparticles as a Catalyst Applied to the Fischer-Tropsch Synthesis. *Catal. Lett.* **2009**, *127*, 204–212. [CrossRef]
101. Sineva, L.V.; Asalieva, E.Y.; Mordkovich, V.Z. The role of zeolite in the Fischer–Tropsch synthesis over cobalt–zeolite catalysts. *Russ. Chem. Rev.* **2015**, *84*, 1176–1189. [CrossRef]
102. Gupta, P.K.; Mahato, A.; Oraon, P.; Gupta, G.K.; Maity, S. Coal fly ash-derived mesoporous SBA as support material for production of liquid hydrocarbon through Fischer-Tropsch route. *Asia-Pac. J. Chem. Eng.* **2020**, *15*, 2471. [CrossRef]
103. Feng, Z.; Peng, W.; Wang, Z.; Guo, H.; Li, X.; Yan, G.; Wang, J.-X. Review of silicon-based alloys for lithium-ion battery anodes. *Int. J. Min. Met. Mater.* **2021**, *28*, 1549–1564. [CrossRef]
104. Xiao, Z.; Wang, C.; Song, L.; Zheng, Y.; Long, T. Research progress of nanosiliconbased materials and siliconcarbon composite anode materials for lithiumion batteries. *J. Solid State Electrochem.* **2022**, *26*, 1137. [CrossRef]
105. Meng, G.; Yang, H.; Zheng, J.; Wang, C. Nanoscale Silicon as Anode for Li-ion Batteries: The Fundamentals, Promises, and Challenges. *Nano Energy* **2015**, *17*, 366–383. [CrossRef]
106. Rahman, M.A.; Song, G.; Bhatt, A.I.; Wong, Y.C.; Wen, C. Nanostructured silicon anodes for high-performance lithiumion batteries. *Adv. Funct. Mater.* **2016**, *26*, 647–678. [CrossRef]
107. Ryu, J.; Hong, D.; Lee, H.W.; Park, S. Practical considerations of Si-based anodes for lithium-ion battery applications. *Nano Res.* **2017**, *10*, 3970–4002. [CrossRef]
108. Liu, X.; Zhang, Q.; Zhu, Y.; Zhao, J.; Chen, J.; Ye, H.; Wei, H.; Liu, Z. Trash to Treasure: Harmful Fly Ash Derived Silicon Nanoparticles for Enhanced Lithium-Ion Batteries. *Silicon* **2022**, *14*, 7983–7990. [CrossRef]
109. Xing, A.; Zhang, J.; Wang, R.; Wang, J.; Liu, X. Fly ashes as a sustainable source for nanostructured Si anodes in lithium-ion batteries. *SN Appl. Sci.* **2019**, *1*, 181. [CrossRef]

Disclaimer/Publisher's Note: The statements, opinions and data contained in all publications are solely those of the individual author(s) and contributor(s) and not of MDPI and/or the editor(s). MDPI and/or the editor(s) disclaim responsibility for any injury to people or property resulting from any ideas, methods, instructions or products referred to in the content.

Article

Effect of Cu on the Formation of Reversed Austenite in Super Martensitic Stainless Steel

Wen Jiang [1,*] and Kunyu Zhao [2]

1 School of Chemistry and Chemical Engineering, Kunming University, Kunming 650214, China
2 Department of Materials Science and Engineering, Kunming University of Science and Technology, Kunming 650093, China
* Correspondence: jiangwen0730@126.com

Abstract: We investigated the effect of Cu on the formation of reversed austenite in super martensitic stainless steel by using X-ray diffraction (XRD), a transmission electron microscope (TEM) and an energy-dispersive spectrometer (EDS). Our results showed that the microstructure of the steels comprised tempered martensite and diffused reversed austenite after the steels were quenched at 1050 °C and tempered at 550–750 °C. The volume fraction of reversed austenite in the steel with 3 wt.% of Cu (3Cu) was more than that with 1.5 wt.% of Cu (1.5Cu). The transmission electron microscope results revealed that the reversed austenite in 1.5Cu steel mainly had the shape of a thin strip, while that in 3Cu steel had a block shape. The nucleation points and degree of Ni enrichment of reversed austenite in 3Cu steel were higher than those in 1.5Cu steel. The reversed austenite was more likely to grow in ε-Cu enriched regions. Therefore, Cu can promote reversed austenite nucleation and growth. The mechanical properties of 3 Cu steel are obviously better than those of 1.5Cu steel when tempered at 550–650 °C.

Keywords: Cu; Ni; reversed austenite; super martensitic stainless steel

Citation: Jiang, W.; Zhao, K. Effect of Cu on the Formation of Reversed Austenite in Super Martensitic Stainless Steel. *Materials* **2023**, *16*, 1302. https://doi.org/10.3390/ma16031302

Academic Editor: Francesco Iacoviello

Received: 4 January 2023
Revised: 30 January 2023
Accepted: 31 January 2023
Published: 3 February 2023

Copyright: © 2023 by the authors. Licensee MDPI, Basel, Switzerland. This article is an open access article distributed under the terms and conditions of the Creative Commons Attribution (CC BY) license (https://creativecommons.org/licenses/by/4.0/).

1. Introduction

Reversed austenite as an important tempered microstructure in iron and steel has been studied by researchers; the effects of alloying elements on the formation of reversed austenite have received considerable scholarly attention. Ni is the key element that promotes α-γ phase transition, while Ni gathers and provides the conditions for the reversed austenite nucleation [1–6]. In addition, N, Si, and other elements impact the phase transition process [7–11]. As an austenitizing element, the effect of Cu on the formation of reversed austenite is negligible.

Over the years, Cu has been used as an alloying element added to steels, and the main functions of Cu in steels are as follows. (1) Solid solution strengthening: After quenching the steel, Cu is dissolved in the steel matrix, which plays the role of solid solution strengthening. The solution-strengthening effect increases in proportion to the increase in carbon content. Cu in an austenitic matrix can be precipitated as finely dispersed Cu-rich phase particles during the aging process, which has an excellent dispersion-strengthening effect. Cu atoms segregated to stacking faults can also pin dislocations and induce a matrix-strengthening effect [12–15]. (2) Improvement of formability and processability of steels: Cu can improve the anisotropy of steel and reduce the work hardening index. (3) Improvement of corrosion resistance of steels: The enhanced corrosion resistance of steel is attributable to the formation of a dense copper layer on the surface of the steel during the corrosion process. However, some researchers believe it is related to anodic polarization. Studies have shown that only 0.1 wt.% of Cu can significantly improve the atmospheric corrosion resistance of steel [16–19]. To date, few studies have reported the effect of Cu on the formation of reversed austenite in super martensitic stainless steel. Thus, this study compared the solid solution and precipitation behavior and investigated their effect on microstructures and mechanical properties of Cu-free and Cu-containing super martensitic stainless steel [20–22]. By comparing super martensitic stainless steels with different Cu content, the effects of Cu on the formation of reversed austenite and mechanical properties were investigated.

2. Materials and Methods

Tested steel with extra low impurity contents was designed and melted in a vacuum-induction melting furnace (rated capacity: 25 kg, rated power: 100 kw, limited vacuum degree: 6 × 10^{-3} Pa, boosting rate: 0.05 Pa/min, maximum temperature: 1700 °C, cooling water pressure: 0.35 MPa). The ingots were hot forged into round bars with a 15 mm diameter. To obtain the complete lath martensite, the samples were quenched in oil at a temperature of 1050 °C for 0.5 h using a vertical-type furnace. Additionally, samples were tempered at temperatures of 550, 600, 650, 700, and 750 °C for 2 h. The chemical compositions of the test steels are shown in Table 1. An etching solution consisting of 1 g of FeCl$_3$, 10 mL of hydrochloric acid and 120 mL of distilled water was used to etch the specimens for 10 s. The volume fraction of retained/reversed austenite at different temperatures was determined using PHILIPS APD-10 full-automatic X-ray diffraction instrument (XRD, Amsterdam, Netherlands) with Co radiation from 40 to 120° at a step interval of 0.02°. According to the integral intensity of diffraction peaks, the volume fraction of retained/reversed austenite was calculated by 6-line method and Formula (1), where $K = I_{0A111}/I_{0F110}$, I_{0A111} and I_{0F110} are the integral strengths of pure austenite and pure ferrite, respectively.

$$\phi_A = 1/(1 + K\frac{I_{F110}}{I_{A111}}) \tag{1}$$

Table 1. Chemical composition (wt%) of the test steel.

Steel Grade	C	Mn	Si	Cr	Ni	Mo	W	Cu
1.5Cu	0.021	0.4	0.27	14.78	6.5	2.04	0.8	1.44
3Cu	0.02	0.38	0.26	14.78	6.6	2.05	0.88	2.74

The morphology and distribution of austenite in the martensitic matrix and the element concentration distributions of the microstructure were investigated using FEI high resolution transmission electron microscope (HRTEM, Hillsboro, OR, USA) after jet polishing with thin foil in a solution of 6% perchloric acid and 94% anhydrous ethanol.

3. Results

3.1. Effect of Cu on the Quenched Microstructure

At 1050 °C, the austenite grains of 1.5Cu and 3Cu test steels have reached uniformity and there is no excessive grain phenomenon. The microstructures of the two steels are both lath martensite at a quenching temperature of 1050 °C (Figure 1). There are several martensite lath bundles with different orientations in an original austenite grain. The volume fraction of austenite was measured with XRD to confirm the presence of austenite in the quenched microstructure of the two test steels. Figure 2 shows that the retained austenite is found in two steels. All the austenite diffraction peaks in the XRD patterns of 3Cu were stronger than those of 1.5Cu. After the volume fraction of austenite was measured, the volume fraction of retained austenite in 3Cu steel after quenching was 13.32%, while that in 1.5Cu steel was 7.24%. This result is because Cu will be completely dissolved in austenite when the steels are quenched at a higher temperature of 1050 °C. With the increase in Cu content, the dissolved Cu content in the austenite increased. Cu stabilized austenite as an austenite-forming element. Thus, the martensite transformation was delayed, and the volume fraction of retained austenite increased. In addition, Mn- or Si-alloying elements in the steel enhanced retained austenite. Thus, more austenites were retained in 3Cu steel than in 1.5Cu steel.

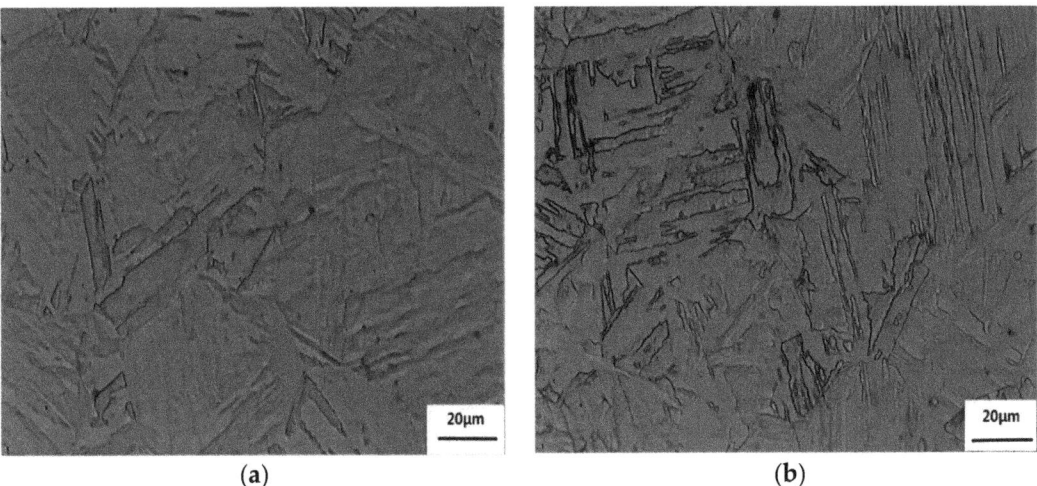

Figure 1. Microstructures of (**a**) 1.5Cu and (**b**) 3Cu steels quenched at 1050 °C.

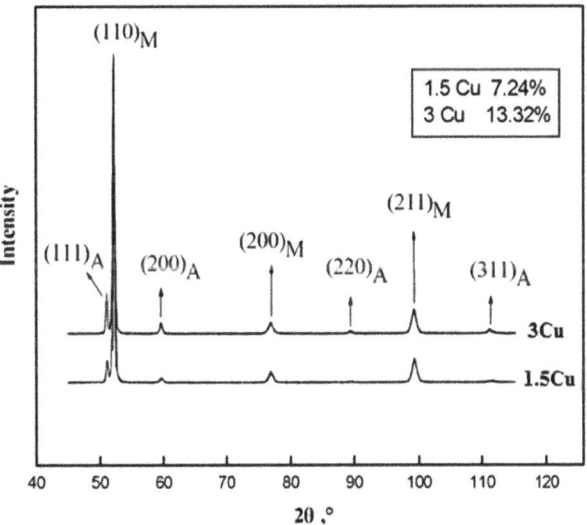

Figure 2. XRD patterns of retained austenite after quenching treatment of 1.5Cu and 3Cu steels. (M is martenite, and A is austenite).

3.2. Effect of Cu on the Tempered Microstructure

Figure 3 shows the microstructure of 1.5Cu and 3Cu tempered at 650 °C for 2 h after quenching at 1050 °C. The microstructure of the two test steels changed from thick lath martensite to fine-tempered martensite. After the test steels were tempered, the saturation of the supersaturated α-Fe decreased continuously due to the decomposition of the thick lath martensite and the precipitation of the carbides and intermetallic compound. At the same time, a few retained austenites were decomposed to form supersaturated ferrite and carbides, which were the tempered martensites.

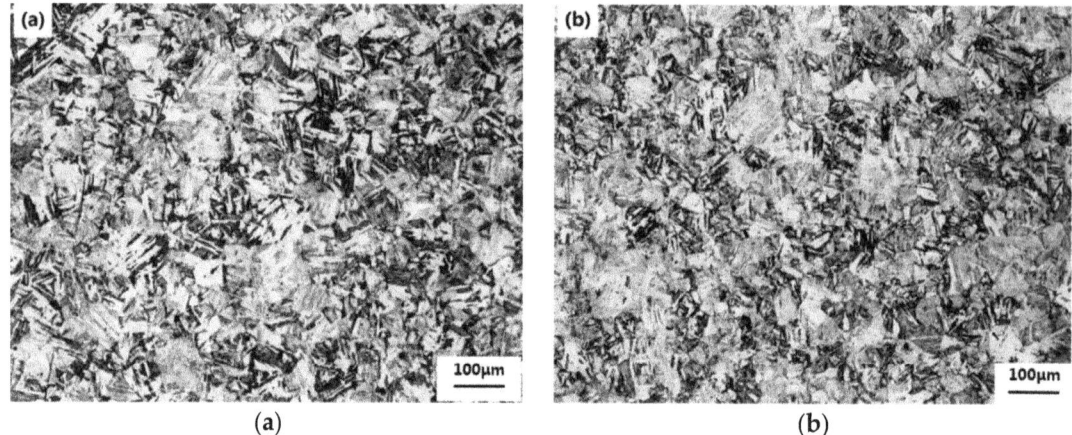

Figure 3. Microstructures of (a) 1.5Cu and (b) 3Cu tempered at 650 °C for 2 h after quenching at 1050 °C.

In addition, another type of austenite microstructure, namely, reserved austenite, was discovered in the two test steels. Reserved austenite was produced during the reversion of α to γ phase in the martensitic matrix. The variation of volume fractions of reversed austenite at different temperatures between 550 and 750 °C after quenching at 1050 °C are shown in Table 2 and Figure 4. A similar trend was observed in the amount of reserved austenite after the two steels were tempered. The volume fraction of reserved austenite of the two steels first increased and then decreased with the increasing tempering temperature and then reached a maximum value at 650–700 °C. The maximum values of the reserved austenite content of the two steels were 31.19% and 55.9%, respectively. Ni and Cu as the forming elements of austenite, can, in addition, increase the formation rate and decrease the formation temperature of reversed austenite. This is probably the main reason for the maximum amount of reversed austenite in 3Cu at 650 °C. Subsequently, the volume fraction of reserved austenite decreased due to the transformation from γ to α during the cooling process [23]. At all tempering temperatures, the volume fraction in 3Cu steel was greater than that in 1.5Cu steel. The 3Cu had a larger volume fraction of reversed austenite steel because Cu is an austenite-forming element. As a result, the greater the content of Cu, the easier the formation of austenite. In addition, the segregation of Cu reduces the faulting energy of the grain boundary, which is conducive to the diffusion of Ni. Earlier studies [24,25] have shown that Ni diffusion contributes to the formation of reversed austenite and increases its volume fraction. According to the above analysis, the increase in Cu content can promote the diffusion of Ni and the formation of reversed austenite. Thus, Cu and Ni can facilitate the formation of reversed austenite.

Table 2. The volume fraction of austenite at different heat treatment.

Steel Grade	1050 °C Quenching	550 °C Tempering	600 °C Tempering	650 °C Tempering	700 °C Tempering	750 °C Tempering
1.5Cu	7.24	12.48	18.63	30.93	31.19	24.33
3Cu	13.32	34.46	39.55	55.90	37.87	33.18

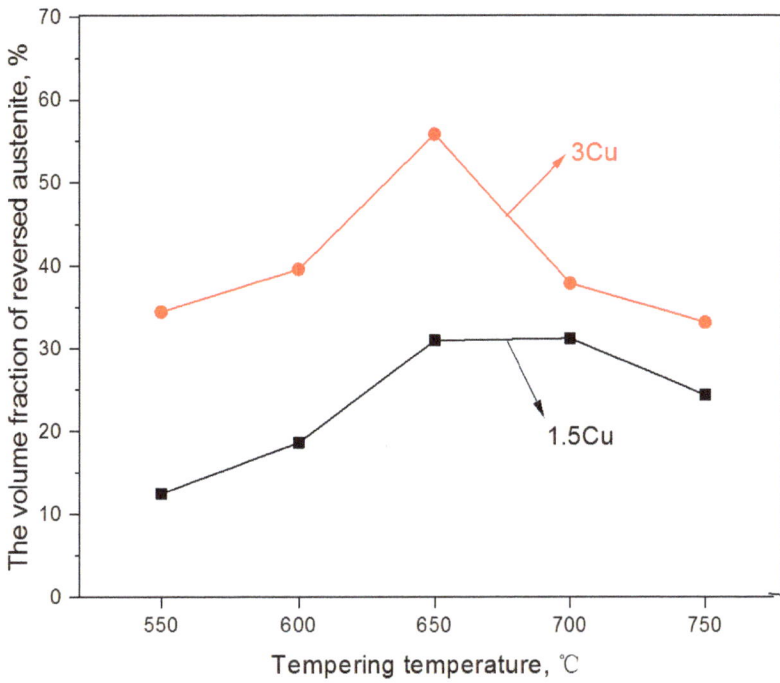

Figure 4. The curve of the volume fraction of reversed austenite and tempering temperature.

3.3. Enrichment of Cu and Ni in Reversed Austenite

According to the above analysis, the volume fraction of reversed austenite was greater in 1.5Cu and 3Cu test steels at a tempering temperature of 650–700 °C. To study the enrichment degree of Cu and Ni in the reversed austenite, TEM-EDS was used to analyze elements in the several forming regions of the reversed austenite at a tempering temperature of 650 °C. The average concentrations of Ni and Cu in the reversed austenite and in the matrix of two steels are shown in Table 3. As shown in Table 3, the Ni elemental concentration in reversed austenite in the two test steels was about three times higher than that in the matrix, indicating that Ni was enriched in the reversed austenite in the two steels. Furthermore, the content of Cu in the reserved austenite was more than that in the matrix, meaning Cu was also enriched in reversed austenite.

Table 3. EDS analysis result of austenite and matrix at a tempering temperature of 650 °C.

Element (wt.%)	1.5Cu			3Cu		
	Austenite	Matrix	Δ	Austenite	Matrix	Δ
Ni	9.93	3.10	6.83	11.96	4.76	7.20
Cu	1.89	1.33	0.56	3.79	2.29	1.50

The data of Ni and Cu presented in Table 3 were represented in a bar chart (Figure 5). The concentration of Ni in 3Cu was higher than that in 1.5Cu. The average Ni concentration in 3Cu was approximately 11.96%, and the Ni concentration in 1.5Cu was approximately 9.93%. These results showed that increased Cu content could increase the enrichment concentration of Ni in reversed austenite, thus making the formation of reversed austenite easier in the Ni-enrichment area. In addition, from the bar chart of Cu element distribution shown in Figure 5, the difference in Cu concentration between reversed austenite and the matrix in 1.5Cu steel was 0.56%. However, the difference in Cu concentration between reversed austenite and the matrix in 3Cu steel was 1.5%. In other words, the difference in Cu concentration between reversed austenite and the matrix in 3Cu was two times greater than that in 1.5Cu steel, indicating that the enrichments of Ni and Cu in 3Cu steel were both higher than those in 1.5Cu steel. The enrichment of Cu was easier in the region with higher Ni concentration. As a result, the enrichment degree of Cu increased. Therefore, from the above analysis of the concentration of two elements, Cu and Ni contributed to the enrichment of each other in reversed austenite.

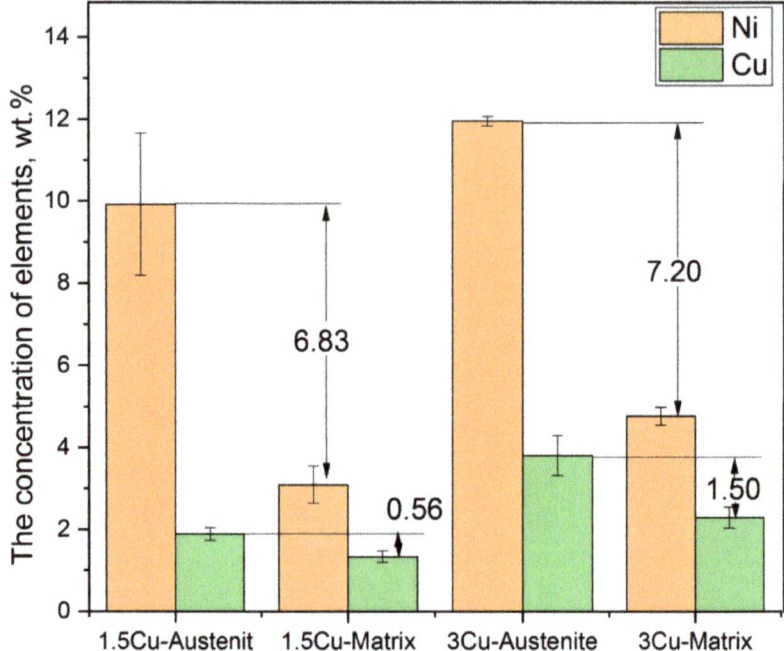

Figure 5. The concentration of Ni and Cu in reversed austenite and matrix in two test steels.

To determine the effect of Cu on the reversed austenite formation, TEM was used to examine the microstructure of 1.5Cu and 3Cu test steels after they were quenched at 1050 °C for 30 min and tempered at 650 °C for 2 h. Figure 6 shows TEM bright field image. Figure 6a,b show dark long strips or block reversed austenite distributes in the boundary and interior of lath martensite. The reversed austenite in 1.5Cu steel was mainly thin and long along the boundary of martensite lath. However, the reversed austenite in 3Cu steel was blocked and was shorter and wider than that in 1.5Cu steel. The number of nucleation points of the reversed austenite of 3Cu steel was significantly more than that of 1.5Cu steel in the same photographic area. This phenomenon indicates that the addition of Cu in 3Cu steel can promote the diffusion of Ni and increase its enrichment degree and the number of Ni enrichment areas, thereby increasing the number of reversed austenite nucleation points. Therefore, the addition of Cu facilitated the formation of reversed austenite.

The different number of nucleation points was mainly attributed to the different shapes of reversed austenite in the two test steels. More nucleation points enhance the growth of reversed austenite grain to meet another growing austenite grain in the growth process. When the two grains meet, then the two grains will stop growing in the original direction and grow in other directions; thus, the shape of reversed austenite becomes a block. The nucleation number and shape of reversed austenite were relatively smaller and finer, respectively, with the increasing Cu content, which improved the mechanical properties of the test steel. As shown in Figure 6a,b, many fine precipitates were observed around the reversed austenite in the two test steels. These precipitates appeared in symmetrical or elliptical shapes, and the grain sizes of 20–50 nm were ε-Cu [12,26]. Compared with 1.5Cu steel, ε-Cu was distributed more densely around the reversed austenite of 3Cu steel (dashed region), indicating that the region with higher Ni enrichment increased the enrichment of Cu and promoted the precipitation of ε-Cu in this region. This phenomenon confirmed that the enrichment of Ni promotes the enrichment and precipitation of Cu in reversed austenite.

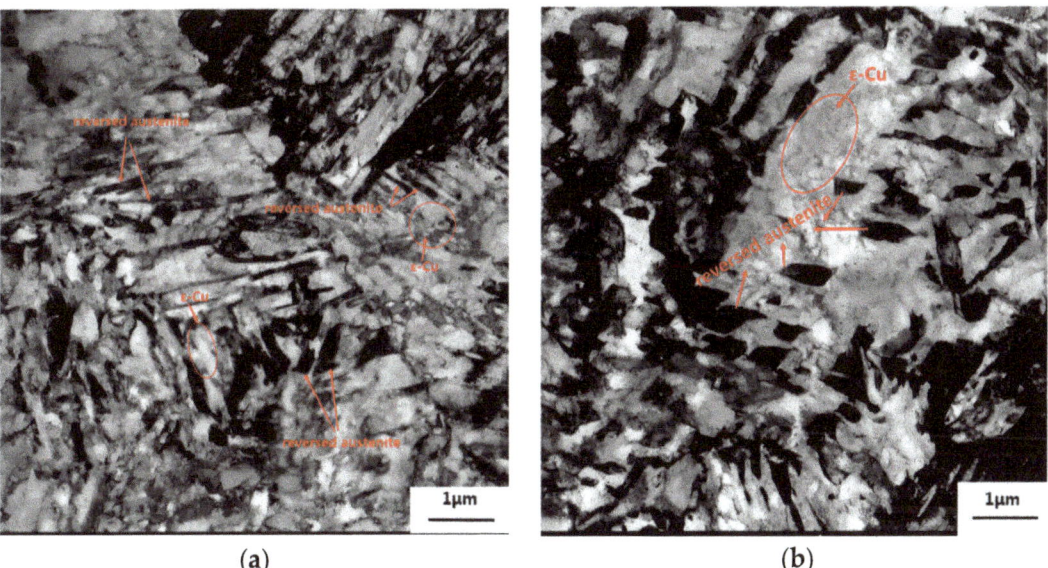

Figure 6. TEM bright field of (**a**) 1.5Cu and (**b**) 3Cu at tempering temperature of 650 °C.

Given the above analysis of the two test steels with the same content of Ni, the volume fraction and nucleation points number of the reversed austenite in 3Cu steel are more than those in 1.5Cu steel, and the enrichment degree of Ni of the reversed austenite in 3Cu steel is greater than that in 1.5Cu, indicating that the addition of Cu promotes the enrichment of Ni. Moreover, Cu was also enriched around the reversed austenite or in the Ni enrichment area. The concentration of Cu and density of ε-Cu were also high in the area with a high concentration of Ni, indicating that Ni also contributed to the enrichment of Cu. Therefore, Cu and Ni synergically promote the formation of reversed austenite formation.

3.4. Effect of Cu on Mechanical Properties

Figure 7 shows the variation of Rockwell hardness (HRC) of 1.5Cu and 3Cu steels at different tempering temperatures. The HRC of the two test steels gradually decreased as the tempering temperature increased, until the latter reached the minimum at 650 °C, and then it slightly increased. During the tempering process, martensite desolated and transformed into reversed austenite with carbide precipitation. With increased tempering temperature, the volume fraction of reversed austenite gradually increased due to the increased driving force of α to γ transformation. Reversed austenite can reduce the hardness of steel as a softening phase [27]. Thus, HRC decreased at a temperature below 650 °C. With the increase in tempering temperature, the concentration of austenitic elements in the reversed austenite decreased gradually due to composition homogenization. The thermal stability of the reversed austenite decreased; as a result, the reversed austenite retransformed into martensite in the process of tempering and cooling, thereby decreasing the volume fraction of the reversed austenite. At the same time, the HRC of the test steels increased. Thus, the optimal tempering temperature for the reversed austenite and HRC of the two steels was 650 °C. The lowest HRC was obtained at 650 °C, consistent with the reversed austenite content test results.

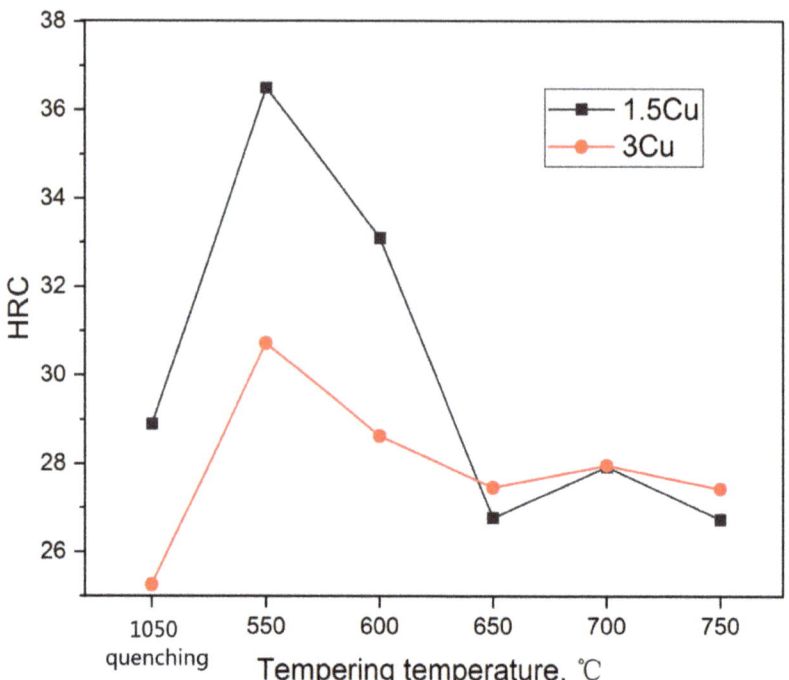

Figure 7. Variation of the hardness of the two test steels with tempering temperature.

The HRC of 3Cu is slightly smaller than that of 1.5Cu after quenching at 1050 °C, which is due to more retained austenite in the 3Cu steel. When the tempering temperature was between 550 °C and 650 °C, the HRC of 1.5Cu steel was significantly greater than that of 3Cu. However, no significant differences were observed in the HRC of the two test steels when the tempering temperature was above 650 °C. This phenomenon is because the 3Cu steel precipitated more fine grains and diffused a copper-rich phase during the tempering process to refine the grain at 550–650 °C. At the same time, there was the role of precipitation hardening [28,29]. Thus, the hardness can be improved to a certain extent. In addition, the increased volume fraction of reversed austenite in 3Cu significantly reduced the hardness [30,31], and the combined effect of the two factors reduced the HRC of 3Cu steel compared with that of 1.5Cu steel. At tempering temperatures above 650 °C, the refinement of ε-Cu relative to the grain weakened, along with the reduction in the reversed austenite, balancing each other and making the HRC of the two test steels remain basically the same. Figure 8 shows the strength-strain curves of the two tested steels tempered at 650 °C for 2 h. It can be seen that 3Cu steel has the highest tensile strength and the best plasticity. Although the yield strength of 3Cu steel is less than that of 1.5Cu steel, 3Cu has the largest strain hardening rate and the longest uniform deformation stage.

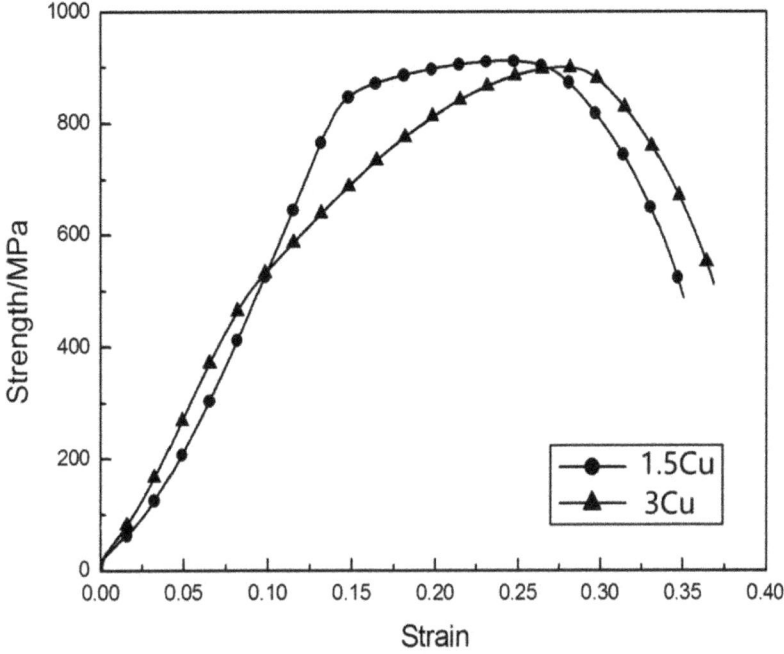

Figure 8. Strength-strain curves of the three tested steels tempered at 650 °C for 2 h.

Table 4 shows the strength and elongation of 1.5Cu steel and 3Cu steel at tempering temperatures of 550–750 °C. As presented in Table 5, the tensile strength of 1.5Cu steel was 903.5–1008.8 MPa, the elongation was 16.66–19.62%, and the sectional shrinkage was 66.04–72.58%. The tensile strength, elongation, and sectional shrinkage of 3Cu steel were 902.5–985.2 MPa, 16.29–26.72%, and 61.51–73.17%, respectively. Overall, the elongation of 3Cu steel was larger than that of 1.5Cu, while the tensile strength of 3Cu steel was lower than that of 1.5Cu. It can be seen from the above data that the mechanical properties of 3Cu steel are obviously better than 1.5Cu steel when tempered at 550–650 °C. This phenomenon is because 3Cu steel has more reversed austenite after tempering.

Table 4. The data of mechanical properties in 1.5Cu and 3Cu steels.

Cu Content (wt.%)	Tempering Temperature	Tensile Strength (MPa)	Elongation to Failure, εu (%)	Product of Strength and Elongation (MPa%)
1.5Cu	550	1008.82	19.38	19,550.93
	600	905.84	19.62	17,772.58
	650	894.35	21.12	18,888.67
	700	924.19	19.15	17,698.24
	750	939.81	16.66	15,657.23
3Cu	550	903.53	26.72	24,142.32
	600	902.48	28.00	25,269.44
	650	912.40	22.58	20,601.99
	700	985.25	17.67	17,409.37
	750	973.70	16.29	15,861.57

Table 5. The diffusion coefficients at different tempering temperatures.

T	$D_{Cu-\alpha}$	$D_{Cu-\gamma}$	$D_{Cu\text{-}self}$
550	2.85×10^{-16}	1.04×10^{-18}	3.18×10^{-14}
600	3.07×10^{-15}	1.01×10^{-17}	1.86×10^{-13}
650	2.55×10^{-14}	7.72×10^{-17}	8.97×10^{-13}
700	1.71×10^{-13}	4.77×10^{-16}	3.68×10^{-12}
750	9.50×10^{-13}	2.47×10^{-15}	1.32×10^{-11}

Figure 9 shows the curves of strength and elongation with tempering temperature. In general, the elongation of 3Cu steel is higher than that of 1.5Cu, while the tensile strength of 3Cu steel is lower than that of 1.5Cu. This is related to more austenite in the 3Cu steel after tempering. As the tempering temperature increased, the tensile strength of the two test steels at first decreased and then increased, whereas the elongation first increased and then decreased. When the tensile strength was minimum, the elongation became maximum. From the previous analysis, the test steel after tempering with the precipitation of reversed austenite. The volume fraction of reversed austenite with the increased tempering temperature first increased, gradually decreased, and peaked at 650 °C. The soft and tough phases and the reversed austenite were diffusely distributed in the martensitic laths and austenite grain boundaries after tempering, which could absorb the deformation work during plastic deformation of the material at room temperature. The plasticity induced by the martensitic phase transformation can significantly improve the plastic toughness of the material [32–34]. Therefore, the strength and elongation of the two test steels changed with the change in austenite content, indicating that the mechanical properties of the test steels were related to the reversed austenite. Reversed austenite has a dual effect on super martensitic properties. On the one hand, an appropriate amount of fine block reversed austenite has a certain strengthening effect and toughening effect, but too much reversed austenite will reduce the strength of steel.

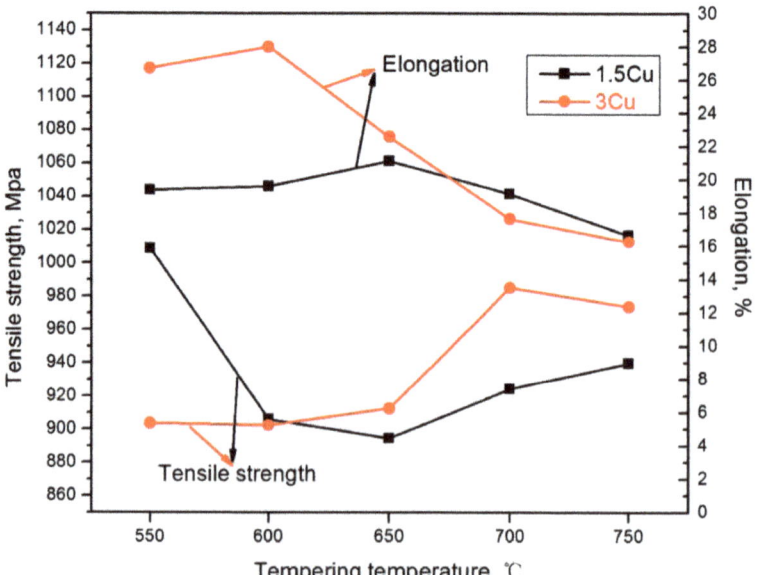

Figure 9. Variation of tensile strength and elongation of the two test steels with tempering temperature.

Because the strength and elongation of the two test steels varied, the strong plastic product (PSE) of the two test steels was calculated (Table 4) to illustrate their comprehensive mechanical properties. The curve graph of the PSE of the two steels is shown in Figure 10. The PSE of 3Cu steel was greater than that of 1.5Cu steel. This result showed that the strength and plasticity could be improved by adding Cu to some extent. The super martensitic stainless steel containing Cu had a high PSE value (15,657–25,270 MPa%). This result further indicated that the synergistic effect of Cu and Ni contributed to the formation of reversed austenite and improved the strength and toughness of the steels.

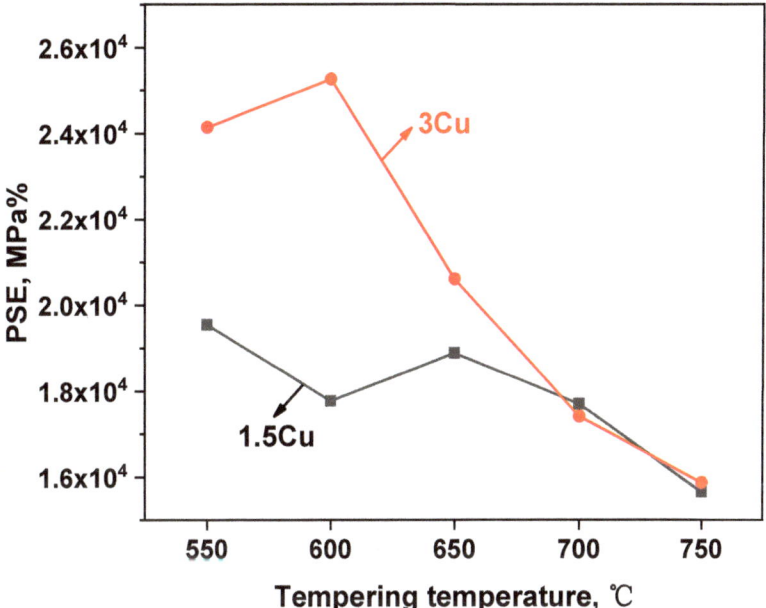

Figure 10. Variation of strength and elongation of the test steels with tempering temperature.

4. Discussion

Given the above analysis, Cu was enriched in the reversed austenite to a certain extent in 1.5Cu and 3Cu steels. The process of Cu enrichment can be explained using diffusion theory. The diffusion of Cu can be expressed as follows [35]:

$$D = D_0 \exp(-Q/RT) \quad (2)$$

where D is diffusion coefficient (cm^2/s), D_0 is frequency factor (cm^2/s), Q is activation energy (kJ/mol), R is the Boltzmann constant, and T is diffusion temperature (K). The diffusion coefficient of Cu in martensite ($D_{Cu-\alpha}$), Cu in austenite ($D_{Cu-\gamma}$), and self-diffusing ($D_{Cu-self}$) are calculated using the following formulas:

$$\begin{aligned} D_{Cu-\alpha} &= 300\exp(-284{,}000/RT) \\ D_{Cu-\gamma} &= 0.19\exp(-272{,}000/RT) \\ D_{Cu-self} &= 0.78\exp(-211{,}000/RT) \end{aligned} \quad (3)$$

Based on Formula (3), the data and curves of $D_{Cu-\alpha}$, $D_{Cu-\gamma}$, and $D_{Cu-self}$ at different tempering temperatures are shown in Table 5 and Figure 10, respectively. As presented in Table 5, values of $D_{Cu-self}$ were larger than those of $D_{Cu-\alpha}$ and $D_{Cu-\gamma}$ at all the tempering temperatures, and the largest value was about two to four orders of magnitude. $D_{Cu-\alpha}$

was 1–2 orders of magnitude larger than $D_{Cu-\gamma}$. The variations in values of the three diffusion coefficients were more significant with increasing tempering temperature. The relationship between the diffusion coefficients and tempering temperature is represented in Figure 11. In the tempering temperature range of 550–750 °C, the three diffusion coefficients exponentially increased in the following order: $D_{Cu-self} > D_{Cu-\alpha} > D_{Cu-\gamma}$. When the tempering temperature was in the range of 550–600 °C, $D_{Cu-self}$ was small and stable, while the self-diffusion rate of Cu was slow. When the tempering temperature was 600–750 °C, the self-diffusion rate of Cu became faster with the increasing temperature, and ε-Cu was gradually formed. Figure 11b is a larger version of Figure 11a, and it shows an increase in $D_{Cu-\alpha}$ starts. Because $D_{Cu-\gamma}$ was relatively small, the Cu atoms in austenite could be seen as stationary compared with the rapid diffusion of Cu atoms in martensite. Then, when the Cu atoms in martensite entered the austenite, its diffusion rate rapidly decreased and became stationary. As the tempering temperature increased, more Cu atoms entered the austenite and enriched the reserved austenite. Therefore, the volume fraction of reversed austenite rapidly increased within the tempering temperature range. In addition, both Ni and Cu played a pivotal role in the enrichment of reversed austenite.

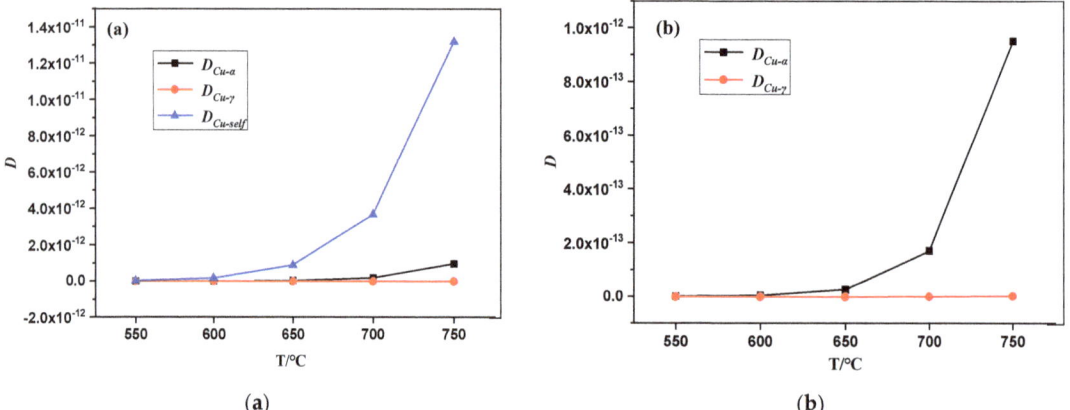

Figure 11. The curves between the diffusion coefficients and tempering temperature. (**a**) The curves of $D_{Cu-self}$, $D_{Cu-\alpha}$, and $D_{Cu-\gamma}$; (**b**) The curves of $D_{Cu-\alpha}$ and $D_{Cu-\gamma}$.

5. Conclusions

1. The matrices of 1.5Cu and 3Cu super martensitic stainless steels were quenched at 1050 °C. A small amount of retained austenite in two test steels and the addition of Cu increased the volume fraction of retained austenite after quenching.
2. The two test steels were quenched at 1050 °C and tempered at 550–750 °C. The tempered microstructures included tempered martensite, reversed austenite, and ε-Cu. The reversed austenite content in both test steels initially increased, then decreased, and finally reached a maximum at 650–700 °C as the tempering temperature increased. The reversed austenite content in 3Cu steel was more than that in 1.5Cu steel, and the distribution in the matrix was also denser.
3. The degree of Ni enrichment and the number of enrichment areas within the reversed austenite in 3Cu steel was higher than that in 1.5Cu steel. The enrichment of Cu elements was also more pronounced in the area of high Ni concentration in the reversed austenite, and more ε-Cu precipitated near this area, indicating that Cu contributed to the nucleation and growth of reversed austenite.
4. The addition of Cu to the test steels facilitated the reversed austenite formation, and the mechanical properties of 3Cu steel are obviously better than those of 1.5Cu steel when tempered at 550–650 °C.

Author Contributions: Methodology, K.Z.; Project administration, formal analysis, resources and writing, W.J. All authors have read and agreed to the published version of the manuscript.

Funding: This research was funded by [Yunnan Province Science and Technology Departemet] grant number [202001BA070001-153].

Institutional Review Board Statement: Not applicable.

Informed Consent Statement: Not applicable.

Data Availability Statement: The data that support the findings of this study are available from the corresponding author, [Jiang W], upon reasonable request.

Conflicts of Interest: The authors declare no conflict of interest.

References

1. Lee, T.-H.; Kim, S.-J. Phase identification in an isothermally aged austenitic 22Cr-21Ni-6Mo-N stainless steel. *Scr. Mater.* **1998**, *39*, 951–956. [CrossRef]
2. Song, Y.Y.; Li, X.Y.; Rong, L.J.; Li, Y.Y.; Nagai, T. Reversed austenite in 0Cr13Ni4Mo martensitic stainless steels. *Mater. Chem. Phys.* **2014**, *143*, 728–734. [CrossRef]
3. Yang, Z.; Liu, Z.; Liang, J.; Su, J.; Yang, Z.; Zhang, B.; Sheng, G. Correlation between the microstructure and hydrogen embrittlement resistance in a precipitation-hardened martensitic stainless steel. *Corros. Sci.* **2021**, *182*, 109260. [CrossRef]
4. Niessen, F.; Apel, D.; Danoix, F.; Hald, J.; Somers, M.A.J. Evolution of substructure in low-interstitial martensitic stainless steel during tempering. *Mater. Charact.* **2020**, *167*, 110494. [CrossRef]
5. Wu, S.; Wang, D.; Zhao, C.; Zhang, Z.; Li, C.; Di, X. Enhanced toughness of Fe–12Cr–5.5Ni–Mo-deposited metals through formation of fine reversed austenite. *J. Mater. Sci.* **2018**, *53*, 15679–15693. [CrossRef]
6. Zhang, Y.; Zhang, C.; Yuan, X.; Li, D.; Yin, Y.; Li, S. Microstructure Evolution and Orientation Relationship of Reverted Austenite in 13Cr Supermartensitic Stainless Steel During the Tempering Process. *Materials* **2019**, *12*, 589. [CrossRef]
7. Misra, R.D.K.; Zhang, Z.; Venkatasurya, P.K.C.; Somani, M.C.; Karjalainen, L.P. The effect of nitrogen on the formation of phase reversion-induced nanograined/ultrafine-grained structure and mechanical behavior of a Cr–Ni–N steel. *Mater. Sci. Eng. A* **2011**, *528*, 1889–1896. [CrossRef]
8. Xie, Y.; Cheng, H.; Tang, Q.; Chen, W.; Chen, W.; Dai, P. Effects of N addition on microstructure and mechanical properties of CoCrFeNiMn high entropy alloy produced by mechanical alloying and vacuum hot pressing sintering. *Intermetallics* **2018**, *93*, 228–234. [CrossRef]
9. Sawada, M.; Adachi, K.; Maeda, T. Effect of V, Nb and Ti Addition and Annealing Temperature on Microstructure and Tensile Properties of AISI 301L Stainless Steel. *ISIJ Int.* **2011**, *51*, 991–998. [CrossRef]
10. Kisko, A.; Hamada, A.S.; Talonen, J.; Porter, D.; Karjalainen, L.P. Effects of reversion and recrystallization on microstructure and mechanical properties of Nb-alloyed low-Ni high-Mn austenitic stainless steels. *Mater. Sci. Eng. A* **2016**, *657*, 359–370. [CrossRef]
11. Challa, V.S.A.; Wan, X.L.; Somani, M.C.; Karjalainen, L.P.; Misra, R.D.K. Strain hardening behavior of phase reversion-induced nanograined/ultrafine-grained (NG/UFG) austenitic stainless steel and relationship with grain size and deformation mechanism. *Mater. Sci. Eng. A* **2014**, *613*, 60–70. [CrossRef]
12. Sen, I.; Amankwah, E.; Kumar, N.S.; Fleury, E.; Oh-ishi, K.; Hono, K.; Ramamurty, U. Microstructure and mechanical properties of annealed SUS 304H austenitic stainless steel with copper. *Mater. Sci. Eng. A* **2011**, *528*, 4491–4499. [CrossRef]
13. Izotov, V.I.; Ilyukhin, D.S.; Getmanova, M.E.; Filippov, G.A. Influence of copper on the structure and mechanical properties of pearlitic steels. *Phys. Met. Metallogr.* **2016**, *117*, 588–593. [CrossRef]
14. Saeidi, N.; Raeissi, M. Promising effect of copper on the mechanical properties of transformation-induced plasticity steels. *Mater. Sci. Technol.* **2019**, *35*, 1708–1716. [CrossRef]
15. Sazegaran, H.; Hojati, M. Effects of copper content on microstructure and mechanical properties of open-cell steel foams. *Int. J. Miner. Metall. Mater.* **2019**, *26*, 588–596. [CrossRef]
16. Zhao, J.; Yang, C.; Zhang, D.; Zhao, Y.; Khan, M.S.; Xu, D.; Xi, T.; Li, X.; Yang, K. Investigation on mechanical, corrosion resistance and antibacterial properties of Cu-bearing 2205 duplex stainless steel by solution treatment. *RSC Adv.* **2016**, *6*, 112738–112747. [CrossRef]
17. Hao, X.; Xi, T.; Xu, Z.; Liu, L.; Yang, C.; Yang, K. Effect of tempering temperature on the microstructure, corrosion resistance, and antibacterial properties of Cu-bearing martensitic stainless steel. *Mater. Corros.* **2021**, *72*, 1668–1676. [CrossRef]
18. Liu, H.; Teng, Y.; Guo, J.; Li, N.; Wang, J.; Zhou, Z.; Li, S. Corrosion resistance and corrosion behavior of high-copper-bearing steel in marine environments. *Mater. Corros.* **2021**, *72*, 816–828. [CrossRef]
19. Wang, Y.; Zhang, X.; Wei, W.; Wan, X.; Liu, J.; Wu, K. Effects of Ti and Cu Addition on Inclusion Modification and Corrosion Behavior in Simulated Coarse-Grained Heat-Affected Zone of Low-Alloy Steels. *Materials* **2021**, *14*, 791. [CrossRef]
20. Ye, D.; Li, J.; Yong, Q.L.; Su, J.; Tao, J.M.; Zhao, K.Y. Microstructure and properties of super martensitic stainless steel microalloyed with tungsten and copper. *Mater. Technol.* **2012**, *27*, 88–91. [CrossRef]

21. Zhu, H.-m.; Luo, C.-p.; Liu, J.-w.; Jiao, D.-l. Effects of Cu addition on microstructure and mechanical properties of as-cast magnesium alloy ZK60. *Trans. Nonferrous Met. Soc. China* **2014**, *24*, 605–610. [CrossRef]
22. Barbosa, B.A.R.S.; Tavares, S.S.M.; Bastos, I.N.; Silva, M.R.; de Macedo, M.C.S. Influence of heat treatments on microstructure and pitting corrosion resistance of 15%Cr supermartensitic stainless steel. *Corros. Eng. Sci. Technol.* **2014**, *49*, 311–315. [CrossRef]
23. Yang, Y.H.; Hui-Bin, W.U.; Cai, Q.W.; Cheng, L. Formation of reversed austenite and its stability in 9Ni steel during tempering. *Trans. Mater. Heat Treat.* **2010**, *31*, 73–77.
24. Jiang, W.; Zhao, K.-y.; Ye, D.; Li, J.; Li, Z.-d.; Su, J. Effect of Heat Treatment on Reversed Austenite in Cr15 Super Martensitic Stainless Steel. *J. Iron Steel Res. Int.* **2013**, *20*, 61–65. [CrossRef]
25. Jiang, W.; Ye, D.; Li, J.; Su, J.; Zhao, K. Reverse Transformation Mechanism of Martensite to Austenite in 00Cr15Ni7Mo2WCu2 Super Martensitic Stainless Steel. *Steel Res. Int.* **2014**, *85*, 1150–1157. [CrossRef]
26. Tan, S.P.; Zhen-Hau, W.A.N.G.; Cheng, S.C.; Liu, Z.D.; Han, J.C.; Fu, W.T. Effect of Cu Content on Aging Precipitation Behaviors of Cu-Rich Phase in Fe-Cr-Ni Alloy. *J. Iron Steel Res. Int.* **2010**, *17*, 6. [CrossRef]
27. Song, Y.Y.; Ping, D.H.; Yin, F.X.; Li, X.Y.; Li, Y.Y. Microstructural evolution and low temperature impact toughness of a Fe–13%Cr–4%Ni–Mo martensitic stainless steel. *Mater. Sci. Eng. A* **2010**, *527*, 614–618. [CrossRef]
28. Rowolt, C.; Milkereit, B.; Springer, A.; Kreyenschulte, C.; Kessler, O. Dissolution and precipitation of copper-rich phases during heating and cooling of precipitation-hardening steel X5CrNiCuNb16-4 (17-4 PH). *J. Mater. Sci.* **2020**, *55*, 13244–13257. [CrossRef]
29. Wang, Z.; Fang, X.; Li, H.; Liu, W. Atom Probe Tomographic Characterization of Nanoscale Cu-Rich Precipitates in 17-4 Precipitate Hardened Stainless Steel Tempered at Different Temperatures. *Microsc. Microanal.* **2017**, *23*, 340–349. [CrossRef]
30. Rodrigues, C.A.D.; Bandeira, R.M.; Duarte, B.B.; Tremiliosi-Filho, G.; Roche, V.; Jorge, A.M. The Influence of Ni Content on the Weldability, Mechanical, and Pitting Corrosion Properties of a High-Nickel-Bearing Supermartensitic Stainless Steel. *J. Mater. Eng. Perform.* **2021**, *30*, 3044–3053. [CrossRef]
31. Wang, X.Y.; Li, M.; Wen, Z.X. The Effect of the Cooling Rates on the Microstructure and High-Temperature Mechanical Properties of a Nickel-Based Single Crystal Superalloy. *Materials* **2020**, *13*, 4256. [CrossRef] [PubMed]
32. Liu, H.; Du, L.X.; Hu, J.; Wu, H.Y.; Gao, X.H.; Misra, R.D.K. Interplay between reversed austenite and plastic deformation in a directly quenched and intercritically annealed 0.04C-5Mn low-Al steel. *J. Alloy. Compd.* **2017**, *695*, 2072–2082. [CrossRef]
33. Zhang, M.; Tong, X.; Xu, S.; Hao, C.; Xia, T. Characterization of Precipitated Phases in the Weld of G520 Steel during Short-Time High-Temperature Tempering and its Influence on Properties. *J. Mater. Eng. Perform.* **2021**, *30*, 5921–5930. [CrossRef]
34. Li, Y.Z.; Wang, M.; Huang, M.X. In-situ measurement of plastic strain in martensite matrix induced by austenite-to-martensite transformation. *Mater. Sci. Eng. A Struct. Mater. Prop. Misrostructure Process.* **2021**, *811*, 141061. [CrossRef]
35. Yong, Q.L. *Second Phases in Structural Steels*; Metallurgical Industry Press: Beijing, China, 2006.

Disclaimer/Publisher's Note: The statements, opinions and data contained in all publications are solely those of the individual author(s) and contributor(s) and not of MDPI and/or the editor(s). MDPI and/or the editor(s) disclaim responsibility for any injury to people or property resulting from any ideas, methods, instructions or products referred to in the content.

Article

Adsorption of Sc on the Surface of Kaolinite (001): A Density Functional Theory Study

Zilong Zhao [1,†], Kaiyu Wang [1,†], Guoyuan Wu [1], Dengbang Jiang [2,*] and Yaozhong Lan [1,*]

[1] School of Materials and Energy, Yunnan University, Kunming 650091, China; zhaozilong@itc.ynu.edu.cn (Z.Z.); w13835318000@163.com (K.W.); wgy66@tom.com (G.W.)
[2] Green Preparation Technology of Biobased Materials National & Local Joint Engineering Research Center, Yunnan Minzu University, Kunming 650500, China
* Correspondence: 041814@ymu.edu.cn (D.J.); yzhlan@ynu.edu.cn (Y.L.)
† These authors contributed equally to this work.

Abstract: The adsorption behavior of Sc on the surface of kaolinite (001) was investigated using the density functional theory via the generalized gradient approximation plane-wave pseudopotential method. The highest coordination numbers of hydrated Sc^{3+}, $Sc(OH)^{2+}$, and $Sc(OH)_2^+$ species are eight, six, and five, respectively. The adsorption model was based on $Sc(OH)_2(H_2O)_5^+$, which has the most stable ionic configuration in the liquid phase. According to the adsorption energy and bonding mechanism, the adsorption of Sc ionic species can be categorized into outer layer and inner layer adsorptions. We found that the hydrated Sc ions were mainly adsorbed on the outer layer of the kaolinite (001)Al-OH and (00−1)Si-O surfaces through hydrogen bonding while also being adsorbed on the inner layer of the deprotonated kaolinite (001)Al-OH surface through coordination bonding. The inner layer adsorption has three adsorption configurations, with the lying hydroxyl group (O_l) position having the lowest adsorption energy (−653.32 KJ/mol). The adsorption energy for the inner layer is lower compared to the outer layer, while the extent of deprotonation is limited. This is because the deprotonation of the inner adsorption layer is energetically unfavorable. We speculate that Sc ions species predominantly adsorb onto the surface of kaolinite (001) in an outer layer configuration.

Citation: Zhao, Z.; Wang, K.; Wu, G.; Jiang, D.; Lan, Y. Adsorption of Sc on the Surface of Kaolinite (001): A Density Functional Theory Study. *Materials* **2023**, *16*, 5349. https://doi.org/10.3390/ma16155349

Academic Editor: Franz Saija

Received: 23 June 2023
Revised: 12 July 2023
Accepted: 24 July 2023
Published: 29 July 2023

Copyright: © 2023 by the authors. Licensee MDPI, Basel, Switzerland. This article is an open access article distributed under the terms and conditions of the Creative Commons Attribution (CC BY) license (https:// creativecommons.org/licenses/by/ 4.0/).

Keywords: rare earth; scandium; kaolinite; density functional theory; adsorption

1. Introduction

In recent years, rare earth elements have been highly regarded due to their distinctive physical and chemical properties [1,2]. Among the most crucial strategic metals, they find extensive applications across numerous high-tech sectors [3–6]. According to their formation process, rare earth ores are primarily categorized into mineral and weathered types. Mineral-type rare earth ores serve as the primary source of light rare earth elements, typically represented by bastnasite and monazite [7]. On the other hand, weathered rare earth ores are medium in medium and heavy rare earth elements, with a notable example being weathered crust elution-deposited rare earth ore, also referred to as ion-adsorption-type rare earth ore [8]. Despite its low grade (0.05–0.2 wt.% rare earth oxides), weathered rare earth ore holds an 80% share in the global supply of medium and heavy rare earth elements [9].

Scandium (Sc) is a rare and expensive metal widely used in various industries such as electronics, optics, automotive, aerospace, transportation, and the production of advanced materials [10,11]. It falls into the heavy rare earths group due to its similar physical and chemical properties to the rare earth elements and its frequent occurrence in symbiosis with yttrium and lanthanides [12]. Rare earth elements primarily adsorb onto clay minerals in hydrated or hydroxy-hydrated forms [8]. Many scholars have investigated the coordination states of hydrated or hydroxy-hydrated rare earth ions in solution. Qiu et al. [13] employed a density functional theory to study the coordination states and bonding mechanisms of La^{3+} in aqueous environments and found that La^{3+} tends to coordinate with more water molecules. $La(H_2O)_{10}^{3+}$ was found to be the most stable structure for La ions in low pH solutions, whereas $La(OH)(H_2O)_8^{2+}$ was deemed more plausible in high pH solutions. Rudolph et al. [14] investigated the hydration of lanthanide(III) aqua ions in aqueous solutions and observed that light rare earth ions formed nona-hydrates while heavy rare earth ions formed octa-hydrates.

Typical ion-adsorption type rare earth ores include kaolinite, montmorillonite and illite, and mineralogical analyses have indicated that kaolinite is the main mineral of this type [15]. Kaolinite ($Al_2O_3 \cdot 2SiO_2 \cdot 2H_2O$) is a 1:1 layered silicate characterized by the alternation of silicon–oxygen tetrahedra and aluminum–oxygen octahedra stacked along the c-axis [16]. Kaolinite layers are held together by interlayer hydrogen bonding, which restricts the diffusion of water molecules into the interlayer space. Therefore, kaolinite has a low coefficient of expansion and is prone to crack along the (001) plane [17–19]. As rare earth species in ion-adsorption-type rare earth ores are mainly adsorbed as ions on the surface of clay minerals, their separation can be enabled through ion exchange in electrolyte solutions (e.g., NaCl, NH_4Cl, or $(NH_4)_2SO_4$ [20,21]), without the need for conventional mineral processing methods such as flotation or magnetic separation [22,23]. The current leaching process for ion-adsorbed-type rare earth ores mainly uses $(NH_4)_2SO_4$ as the leaching agent and NH_4HCO_3 as the precipitant in the in situ leaching process. The in situ leaching process is less expensive and causes limited environmental damage than pool and heap leaching processes but suffers from various disadvantages such as poor leaching efficiency, ammonia and nitrogen pollutant emission, and difficulty in effectively recovering rare earth ions [24]. These issues result in a large amount of rare earths remaining in the waste residue, requiring secondary or tertiary leaching to recover the rare earth resources. Various factors contribute to the presence of residual rare earths in the waste residue, including blind spots in the leaching process, capillary phenomena, incomplete weathering, and desorption phenomena [25]. To address these problems, it is necessary to study the adsorption mechanism of rare earths at the molecular level in order to gain a deeper understanding of the adsorption mechanism and surface interactions. This will provide guidance for improving the efficiency and selectivity of metallurgical residue recovery and contribute to the development of new environmentally friendly and efficient leaching agents.

With the development of quantum chemical calculations, density functional theory (DFT) has found extensive applications in the field of mineral processing [26–28]. In recent years, this method has also been widely employed to investigate the adsorption of metal ions on the surfaces of various clay minerals, including kaolinite. Chen et al. [29] conducted density functional theory (DFT) calculations to investigate the adsorption of Hg^{2+} on the surface of kaolinite (001). The findings revealed that Hg^{2+} exhibits a maximum hydration coordination number of six in the liquid phase; they also determined its optimal adsorption site and bonding mechanism on the surface of kaolinite. Qiu et al. [30] conducted a study on the adsorption mechanism of $Lu(OH)_2^+$ on the surface of kaolinite. Their findings indicated that hydrated $Lu(OH)_2^+$ was more stable than hydrated Lu^{3+}. Moreover, they reported that the adsorption energy of hydrated $Lu(OH)_2^+$ was higher for inner layer adsorption than for outer layer adsorption and that $Lu(OH)_2^+$ was preferentially adsorbed on the deprotonated Al-OH surface. These results suggest that the primary adsorption mechanism of rare earth ions on kaolinite involves outer layer adsorption, whereas the inner layer adsorption mechanism may dominate in high pH environments. Peng et al. [31] investigated the adsorption behavior of $Y(OH)_{3-n}^{n+}$ (n = 1–3) on kaolinite surfaces at different degrees of deprotonation using DFT calculations. They found that the deprotonation reaction of the Al-OH surface was energetically unfavorable, with an energy of 145.5 KJ/mol. Furthermore, they observed that the surface activity and adsorption energy of rare earth ions increased as the degree of deprotonation of the adsorbed species on the kaolinite (001) surface increased. Consequently, the adsorption energies of $Y(OH)_{3-n}^{n+}$ on kaolinite (001) and (00−1) surfaces gradually decreased with increasing degree of hydrolysis of Y^{3+} ions. Yan et al. [32] conducted DFT calculations to investigate the adsorption of $Lu(OH)^{2+}$ and $Al(OH)^{2+}$ on kaolinite (001) and (00−1) surfaces. Their findings revealed that the coordination numbers of these two hydrated rare earth ions are seven and five, respectively. Additionally, they observed that both hydrated ions could be adsorbed on kaolinite Al-OH and Si-O surfaces through hydrogen bonding and on the deprotonated Al-OH surface through coordination bonding. They further noted that hydrated $Al(OH)^{2+}$ ions exhibit easier adsorption on kaolinite surfaces compared to hydrated $Lu(OH)^{2+}$, primarily due to their smaller radius and lower coordination numbers.

In this study, we investigated the hydration structures of Sc^{3+}, $Sc(OH)^{2+}$, and $Sc(OH)_2^+$ in a liquid-phase environment via DFT calculations. In addition, the optimal hydrated configuration of $Sc(OH)_2(H_2O)_5^+$ in a liquid-phase environment was investigated for outer layer adsorption on the (001)Al-OH and (00−1)Si-O surfaces of kaolinite, as well as its monodentate inner layer adsorption on the deprotonated (001)Al-OH surface. Furthermore, the partial density of states (PDOS) and Mulliken charge population were also analyzed.

2. Theoretical Methods and Models

2.1. Calculation Methods and Parameters

DFT calculations were carried out based on the plane wave pseudopotential implemented using the Cambridge Sequential Total Energy Package (CASTEP) 19.1 software package [33–36]. The Perdew–Burke–Ernzerhof (PBE) functional in the generalized gradient approximation allows a more accurate description of the hydrogen bonding system consisting of water and has been used to analyze the exchange-correlation potential [37]. The ultrasoft pseudopotential (USP) was utilized to model the interaction between the ionic core and valence electron [38], and the Grimme method in DFT-D dispersion correction was employed to describe systemic weak interactions [39]. The plane-wave cut-off energy was set at 480 eV, and the model was geometrically optimized using the Broyden–Fletcher–Goldfarb–Shanno (BFGS) algorithm [40]. The convergence tolerance values were set as follows: maximum atomic force, 0.03 eV/Å; maximum atomic displacement, 0.001 Å; energy, 1.0×10^{-5} eV/atom; and maximum stress, 0.05 GPa. The K-point grids of Brillouin zone integrations used for the kaolinite bulk phase and surface are (3 × 2 × 2) and (3 × 2 × 1), respectively. The SCF converged to an accuracy of 2.0×10^{-6}. The dipole moment correction was taken into account in the optimization and property evaluation of the surface.

2.2. Construction of Computational Models

The chemical formula of kaolinite is $Al_4Si_4(OH)_8$, and we used the initial input cell structure of kaolinite with the space group C1, which was determined through low-temperature (1.5 K) neutron powder diffraction by Bish [41] in the USA [42,43]. The geometry of bulk kaolinite was optimized using the following lattice constants: a = 5.19 Å, b = 8.98 Å, c = 7.35 Å, α = 91.48°, β = 105.01°, and γ = 89.91°, within 2% error of experimentally measured cell parameters. We chose the (001) cleavage plane of kaolinite as the adsorption surface and set the vacuum layer thickness as 15 Å. The kaolinite surface was subjected to a 2 × 2 × 1 supercell. To optimize the two surfaces, the bottom three layers of atoms are fixed, and only the top three layers are relaxed. The optimized kaolinite surface is shown in Figure 1. The kaolinite (001) surface exposes the Al-OH layer, while the kaolinite (00−1) surface exposes the Si-O layer. After optimization, the Al-OH surface exhibits three distinct types of hydroxyl groups: tilted hydroxyl groups (O_tH), lying hydroxyl groups (O_lH), and upright hydroxyl groups (O_uH) [44].

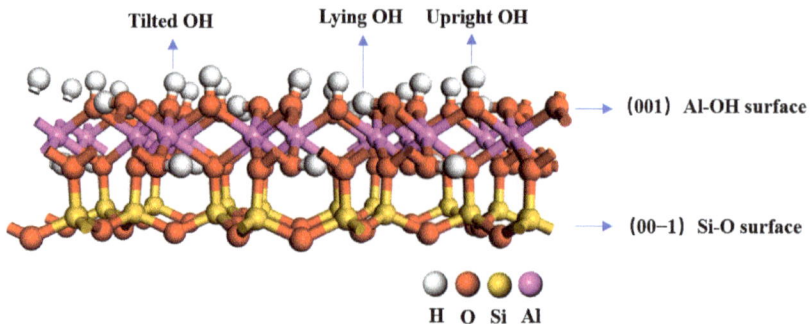

Figure 1. Structure of the kaolinite (001) surface.

Due to the uncertain number of coordinated water molecules, models of all possible hydration and hydroxyl hydration configurations of Sc^{3+}, viz., $[Sc(OH)_m(H_2O)_n]^{(3-m)+}$ ($0 \leq m \leq 2$, $1 \leq n \leq 10$) were constructed and optimized to determine the most stable ionic configuration for adsorption analysis. The binding energy between Sc^{3+} and the H_2O molecule is calculated as (1):

$$Sc^{3+} + m(OH)^- + nH_2O \rightarrow Sc(OH)_m(H_2O)_n^{(3-m)+}$$
$$E_{binding} = E_{Sc/H} - E_{Sc} - mE_{OH} - nE_{H_2O} \tag{1}$$

where $E_{binding}$ represents the binding energy of the Sc ion hydrate, $E_{Sc/H}$ denotes the total energy of the Sc ion hydrate, E_{Sc} refers to the energy of the Sc ion, m and n indicate the number of H_2O and -OH species in the hydrate, and E_{OH} and E_{H_2O} represent the energies of the -OH species and H_2O molecules, respectively.

According to the different adsorption mechanisms of rare earth ions on kaolinite surfaces, rare earth adsorption can be categorized into outer layer and inner layer adsorption types [45]. The outer layer adsorption is due to physical adsorption, and the rare earth hydrates are bound to the kaolinite surface via hydrogen bonds, and the bonding force is weak. The inner layer adsorption is due to chemisorption, involving the formation of coordination bonds between rare earth hydrates and oxygen atoms on the deprotonated kaolinite surface, resulting in a strong bonding force [46]. In this study, rare earth hydrates were placed directly over the central Al atom on the surface of (001)Al-OH and over the central silicon ring on the surface of (00−1)Si-O as initial outer adsorption structures. The inner layer adsorption occurs exclusively on the (001) Al-OH surface, where a water molecule of the Sc^{3+} hydrate with a saturated coordination structure is removed and placed directly above deprotonated oxygen atoms with three different configurations (O_tH, O_lH, and O_uH) around the Al atom at the center of the (001) Al-OH surface as the initial inner layer adsorption structure. The adsorption energy of the Sc^{3+} hydrate adsorbed on the outer and inner layers of the kaolinite surface are calculated as (2) and (3), respectively:

$$S \equiv Al(OH)_3 + [Sc(OH)_2(H_2O)_n]^+ \rightarrow$$
$$[S \equiv Al(OH)_3Sc(OH)_2(H_2O)_m]^+ + (n-m)H_2O \tag{2}$$
$$E_{ads} = E_{Sc/S} - E_{Sc} - E_S$$

$$S \equiv Al(OH)_3 + [Sc(OH)_2(H_2O)_n]^+ + OH^- \rightarrow$$
$$[S \equiv Al(OH)_2(O)Sc(OH)_2(H_2O)_m]^+ + (n-m+1)H_2O \tag{3}$$
$$E_{ads} = E_{Sc/S} - E_s - E_{Sc} + (n-m+1)E_{H_2O} - E_{OH^-}$$

where $S \equiv Al(OH)_3$ denotes the surface of kaolinite (001), n and m are the numbers of aqueous ligands of Sc ion hydrate before and after adsorption. E_{ads} represents the adsorption energy of the system, $E_{Sc/S}$ denotes the total energy of the system after adsorption, E_s is the total energy of the kaolinite (001) surface, E_{Sc} is the energy of the Sc ion hydrate and, E_{H_2O} is the energy of the water molecule. A negative value of the adsorption energy (E_{ads}) indicates an exothermic reaction and a larger negative value signifies greater adsorption stability.

3. Results and Discussion

3.1. Geometric Configuration of $[Sc(OH)_m(H_2O)_n]^{(3-m)+}$

The equilibrium geometry of the DFT-optimized $Sc(H_2O)_{1-9}^{3+}$ structures are shown in Figure 2. Sc^{3+} forms a coordination bond with the O atom of the H_2O molecule. When the number of coordinated H_2O was greater than eight, one of the H_2O ligands in $Sc(H_2O)_9^{3+}$ broke away from the first hydration layer of Sc^{3+} and became conformationally unstable; this observation suggests that Sc^{3+} can have a maximum coordination number of eight. The equilibrium geometrical parameters and binding energies of $Sc(H_2O)_{1-9}^{3+}$ are presented in Table 1. As the number of aqueous ligands increases, the steric hindrance around Sc^{3+} increases, and the average bond length, $R(Sc-O_w)_{avg}$ between Sc^{3+} and O_w in aqueous ligands increases. Sc^{3+} gains electrons from the coordinated water molecules, and its charge tends to decrease as a result, as with the hydrate of the La ion [13]. As more water molecules are included, the binding energy of the rare earth hydrates decreases. A lower binding energy indicates a higher stability of the system. The adsorption of hydrated Sc^{3+} exhibits the most stable configuration with $Sc(H_2O)_8^{3+}$, with a binding energy of −2629.44 KJ/mol.

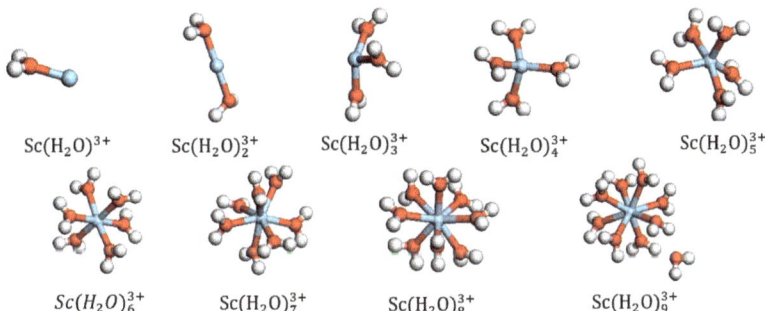

Figure 2. Equilibrium geometries of $Sc(H_2O)_{1-9}^{3+}$ (blue ball represents Sc).

Table 1. Geometric structural parameters and binding energies of the equilibrium structure of $Sc(H_2O)_{1-9}^{3+}$.

n [a]	$R(Sc-O_w)_{min}$ /Å	[b] $R(Sc-O_w)_{max}$ /Å	$R(Sc-O_w)_{avg}$ /Å	$E_{binding}$ /KJ·mol^{-1}	Sc Charge /e
1	1.94309	1.94309	1.94309	−709.44	2.52
2	2.20016	2.20044	2.2003	−1208.64	2.3
3	2.03104	2.03233	2.03155	−1607.04	2.19
4	2.06466	2.07317	2.06876	−1935.36	2.13
5	2.08986	2.1409	2.11343	−2184	2.08
6	2.13676	2.14966	2.14483	−2404.8	2.07
7	2.17566	2.24127	2.20447	−2528.64	2.04
8	2.18095	2.35703	2.26569	−2629.44	2.03

[a] Number of coordinated water molecules. [b] Distance of Sc from the oxygen atom of the coordinated water.

The water molecule in $Sc(H_2O)_{1-9}^{3+}$ was replaced with a hydroxyl group to obtain the geometric configurations of mono- and di-hydroxy hydrates. Upon DFT optimization, it was observed that as the number of coordinated water ligands exceeded six, one of the H_2O molecules in the initial $Sc(OH)(H_2O)_7^{2+}$ structure detached; similarly, one of the H_2O molecules in $Sc(OH)_2(H_2O)_6^+$ detached when the number of coordinated water molecules exceeded five. The most stable coordinated structures of the mono- and di-hydroxy hydrates are $Sc(OH)(H_2O)_6^{2+}$ and $Sc(OH)_2(H_2O)_5^+$, respectively, as shown in Figure 3. As shown in Table 2, as the hydroxyl group replaces the coordinated water molecule, there is a slight increase in the length of the Sc-Ow bond, which increases slightly, and the binding energy decreases further. Thus, the binding energy of $Sc(OH)_2(H_2O)_5^+$ (−3661.44 KJ/mol) is lower than that of $Sc(OH)(H_2O)_6^{2+}$ (−3180.48 KJ/mol) and $Sc(H_2O)_8^{3+}$ (−2629.44 KJ/mol). Therefore, it is inferred that $Sc(OH)_2(H_2O)_5^+$ is the most stable structure of the Sc hydrate, and this structure is used for further calculations and analyses of its adsorption on the surface of kaolinite (001).

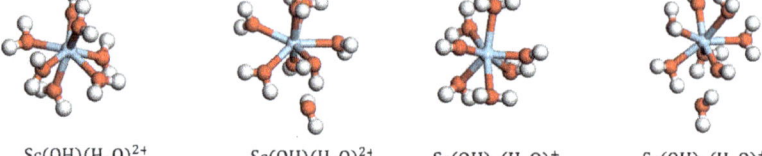

$Sc(OH)(H_2O)_6^{2+}$ $Sc(OH)(H_2O)_7^{2+}$ $Sc(OH)_2(H_2O)_5^+$ $Sc(OH)_2(H_2O)_6^+$

Figure 3. Equilibrium geometries of $Sc(OH)(H_2O)_n^{2+}$ and $Sc(OH)_2(H_2O)_n^+$.

Table 2. Equilibrium geometrical parameters and binding energies of $Sc(OH)(H_2O)_6^{2+}$ and $Sc(OH)_2(H_2O)_5^+$.

Sample	$R(Sc-O_w)_{min}$ /Å	$R(Sc-O_w)_{max}$ /Å	$R(Sc-O_w)_{avg}$ /Å	$E_{binding}$ /KJ·mol^{-1}	Charge on Sc/e
$Sc(OH)(H_2O)_6^{2+}$	1.85728	2.31498	2.21025	−3180.48	1.77
$Sc(OH)_2(H_2O)_5^+$	2.34878	2.34878	2.22228	−3661.44	1.71

3.2. Outer Layer Adsorption of $Sc(OH)_2(H_2O)_5^+$ on the (001)Al-OH Surface

The equilibrium adsorption geometry of $Sc(OH)_2(H_2O)_5^+$ on the outer layer of the kaolinite (001)Al-OH surface is depicted in Figure 4. As shown, the Al-OH surface contains a large number of hydroxyl groups with a high steric hindrance, and one of the coordinated water molecules in the Sc hydrate was squeezed out upon its approach to the surface. Hydrogen bonds are established between the water molecules in $Sc(OH)_2(H_2O)_5^+$ and the hydroxyl groups on the surface of kaolinite. DFT optimizations resulted in a total of five hydrogen bonds between the adsorbate and adsorbent with bond lengths of 1.299, 1.598, 1.842, 2.035, and 2.431 Å. The equilibrium geometrical parameters and adsorption energies of $Sc(OH)_2(H_2O)_5^+$ bound to the Al-OH surface are provided in Table 3. Following the adsorption of $Sc(OH)_2(H_2O)_5^+$, the average Sc-Ow bond length decreased from 2.22 to 2.14 Å. This reduction can be attributed to the compression of water ligands near the surface and the detachment of one of the coordinated water molecules, resulting in a tighter bond between Sc^{3+} and the surrounding water ligands. The adsorption energy of $Sc(OH)_2(H_2O)_5^+$ is −522.24 KJ/mol.

Table 3. Equilibrium geometrical parameters and adsorption energies for the adsorption structures of $Sc(OH)_2(H_2O)_5^+$ on the outer layer of the (001)Al-OH surface.

Name	State	N	$R(Sc-O_w)$ /Å	$R(Sc-O_w)_{avg}$ /Å	E_{ads} /KJ·mol^{-1}
$Sc(OH)_2(H_2O)_5^+$	Before	7	1.90, 1.98, 2.30, 2.31, 2.34, 2.35, 2.38	2.22	−522.24
	After	6	1.90, 2.02, 2.12, 2.22, 2.26, 2.32	2.14	

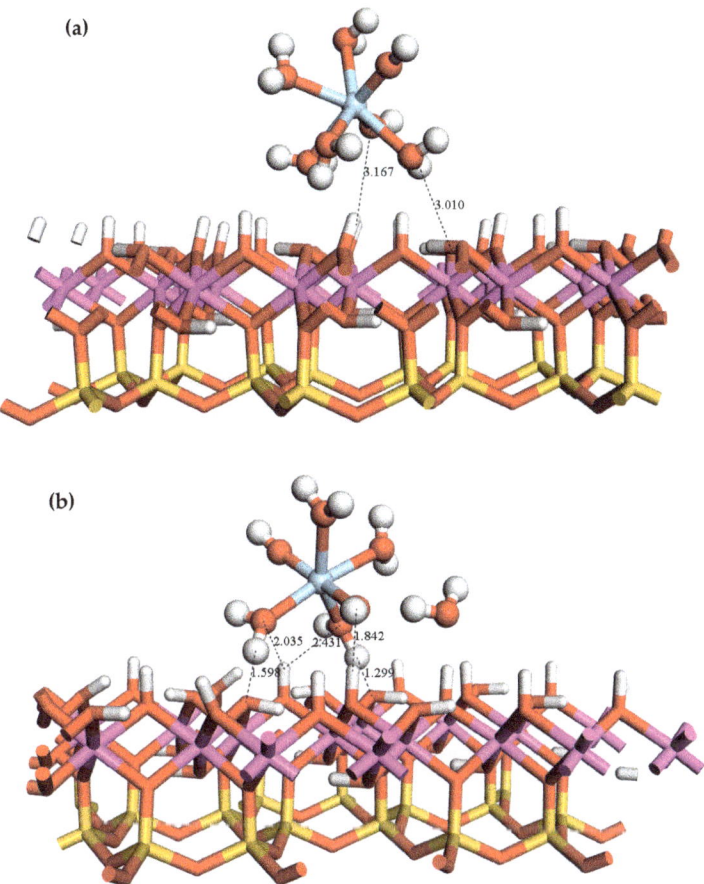

Figure 4. Equilibrium geometry of $Sc(OH)_2(H_2O)_5^+$ adsorbed on the outer layer of the (001)Al-OH surface (**a**) before adsorption; (**b**) after adsorption.

Figure 5 shows the PDOS of the Sc ion and kaolinite surface before and after the adsorption of $Sc(OH)_2(H_2O)_5^+$ on the outer layer of the (001)Al-OH surface. The PDOS of the Sc ion is overall shifted towards lower energy after adsorption. The non-localization of 3d orbitals above the Fermi energy level is enhanced, and the density of states peak shifts from 4.6–6.9 eV to 2.6–7.5 eV. The 3p orbitals change from −(26.7–25.3) eV to −(29.2–27.8) eV. The Al-OH surface has a slight shift in the density of states towards lower energies after adsorption, and the 2p orbital near the Fermi energy level changes from −9.6–0.74 eV to −9.9–0.75 eV. The system becomes more stable, and there is a significant enhancement in the localization of the 2s and 2p orbitals in the conduction band.

The Mulliken atomic population analysis of the Sc hydrate before and after its adsorption is presented in Table 4. The data indicate that, upon adsorption, the 3s and 3p orbitals of Sc gain 0.02e and 0.04e, whereas the 3d orbital loses 0.11e. Due to the presence of hydrogen bonding interactions rather than coordination bonds, the total charge change is minimal (−0.05e), which is also consistent with the subtle changes observed in the PDOS plot. This indicates that the charge transfer from $Sc(OH)_2(H_2O)_5^+$ to the Al-OH surface is small.

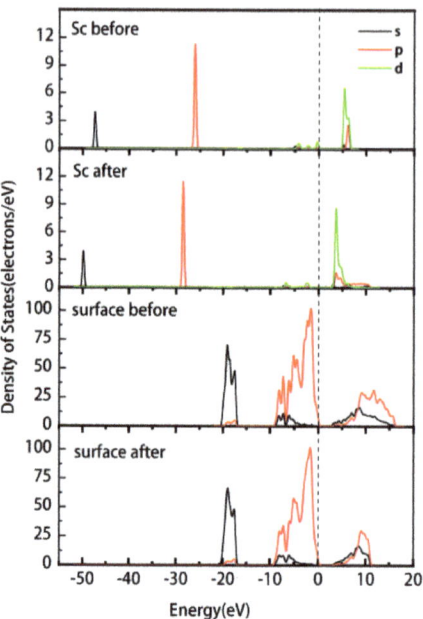

Figure 5. PDOS of Sc and the (001) Al-OH surface before and after the outer layer adsorption of the Sc hydrate.

Table 4. Mulliken atomic population of Sc adsorbed on the outer layer of the (001) Al-OH surface.

State	3s	3p	3d	Total	Charge/e
Before adsorption	2.14	5.94	1.21	9.29	1.71
After adsorption	2.16	5.98	1.1	9.24	1.76
Δcharge [a]	0.02	0.04	−0.11	−0.05	0.05

[a] Amount of change in charge before and after adsorption.

3.3. Adsorption of $Sc(OH)_2(H_2O)_5^+$ on the Outer Layer of the (00−1)Si-O Surface

Figure 6 illustrates the adsorption configuration of $Sc(OH)_2(H_2O)_5^+$ on the outer layer of the kaolinite (00−1)Si-O surface. Near the (00−1)Si-O surface, one of the coordinated water molecules in $Sc(OH)_2(H_2O)_5^+$ detaches from the hydrate. This is in contrast to the (001)Al-OH surface, which does not have hydroxyl groups but is abundant in saturated oxygen atoms. The optimized surface oxygen atoms (O_s) on the outer layer form four hydrogen bonds with the hydrogen atoms (H_w) in the H_2O ligands of the Sc hydrate and the H-bond lengths are 1.841, 1.848, 1.885, and 1.967 Å. The corresponding structural parameters and adsorption energies are presented in Table 5.

After adsorption occurs, the average Sc-O_w bond length experiences a slight reduction from 2.22 to 2.17 Å. This decrease can be attributed to the detachment of one coordinated water ligand and the steric hindrance present at the surface. The adsorption energy of $Sc(OH)_2(H_2O)_5^+$ on the Si-O surface is significantly lower (−648.96 KJ/mol) compared to the Al-OH surface (−522.24 KJ/mol). This is possibly due to the steric hindrance created by the hydroxyl groups on the Al-OH surface, which prevent $Sc(OH)_2(H_2O)_5^+$ from approaching the surface. Based on the analysis, it can be concluded that the outer layer adsorption of $Sc(OH)_2(H_2O)_5^+$ is more likely to occur on the Si-O surface of kaolinite.

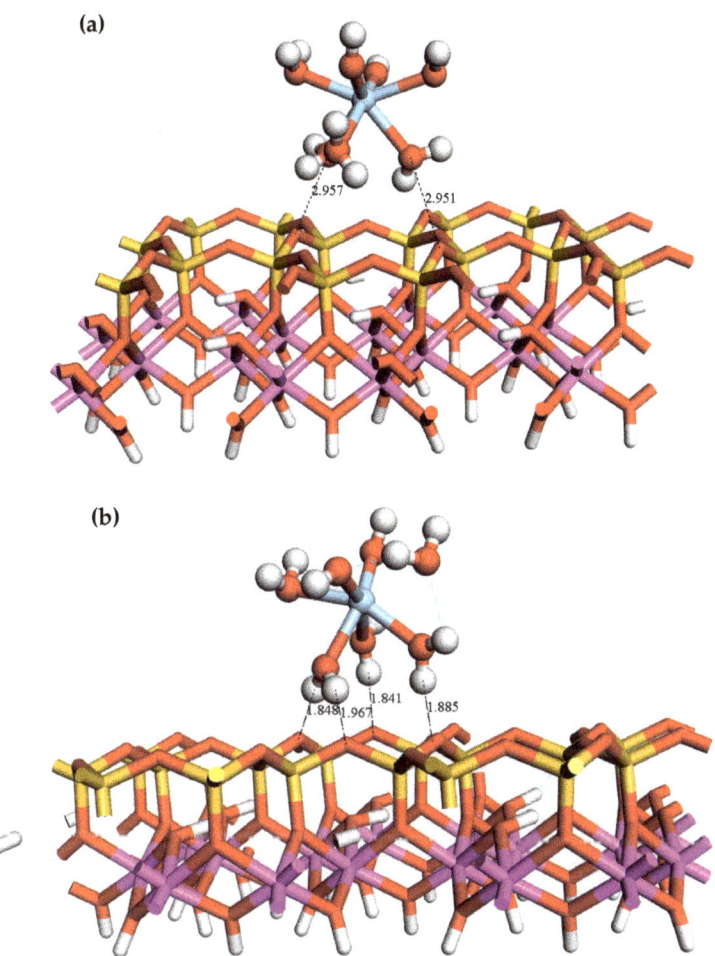

Figure 6. Equilibrium geometry for the adsorption of $Sc(OH)_2(H_2O)_5^+$ on the outer layer of the (00−1)Si-O surface (**a**) before adsorption; (**b**) after adsorption.

Table 5. Equilibrium geometries and adsorption energies for the adsorption of $Sc(OH)_2(H_2O)_5^+$ on the outer layer of the (00−1)Si-O surface.

Name	State	N	$R(Sc-O_w)$ /Å	$R(Sc-O_w)_{avg}$ /Å	Eads /KJ·mol^{-1}
$Sc(OH)_2(H_2O)_5^+$	Before	7	1.90, 1.98, 2.30, 2.31, 2.34, 2.35, 2.38	2.22	−648.96
	After	6	1.95, 1.99, 2.19, 2.23, 2.29, 2.34	2.17	

Figure 7 displays the PDOS of the Sc ion and the surface before and after adsorption. The change in Sc at the Fermi energy level was not significant; the energies of the 3d and 2p orbitals in the conduction band changed from 4.7–6.8 eV before adsorption to 4.4–11.6 eV after adsorption. Thus, the non-localization of the electrons was enhanced, and the overall shift toward lower energies was small. The Si-O surface has enhanced electron localization in the 2s and 2p orbitals in the conduction band. The energy of the 2p orbital at the Fermi level shifts from −8.6–0.5 eV to −9.6–0.25 eV, indicating a slight reduction in the reactivity of the Si-O surface. The Mulliken atomic population of Sc (Table 6) indicates that the 3s and 3p orbitals of Sc gain 0.02e each, the 3d orbitals lose 0.15e, and the total charge increases from 1.71e to 1.82e. The adsorption charge transfer of $Sc(OH)_2(H_2O)_5^+$ on the (00−1) Si-O surface is more significant compared to the data in Table 4. Therefore, it can be inferred that outer layer adsorption is more likely to occur on this surface.

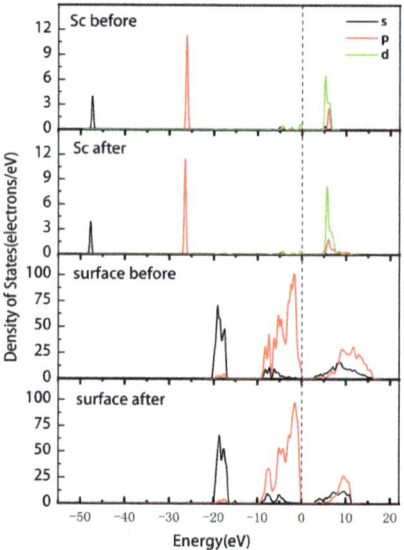

Figure 7. PDOS of Sc and the (00−1)Si-O surface before and after the outer layer adsorption of the Sc hydrate.

Table 6. Mulliken atomic population of Sc adsorbed on the outer layer of the (00−1) Si-O surface.

State	3s	3p	3d	Total	Charge/e
Before adsorption	2.14	5.94	1.21	9.29	1.71
After adsorption	2.16	5.96	1.06	9.18	1.82
Δcharge	0.02	0.02	−0.15	−0.11	0.11

3.4. Inner Layer Adsorption on the (001)Al-OH Surface

The adsorption behavior of rare earth ions on the surface of kaolinite in a liquid phase is pH-dependent. If the pH is higher than the point of zero charge of kaolinite, the surface hydroxyl groups lose their protons and become negatively charged. In such scenarios, the rare earth hydrate is bound to the deprotonated Al-OH surface through inner layer adsorption, specifically by forming coordination bonds with the oxygen atoms. Three forms of hydroxyl groups are present on the optimized Al-OH surface: tilted (O_tH), lying (O_lH), and upright (O_uH) hydroxyl groups. The geometrical configurations of the inner layer adsorption forms of the Sc ions on these three deprotonated oxygen atoms are shown in Figure 8. When adsorbed in the inner layer, the Sc ion in $Sc(OH)_2(H_2O)_5^+$ forms a monodentate adsorption structure by coordinating with the deprotonated oxygen atom (Os) on the Al-OH surface. Due to the steric hindrance imposed by the kaolinite surface, two water ligands are displaced from the coordination sphere of Sc and released as free water molecules after adsorption in all three configurations, and the coordination number of Sc decreases.

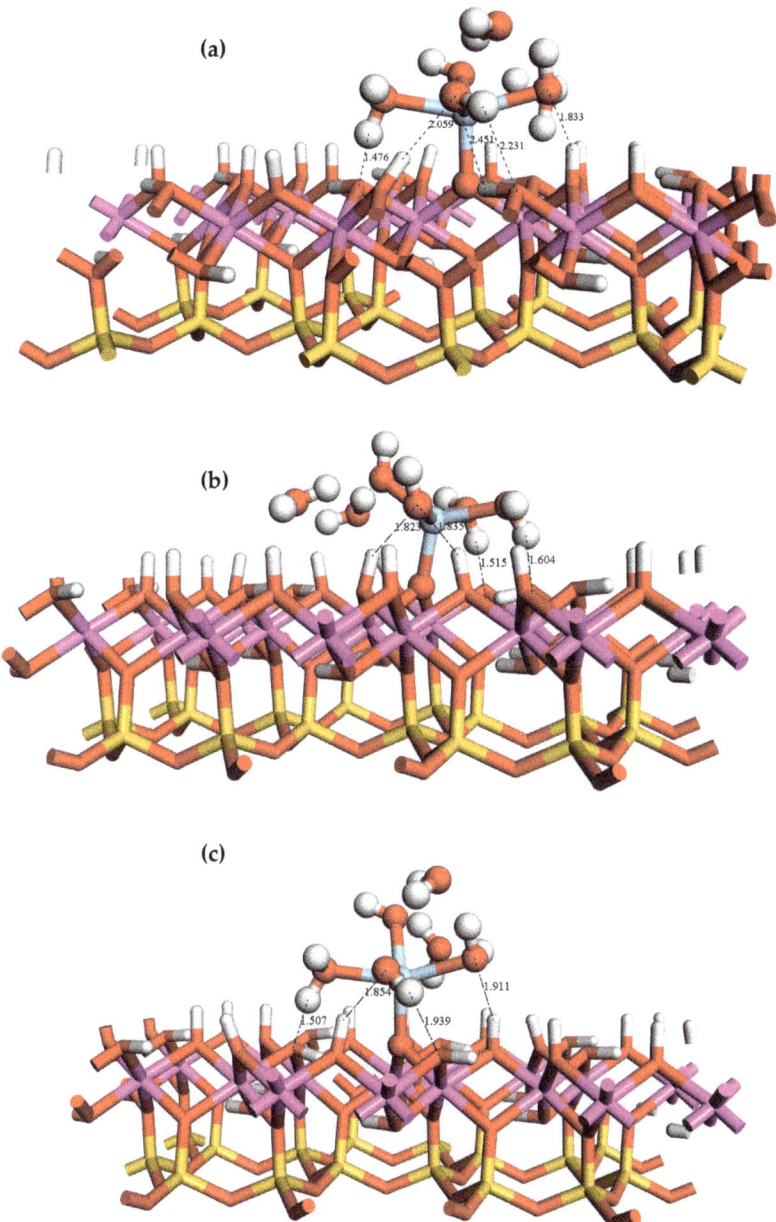

Figure 8. Equilibrium geometries of Sc(OH)$_2$(H$_2$O)$_5^+$ adsorbed on the inner layer O$_u$ (**a**), O$_l$ (**b**), and O$_t$ (**c**) sites of the (001) Al-OH surface.

The equilibrium geometrical parameters and adsorption energies of the three adsorption configurations of the Sc hydrate are presented in Table 7. Unlike the outer layer adsorption case, inner layer adsorption involves both hydrogen bonding and coordination bonding mechanisms. The coordination bond between the Sc ion and the surface oxygen atom (Os) has a shorter bond length compared to the average Sc-O$_w$ bond length.

Table 7. Equilibrium geometries and adsorption energies of three configurations of Sc adsorbed on the inner layer of the (001)Al-OH surface.

Location	N	$R(Sc-O_w)_{avg}$ /Å	$R(Sc-O_s)$ [a] /Å	Eads /KJ·mol^{-1}
O_u	4	2.13	1.95	−643.68
O_l	4	2.13	1.94	−653.32
O_t	4	2.17	2.00	−595.68

[a] Distance between Sc and the surface deprotonated oxygen.

$Sc(OH)_2(H_2O)_5^+$ has the lowest adsorption energy on O_l, indicating that O_l is the best adsorption site. Upon adsorption at the O_l site, one hydroxyl group and two coordinated water molecules in the rare earth hydrate near the kaolinite surface form four hydrogen bonds with the oxygen atom near the kaolinite surface. The bond lengths of these hydrogen bonds are 1.515, 1.604, 1.823, and 1.835 Å. The inner layer adsorption mode exhibits significantly lower adsorption energy compared to the outer layer adsorption mode, indicating that the formation of coordination bonds in the inner layer adsorption enhances the stability of Sc adsorption.

Further, we analyzed the adsorption mechanism by calculating the PDOS and Mulliken atomic population of $Sc(OH)_2(H_2O)_5^+$ adsorbed at the O_l site. The results of this analysis are shown in Figure 9 and Table 8. The overall shift in the density of states peaks of Sc and surface oxygen O_l toward lower energies following the adsorption of the Sc hydrate indicates that the formation of coordination bonds between the adsorbed Sc and O_l results in a lower energy and more stable system. The 2p orbital of O_l moves away from the Fermi level after adsorption, resulting in a new peak at 4.6 eV. The overlapping of the 2p orbital of Ol and the 3d orbital of Sc in the energy range of −8 to −0.3 eV suggests the presence of bonding states in this region.

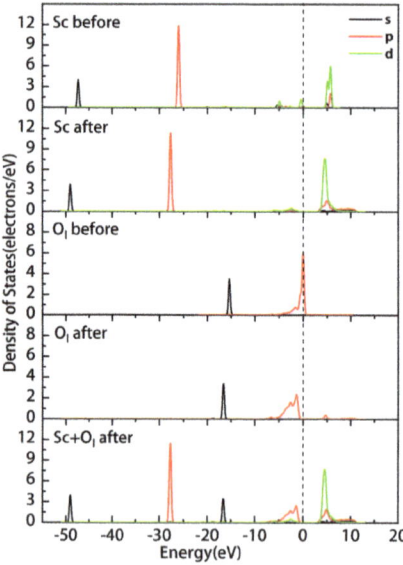

Figure 9. Partial density of states of Sc and Ol before and after the adsorption of the Sc hydrate on the inner layer.

Table 8. Mulliken atomic populations of Sc and O_l before and after adsorption, along with the Mulliken bond population of the Sc-O_l bond.

State	Sc					O_l				Sc-O_l [a]
	s	p	d	Total	Charge	s	p	Total	Charge	
Before	2.14	6.01	1.25	9.40	1.60	1.91	4.96	6.87	−0.87	
After	2.16	5.94	1.12	9.22	1.78	1.86	5.12	6.98	−0.98	0.48
Δcharge	0.02	−0.07	−0.13	−0.18	0.18	−0.05	0.16	0.11	−0.11	

[a] Bond population value for coordination bond formed by Sc and surface oxygen.

Analyses of the Mulliken charge of Sc and O_l before and after adsorption reveal that the Mulliken bond population value of the Sc-O bond is 0.48 and that the coordinative bonds formed on the surface are more covalent. After adsorption, the 4s orbital of Sc gains a charge of 0.02e, while the 3p and 3d orbitals lose charges of 0.13e and 0.18e, respectively, and the charge changes from 1.60 to 1.78 (i.e., 0.18 electrons are lost). Further, the 2s orbital of O_l loses 0.05e, the 2p orbital gains 0.16e, and the charge changes from −0.87 to −0.98 (0.11 electrons are gained). After the formation of the coordination bond, Sc transfers electrons to O_l.

The outer layer adsorption of $Sc(OH)_2(H_2O)_5^+$ on the (001)Al-OH and (00−1)Si-O surfaces of kaolinite results in hydrogen bonding interactions. The adsorption energy of Sc on the outer layer of the (001)Al-OH surface is −522.24 KJ/mol, while on the Si-O surface, it is −648.96 KJ/mol. The steric hindrance of the hydroxyl groups on the (001)Al-OH surface prevents the rare earth ion from approaching. On the other hand, the oxygen atoms on the surface of (00−1)Si-O are strongly electronegative. Thus, $Sc(OH)_2(H_2O)_5^+$ is more stable when adsorbed on the outer layer of the (00−1)Si-O surface. At high pH values, the deprotonation of the hydroxyl group on the (001)Al-OH surface enables the formation of coordination bonds between the oxygen atom on the deprotonated surface (O_s) and Sc ions, referred to as inner layer adsorption. The adsorption of $Sc(OH)_2(H_2O)_5^+$ on the deprotonated (001) surface results in three configurations, with the lowest adsorption energy (−653.32 KJ/mol) at the lying oxygen site (O_l), which is significantly lower than the outer layer adsorption energy. Although the energy of the inner layer adsorption mode is lower, the energetic disadvantage of the surface deprotonation process and the fact that most of the rare earth species can be desorbed by NH_4^+ ion exchange. Hence, the outer layer adsorption mode is the dominant mechanism for rare earth elements to bind with the kaolinite surface.

4. Conclusions

This study investigates the structures and bonding mechanisms of hydrated Sc^{3+}, $Sc(OH)^{2+}$, and $Sc(OH)_2^+$ species adsorbed onto the surfaces of kaolinite (001)Al-OH and (00−1)Si-O using the plane-wave pseudopotential DFT method. Hydrated $Sc(OH)^{2+}$ was found to have a more stable structure and lower adsorption energy than hydrated Sc^{3+} and hydrated $Sc(OH)_2^+$ species. The best adsorption configuration for $Sc(OH)^{2+}$ in the liquid phase is $Sc(OH)_2(H_2O)_5^+$.

The adsorption energies of hydrated $Sc(OH)^{2+}$ on the outer layers of (001) and (00−1) surfaces of kaolinite are −522.24 and −648.96 KJ/mol, respectively. The adsorption process is primarily driven by hydrogen bonding interactions between the water ligands in the Sc hydrate and the surface of kaolinite.

Inner layer adsorption of hydrated $Sc(OH)^{2+}$ on the deprotonated hydroxyl group of the kaolinite (001) surface occurred in three adsorption configurations (O_u, O_l, and O_t), with the lowest adsorption energy at the O_l site (−653.32 KJ/mol). In addition to hydrogen bonds, coordination bonding (Sc-O_S) was also observed between Sc and oxygen atoms of the deprotonated hydroxyl groups on the surface, resulting in higher adsorption energy than that on the outer layer. According to the PDOS and Mulliken population analysis, the coupling between the 3d orbital of Sc and the 2p orbital of O_S to form a bonding state is the main contributor to the Sc-O_S coordination bond.

Although the inner layer adsorption exhibits lower adsorption energy compared to the outer layer adsorption, the presence of hydrated Sc ions primarily occurs in the outer layer adsorption mode on the kaolinite (001)Al-OH surface. The results indicate that inner layer adsorption exhibits greater stability compared to outer layer adsorption. However, the extent of deprotonation of the (001) surface in high-pH solutions is constrained by the energy required to deprotonate the inner layer hydroxyl groups. Most rare earth ions can be desorbed through ion exchange with NH_4^+. It can therefore be concluded that the main adsorption mode for rare earth ions is on the outer layer of the kaolinite surface. To prevent the deprotonation of the hydroxyl groups on the (001) surface of kaolinite, a leaching process at a lower pH can be employed to facilitate the exchange of rare earth ions.

Author Contributions: Conceptualization, D.J. and Y.L.; methodology, D.J., G.W. and Z.Z.; Formal analysis, K.W.; validation, D.J., G.W. and Z.Z.; investigation, Y.L.; resources, D.J.; data curation, K.W. and Z.Z.; writing—original draft preparation, Z.Z.; writing—review and editing, D.J. All authors have read and agreed to the published version of the manuscript.

Funding: This research was funded by the National Natural Science Foundation of China, grant number U2002215, Yunnan Provincial Science and Technology Department, grant number 202001AU070007.

Institutional Review Board Statement: Not applicable.

Informed Consent Statement: Not applicable.

Data Availability Statement: The data presented in this study are available upon request from the corresponding authors.

Conflicts of Interest: The authors declare no conflict of interest.

References

1. Cheisson, T.; Schelter, E.J. Rare earth elements: Mendeleev's bane, modern marvels. *Science* **2019**, *363*, 489–493. [CrossRef]
2. Omodara, L.; Pitkäaho, S.; Turpeinen, E.-M.; Saavalainen, P.; Oravisjärvi, K.; Keiski, R.L. Recycling and substitution of light rare earth elements, cerium, lanthanum, neodymium, and praseodymium from end-of-life applications—A review. *J. Clean. Prod.* **2019**, *236*, 117573. [CrossRef]
3. Dutta, T.; Kim, K.-H.; Uchimiya, M.; Kwon, E.E.; Jeon, B.-H.; Deep, A.; Yun, S.-T. Global demand for rare earth resources and strategies for green mining. *Environ. Res.* **2016**, *150*, 182–190. [CrossRef] [PubMed]
4. Wang, X.; Zhu, Y.; Li, H.; Lee, J.-M.; Tang, Y.; Fu, G. Rare-Earth Single-Atom Catalysts: A New Frontier in Photo/Electrocatalysis. *Small Methods* **2022**, *6*, 2200413. [CrossRef] [PubMed]
5. Zeng, Z.; Xu, Y.; Zhang, Z.; Gao, Z.; Luo, M.; Yin, Z.; Zhang, C.; Xu, J.; Huang, B.; Luo, F.; et al. Rare-earth-containing perovskite nanomaterials: Design, synthesis, properties and applications. *Chem. Soc. Rev.* **2020**, *49*, 1109–1143. [CrossRef]
6. Lincheng, X.; Yue, W.; Yong, Y.; Zhanzhong, H.; Xin, C.; Fan, L. Optimisation of the electronic structure by rare earth doping to enhance the bifunctional catalytic activity of perovskites. *Appl. Energy* **2023**, *339*, 120931. [CrossRef]
7. Wang, L.; Huang, X.; Yu, Y.; Zhao, L.; Wang, C.; Feng, Z.; Cui, D.; Long, Z. Towards cleaner production of rare earth elements from bastnaesite in China. *J. Clean. Prod.* **2017**, *165*, 231–242. [CrossRef]
8. Zhang, Z.; He, Z.; Xu, Z.; Yu, J.; Zhang, Y.; Chi, R. Rare Earth Partitioning Characteristics of China Rare Earth Ore. *Chin. Rare Earths* **2016**, *37*, 121–127.
9. Borst, A.M.; Smith, M.P.; Finch, A.A.; Estrade, G.; Villanova-de-Benavent, C.; Nason, P.; Marquis, E.; Horsburgh, N.J.; Goodenough, K.M.; Xu, C.; et al. Adsorption of rare earth elements in regolith-hosted clay deposits. *Nat. Commun.* **2020**, *11*, 4386. [CrossRef]
10. Ochsenkühn-Petropoulou, M.T.; Hatzilyberis, K.S.; Mendrinos, L.N.; Salmas, C.E. Pilot-Plant Investigation of the Leaching Process for the Recovery of Scandium from Red Mud. *Ind. Eng. Chem. Res.* **2002**, *41*, 5794–5801. [CrossRef]
11. Wang, W.; Pranolo, Y.; Cheng, C.Y. Metallurgical processes for scandium recovery from various resources: A review. *Hydrometallurgy* **2011**, *108*, 100–108. [CrossRef]
12. Zhang, N.; Li, H.-X.; Liu, X.-M. Recovery of scandium from bauxite residue—Red mud: A review. *Rare Met.* **2016**, *35*, 887–900. [CrossRef]
13. Qiu, S.; Yan, H.; Qiu, X.; Wu, H.; Zhou, X.; Wu, H.; Li, X.; Qiu, T. Adsorption of La on kaolinite (001) surface in aqueous system: A combined simulation with an experimental verification. *J. Mol. Liq.* **2022**, *347*, 117956. [CrossRef]
14. Rudolph, W.W.; Irmer, G. On the Hydration of the Rare Earth Ions in Aqueous Solution. *J. Solut. Chem.* **2020**, *49*, 316–331. [CrossRef]

15. Zhang, Z.; Zheng, G.; Takahashi, Y.; Wu, C.; Zheng, C.; Yao, J.; Xiao, C. Extreme enrichment of rare earth elements in hard clay rocks and its potential as a resource. *Ore Geol. Rev.* **2016**, *72*, 191–212. [CrossRef]
16. White, C.E.; Provis, J.L.; Proffen, T.; Riley, D.P.; van Deventer, J.S.J. Density Functional Modeling of the Local Structure of Kaolinite Subjected to Thermal Dehydroxylation. *J. Phys. Chem. A* **2010**, *114*, 4988–4996. [CrossRef]
17. Wang, Q.; Kong, X.-P.; Zhang, B.-H.; Wang, J. Adsorption of Zn(II) on the kaolinite(001) surfaces in aqueous environment: A combined DFT and molecular dynamics study. *Appl. Surf. Sci.* **2017**, *414*, 405–412. [CrossRef]
18. Wu, H.; Yan, H.; Zhao, G.; Qiu, S.; Qiu, X.; Zhou, X.; Qiu, T. Influence of impurities on adsorption of hydrated Y3+ ions on the kaolinite (001) surface. *Colloids Surf. A Physicochem. Eng. Asp.* **2022**, *653*, 129961. [CrossRef]
19. Šolc, R.; Gerzabek, M.H.; Lischka, H.; Tunega, D. Wettability of kaolinite (001) surfaces—Molecular dynamic study. *Geoderma* **2011**, *169*, 47–54. [CrossRef]
20. Chi, R.A.; Tian, J.; Luo, X.P.; Xu, Z.G.; He, Z.Y. The basic research on the weathered crust elution-deposited rare earth ores. *Nonferrous Met. Sci. Eng.* **2012**, *3*, 1–13.
21. He, Z.; Zhang, Z.; Yu, J.; Zhou, F.; Xu, Y.; Xu, Z.; Chen, Z.; Chi, R. Kinetics of column leaching of rare earth and aluminum from weathered crust elution-deposited rare earth ore with ammonium salt solutions. *Hydrometallurgy* **2016**, *163*, 33–39. [CrossRef]
22. Zhao, L.-S.; Wang, L.-N.; Chen, D.-S.; Zhao, H.-X.; Liu, Y.-H.; Qi, T. Behaviors of vanadium and chromium in coal-based direct reduction of high-chromium vanadium-bearing titanomagnetite concentrates followed by magnetic separation. *Trans. Nonferrous Met. Soc. China* **2015**, *25*, 1325–1333. [CrossRef]
23. Wang, P.-P.; Qin, W.-Q.; Ren, L.-Y.; Wei, Q.; Liu, R.-Z.; Yang, C.-R.; Zhong, S.-P. Solution chemistry and utilization of alkyl hydroxamic acid in flotation of fine cassiterite. *Trans. Nonferrous Met. Soc. China* **2013**, *23*, 1789–1796. [CrossRef]
24. Wang, G.; Lai, Y.; Peng, C. Adsorption of rare earth yttrium and ammonium ions on kaolinite surfaces: A DFT study. *Theor. Chem. Acc.* **2018**, *137*, 53. [CrossRef]
25. Long, P.; Wang, G.-S.; Tian, J.; Hu, S.-L.; Luo, S.-H. Simulation of one-dimensional column leaching of weathered crust elution-deposited rare earth ore. *Trans. Nonferrous Met. Soc. China* **2019**, *29*, 625–633. [CrossRef]
26. Huang, H.; Qiu, T.; Ren, S.; Qiu, X. Research on flotation mechanism of wolframite activated by Pb(II) in neutral solution. *Appl. Surf. Sci.* **2020**, *530*, 147036. [CrossRef]
27. Blanchard, M.; Wright, K.; Gale, J.D.; Catlow, C.R.A. Adsorption of As(OH)$_3$ on the (001) Surface of FeS2 Pyrite: A Quantum-mechanical DFT Study. *J. Phys. Chem. C* **2007**, *111*, 11390–11396. [CrossRef]
28. Wang, F.S.; Gao, Z.; Liu, S.G. Model of muti-pressure craft model based on transient liquid-phase bonding. *J. Lanzhou Petrochem. Coll. Technol.* **2008**, *8*, 25–27.
29. Chen, G.; Li, X.; Zhou, L.; Xia, S.; Yu, L. Mechanism insights into Hg(II) adsorption on kaolinite(001) surface: A density functional study. *Appl. Surf. Sci.* **2019**, *488*, 494–502. [CrossRef]
30. Qiu, S.; Wu, H.; Yan, H.; Li, X.; Zhou, X.; Qiu, T. Theoretical investigation of hydrated [Lu(OH)$_2$]+ adsorption on kaolinite(001) surface with DFT calculations. *Appl. Surf. Sci.* **2021**, *565*, 150473. [CrossRef]
31. Peng, C.; Zhong, Y.; Wang, G.; Min, F.; Qin, L. Atomic-level insights into the adsorption of rare earth Y(OH)3-nn+ (n = 1–3) ions on kaolinite surface. *Appl. Surf. Sci.* **2019**, *469*, 357–367. [CrossRef]
32. Yan, H.; Yang, B.; Zhou, X.; Qiu, X.; Zhu, D.; Wu, H.; Li, M.; Long, Q.; Xia, Y.; Chen, J.; et al. Adsorption mechanism of hydrated Lu(OH)$_2$+ and Al(OH)$_2$+ ions on the surface of kaolinite. *Powder Technol.* **2022**, *407*, 117611. [CrossRef]
33. Clark, S.J.; Segall, M.D.; Pickard, C.J.; Hasnip, P.J.; Probert, M.I.J.; Refson, K.; Payne, M.C. First principles methods using CASTEP. *Z. Für Krist.-Cryst. Mater.* **2005**, *220*, 567–570. [CrossRef]
34. Hohenberg, P.; Kohn, W. Inhomogeneous electron gas. *Phys. Rev.* **1964**, *136*, B864–B871. [CrossRef]
35. Kohn, W.; Sham, L.J. Self-Consistent Equations Including Exchange and Correlation Effects. *Phys. Rev.* **1965**, *140*, A1133–A1138. [CrossRef]
36. Segall, M.D.; Philip, J.D.L.; Probert, M.J.; Pickard, C.J.; Hasnip, P.J.; Clark, S.J.; Payne, M.C. First-principles simulation: Ideas, illustrations and the CASTEP code. *J. Phys. Condens. Matter* **2002**, *14*, 2717. [CrossRef]
37. Perdew, J.P.; Burke, K.; Ernzerhof, M. Generalized Gradient Approximation Made Simple. *Phys. Rev. Lett.* **1996**, *77*, 3865–3868. [CrossRef] [PubMed]
38. Vanderbilt, D. Soft self-consistent pseudopotentials in a generalized eigenvalue formalism. *Phys. Rev. B* **1990**, *41*, 7892–7895. [CrossRef]
39. Grimme, S. Semiempirical GGA-type density functional constructed with a long-range dispersion correction. *J. Comput. Chem.* **2006**, *27*, 1787–1799. [CrossRef]
40. Pfrommer, B.G.; Côté, M.; Louie, S.G.; Cohen, M.L. Relaxation of Crystals with the Quasi-Newton Method. *J. Comput. Phys.* **1997**, *131*, 233–240. [CrossRef]
41. Bish, D.L. Rietveld refinement of the kaolinite structure at 1.5 K. *Clays Clay Miner.* **1993**, *41*, 738–744. [CrossRef]
42. White, C.E.; Provis, J.L.; Riley, D.P.; Kearley, G.J.; van Deventer, J.S.J. What Is the Structure of Kaolinite? Reconciling Theory and Experiment. *J. Phys. Chem. B* **2009**, *113*, 6756–6765. [CrossRef] [PubMed]
43. Neder, R.B.; Burghammer, M.; Grasl, T.H.; Schulz, H.; Bram, A.; Fiedler, S. Refinement of the Kaolinite Structure From Single-Crystal Synchrotron Data. *Clays Clay Miner.* **1999**, *47*, 487–494. [CrossRef]
44. Qiu, T.; Qiu, S.; Wu, H.; Yan, H.; Li, X.; Zhou, X. Adsorption of hydrated [Y(OH)$_2$]+ on kaolinite (001) surface: In7sight from DFT simulation. *Powder Technol.* **2021**, *387*, 80–87. [CrossRef]

45. Kremleva, A.; Krüger, S.; Rösch, N. Density Functional Model Studies of Uranyl Adsorption on (001) Surfaces of Kaolinite. *Langmuir* **2008**, *24*, 9515–9524. [CrossRef]
46. Vasconcelos, I.F.; Bunker, B.A.; Cygan, R.T. Molecular Dynamics Modeling of Ion Adsorption to the Basal Surfaces of Kaolinite. *J. Phys. Chem. C* **2007**, *111*, 6753–6762. [CrossRef]

Disclaimer/Publisher's Note: The statements, opinions and data contained in all publications are solely those of the individual author(s) and contributor(s) and not of MDPI and/or the editor(s). MDPI and/or the editor(s) disclaim responsibility for any injury to people or property resulting from any ideas, methods, instructions or products referred to in the content.

Article

The Removal of Inclusions with Different Diameters in Tundish by Channel Induction Heating: A Numerical Simulation Study

Bing Yi [1,2], Guifang Zhang [1,*], Qi Jiang [1,*], Peipei Zhang [1], Zhenhua Feng [1] and Nan Tian [1]

1 Faculty of Metallurgical and Energy Engineering, Kunming University of Science and Technology, Kunming 650093, China; yibing8578@163.com (B.Y.); zpei861@163.com (P.Z.); fengzhenhua666@126.com (Z.F.); tian1852558@163.com (N.T.)
2 Huanan Zhongke Electric Co., Ltd., Electromagnet Center, Yueyang 414000, China
* Correspondence: guifangzhang65@163.com (G.Z.); jiangqi87190500@163.com (Q.J.)

Abstract: The quality of the bloom will be impacted by the non-metallic impurities in the molten steel in the tundish, which will reduce the plasticity and fatigue life of the steel. In this research, a mathematical model of a six-flow double-channel T-shaped induction heating tundish was established, the effects of induction heating conditions on the removal of inclusions in the tundish were investigated, and the impact of various inclusion particle sizes on the removal effect of inclusions under induction heating was explored. The results show that the Residence Time Distribution (RTD) curve produced through numerical simulation and physical simulation is in good agreement. The reduction of inclusion particles in the channel is made affordable by the dual-channel induction heating technique. As the diameter of inclusion particles increases from 10 μm to 50 μm, the probability of inclusion particles being removed from the channel gradually decreases from 70.9% to 56.1%.

Keywords: tundish metallurgy; dual-channel induction heating; inclusion removal; numerical simulation

Citation: Yi, B.; Zhang, G.; Jiang, Q.; Zhang, P.; Feng, Z.; Tian, N. The Removal of Inclusions with Different Diameters in Tundish by Channel Induction Heating: A Numerical Simulation Study. *Materials* 2023, 16, 5254. https://doi.org/10.3390/ma16155254

Academic Editor: Tomasz Trzepieciński

Received: 18 June 2023
Revised: 21 July 2023
Accepted: 24 July 2023
Published: 26 July 2023

Copyright: © 2023 by the authors. Licensee MDPI, Basel, Switzerland. This article is an open access article distributed under the terms and conditions of the Creative Commons Attribution (CC BY) license (https://creativecommons.org/licenses/by/4.0/).

1. Introduction

The quality of the bloom is significantly impacted by the non-metallic impurities in the molten steel in the tundish, and the control of inclusions has always been an important direction of iron and steel metallurgy [1,2]. The number, size, chemical content, shape, and homogeneity of inclusions in the tundish can all be controlled. The behaviors of inclusions mainly include floating, collision condensation, and adsorption [3,4]. Moreover, after turbulence controllers, baffles and dams are added in the tundish, the extended residence time of the molten steel in the tundish is conducive to the full flow of the molten steel in the tundish, so the removal effect of the inclusions in the tundish will be significantly improved [5,6]. However, the temperature of molten steel will gradually decrease with the flow of molten steel, and the viscosity will increase, which is not conducive to the flow and removal of inclusions in the tundish. Wang et al. [7] and Xing et al. [8] reported that the electromagnetic induction heating technology could effectively solve the above problem and can realize the pouring of molten steel at a low degree of superheat, which helps to avoid center segregation and center keyhole and improves the quality of the continuous casting bloom. However, adding a large number of flow control devices, such as dams, will increase the contact area between the molten steel and the refractory material, thereby polluting the molten steel and ultimately affecting the quality of the billet [9].

Based on the principle of electromagnetic mutual induction followed in the metallurgical process, the researchers have developed a dual-channel induction heating tundish, which can compensate for the temperature loss during the flow of molten steel and facilitate the removal of inclusions in molten steel [10]. Mabuchi et al. [11] have found that molten steel will flow obliquely upwards under the influence of buoyancy and electromagnetic force after passing through the induction heating channel, which will promote the full flow of molten steel in the tundish and facilitate the removal of inclusions. Wang et al. [12]

studied the effect of induction heating power on the movement of inclusion particles by combining experiments and simulations, and found that the electromagnetic force acting on molten steel is helpful for the removal of inclusion particles, and the larger the particle size of the inclusions, the more obvious the removal effect. Although the thermophoretic force is not conducive to the removal of inclusion particles, when the induction heating power is 1200 kW, the thermophoretic force suffered by inclusion particles is less than 3%, which can be ignored. Lei et al. [13] studied the physical field of inclusions through the population conservation model or inclusion mass conservation model. The study found that under the action of an electromagnetic field, both the electromagnetic force and induction heating are helpful to the removal of inclusion particles, and the former plays a dominant role. The removal rate of inclusion particles in the tundish increased from 21.4% to 31.05% after using channel induction heating, and the removal rate of inclusion particles in the channel was 1/3 of the removal rate of inclusion particles in the entire tundish. Miki et al. [14] conducted a simulation study on the trajectory of non-metallic inclusions and found that the inclusions' particle size significantly impacted the removal rate. Moreover, when the inclusions collide and grow up from small-grained to large-sized inclusions, it is possible for the small-grained inclusions to float up and then be separated. Additionally, the size of the inclusions in the induction heating tundish has an important influence on its removal effect, which has become a hot spot for metallurgy experts [15,16].

According to the research of the above-mentioned scholars and further literature research, the current research mainly includes the distribution of flow field, temperature field, and electromagnetic force, and the movement of inclusion particles in the two-flow induction heating tundish of the curved channel [17–19]; the influence of the application of double-channel and four-channel bag types on the temperature field of molten steel in the six-flow induction heating tundish, and the influence of different power on the removal of inclusion particles in the channel [20,21], etc. However, few scholars have reported on the influence of inclusions with different particle sizes in different regions of the tundish in the six-fluid induction heating tundish. At the same time, the distribution of inclusions in different regions has a certain influence on optimizing the flow field, improving the crystallinity of molten steel, and improving the quality of the bloom. Therefore, this article investigates the removal effect and trajectory of inclusions in the channel zone and the discharging chamber in the tundish with and without induction heating so as to provide a theoretical reference for tundish induction heating to remove inclusions.

2. Model Building

2.1. Physical Model

2.1.1. Geometry

Figure 1 is a schematic diagram of the three-dimensional geometry of the dual-channel induction heating tundish and the arrangement of the induction coils. The tundish is suitable for pouring 380 × 280 mm^2 billets, the casting speed is 0.68 m/min, and the working fluid level is 900 mm. The distance between the four sides of the iron core and the channel is equal. Relevant dimensions include: the length of the top of the receiving chamber is 3086 mm, the length of the bottom is 2863 mm, the length of the channel is 1630 mm, the diameter of the main channel and the diameter of the sub-channel are both 160 mm, the top length and bottom length of the discharging chamber are 8141 mm and 7824 mm, and the distance between adjacent water outlets is 1500 mm. Moreover, the diameters of the water inlet and outlet are 90 mm and 40 mm, respectively.

The molten steel enters the receiving chamber through the long nozzle and impacts the turbulence controller. Then, the molten steel flows fully into the receiving chamber to form a good flow field. At the same time, most of the inclusions in the molten steel are removed in the receiving chamber. The main removal method is that the inclusion particles float to the surface of the molten steel and are adsorbed by steel slag and adhered to the wall of the receiving chamber. The molten steel enters the channel area after fully flowing in the receiving chamber. At this time, due to induction heating and the Lorentz force, swirling flow is formed within the channel, and the temperature rises rapidly. Due to the effect of the Lorentz force, some inclusion particles will be adsorbed on the inner wall of the channel. The high-temperature molten steel flowing out of the channel forms a density difference with the molten steel in the discharging chamber, so the molten steel flows obliquely upward after flowing out of the channel, which helps the molten steel flow fully in the discharging chamber and promotes the removal of inclusion particles. Finally, the molten steel flows out through the outlet.

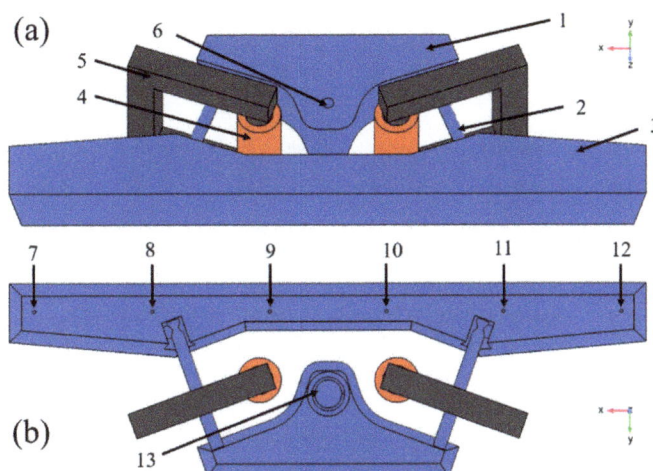

1- The receiving chamber; 2-Channel; 3-The discharging chamber; 4-Coil; 5-Iron core; 6-Inlet; 7-Outlet 1; 8-Outlet 2; 9- Outlet 3; 10- Outlet 4; 11- Outlet 5; 12- Outlet 6; 13-Turbulence controller.

Figure 1. Dual-channel electromagnetic induction heating tundish model: (**a**) upper view; (**b**) bottom view.

2.1.2. Experimental Method

It is the most common method to quantitatively study tundish flow by obtaining the residence time distribution (RTD) curve through the stimulus-response method to calculate the ratio of the dead zone, piston zone, and fully mixed zone of tundish. The theoretical basis of the tundish hydraulic test is the similarity principle, and its basic condition is to ensure that the model is geometrically and dynamically similar to the prototype. The ratio of the geometric size of the model tundish used to that of the prototype tundish is 1:2. In terms of dynamic similarity, the content of the experiment is mainly related to the inertial force, gravity, and viscous force of the fluid in the tundish. According to the knowledge of fluid mechanics, when the molten steel flow in the tundish and the fluid flow in the water model are in the same self-modeling area, as long as the Fr numbers of the model and the prototype are equal, the dynamic conditions required for the experiment can be met. The experimental setup of the water model is shown in Figure 2.

During the water simulation experiment, the temperature of the water is about 20 °C, and its volume expansion coefficient is 2.07×10^{-4} K^{-1}. For the simulated molten steel, its temperature is about 1500 °C, and its volume expansion coefficient is 38.3×10^{-6} K^{-1}. In the tundish water model experiment, the water in the channel is heated to simulate the channel induction heating of the tundish, which makes up for the temperature difference between the inside and outside of the channel, which is about 10 °C. A single channel in the water model uses a 3 kW heating rod. Generally, the temperature difference at both ends of the heating channel in the tundish water model is about 2–3 °C so as to simulate the temperature difference at the channel position in the molten steel.

Figure 2. Six-flow tundish water model experimental device.

2.2. Mathematical Models

2.2.1. Model Assumptions and Control Equations

Based on the COMSOL Multiphysics 6.0 software, this study analyzed in detail the impact of inclusion particles of various diameters being removed from the channel, the wall, and the top surface of the discharging chamber. To reflect the impact of induction heating's presence or absence on the removal of impurities from the molten steel in the tundish, the following assumptions were made for the simulation.

1. Surface slag's impact on flow is not taken into account;
2. Molten steel is an incompressible fluid;
3. The flow field is regarded as a stable flow field, and the inclusions are only affected by gravity, buoyancy, viscous resistance, and electromagnetic force;
4. The collision growth of inclusions is ignored;
5. If the inclusions contact the top slag, it means that they will be separated from the metal melt.

The Navie–Stokes equation, continuity equation, k-ε double equation, mass transfer equation, energy equation, and Maxwell's equations [15,22] are used to establish a 3D numerical model of the tundish. When the inclusions move in the induction heating tundish, they will be affected by gravity, buoyancy, and Saffman lift. The equations used in the model are shown below.

In the calculation of tundish induction heating, gravity, buoyancy, drag force, and other forces are related to the flow rate of the molten steel so as to calculate the flow rate in different regions more accurately. The expression of the momentum equation is shown in Equation (1).

$$\frac{\partial(u_i u_j)}{\partial x_j} = -\frac{\partial P}{\partial x_j} + \frac{\partial}{\partial x_j}\left(\mu_{eff}\frac{\partial u_i}{\partial x_j}\right) + \frac{\partial}{\partial x_i}\left(\mu_{eff}\frac{\partial u_j}{\partial x_i}\right) + \rho g + F_B \quad (1)$$

where x_i and x_j are the space coordinate, m; P is the pressure, Pa; g and ρ are the acceleration of gravity and molten steel's density, respectively, m·s^{-1} and kg·m^{-3}; μ_{eff} is the effective viscosity coefficient, which is the sum of laminar flow viscosity and turbulent flow viscosity, kg·m^{-1}·s^{-1}; F_B is thermal buoyancy, N.

The flow process in the continuous casting process is calculated using the low Reynolds number k-ε turbulence model, and the expression of the turbulent kinetic energy equation is shown in Equation (2).

$$\frac{\partial(\rho u_i k)}{\partial x_i} = \frac{\partial}{\partial x_i}\left[\left(\mu_{eff} + \frac{\mu_t}{\sigma_k}\right)\frac{\partial k}{\partial x_i}\right] + G - \rho\varepsilon \quad (2)$$

where k means turbulent kinetic energy, $m^2 \cdot s^{-2}$; u_i and u_j are the velocity component in the x_i and x_j direction, m/s; ε represents kinetic energy dissipation rate, $m^2 \cdot s^{-3}$; μ_t means turbulent viscosity, $kg \cdot m^{-1} \cdot s^{-1}$; G means the turbulent kinetic energy source term.

The equations of turbulent kinetic energy dissipation rate are shown in Equations (3)–(5).

$$\frac{\partial(\rho u_i \varepsilon)}{\partial x_i} = \frac{\partial}{\partial x_i}\left[\left(\mu_{eff} + \frac{\mu_t}{\sigma_\varepsilon}\right)\frac{\partial \varepsilon}{\partial x_i}\right] + c_1 f_1 G \rho \frac{\varepsilon}{k} - c_2 f_2 \rho \frac{\varepsilon^2}{k} \qquad (3)$$

where $f_1 = 1$, $C_1 = 1.45$, $C_2 = 2.0$, and f_2 can be calculated by Equation (4).

$$f_2 = 1 - 0.3 \exp\left(-Re^2\right) \qquad (4)$$

$$G = \mu_t \frac{\partial u_j}{\partial x_i}\left(\frac{\partial u_i}{\partial x_j} + \frac{\partial u_j}{\partial x_i}\right) \qquad (5)$$

The values of the corresponding coefficients in the formula include $c_\mu = 0.09$, $\sigma_k = 1.00$, and $\sigma_\varepsilon = 1.00$, which are recommended by Launder and Spalding [23].

The expression of the energy equation is Equation (6).

$$\rho\left(\frac{\partial T}{\partial t} + C_p \frac{\partial T}{\partial x_i}\right) = \frac{\partial}{\partial x_i}\left(k_{eff} \frac{\partial T}{\partial x_i}\right) \qquad (6)$$

where k_{eff} is the effective thermal conductivity, $W \cdot m^{-1} \cdot K^{-1}$; T is temperature, K; C_p is the specific heat capacity, $J \cdot kg^{-1} \cdot K^{-1}$.

The differential equation of force balance of inclusions in the tundish is shown in Equation (7). Due to the electromagnetic force, induction heating, and thermophoretic force in the tundish, the inclusion particles are affected by gravity, buoyancy, drag, and Staffman force in the tundish [24,25]. The motion equation of inclusion particles is as follows.

$$\rho_P \frac{\pi}{6} d_P^3 \frac{dv_P}{dt} = F_g + F_b + F_d + F_s + F_p + F_t \qquad (7)$$

where ρ_P is inclusions' density, $kg \cdot m^{-3}$; v_P represents the flow rate of inclusion particles, m/s; d_P is the diameter of inclusion particles, μm; F_g means the gravity of inclusion particles, N; F_b represents the buoyancy force on the inclusion particles, N; F_d is the drag force on the inclusion particles, N; F_s means the Staffman force on the inclusion particles, N; F_p is the electromagnetic pressure force on the inclusion particles, N; F_t is the thermophoretic force on the inclusion particles, N.

2.2.2. Boundary Conditions

Based on the actual situation, the outlet of the inclusion particles is set to the section of the outlet and the top surface in the pouring area, and the wall condition at the outlet is set to freeze. The conditions of the tundish wall and channel wall are set to rebound, and the condition of the primary particle condition is probability 0.5; otherwise, the inclusion particles freeze on the wall. The induction coil used in this study is a pair of single-phase alternating-current windings with a frequency of 50 Hz and a power of 800 kW. The theoretical residence time of the tundish is 1080 s, and double the theoretical residence time, 2160 s, is input into the mathematical model of the motion trajectory of the inclusion particles, providing sufficient time for the removal of the inclusion particles and solving the state of inclusion particles at different times through a transient solver.

Table 1 indicates the properties of molten steel together with parameters in mathematical modeling.

Table 1. Steel physical parameters and simulation parameters.

Parameter	Value	Parameter	Value
Inlet velocity, m·s^{-1}	0.9	Thermal conductivity, W·m^{-1}·K^{-1}	23.5
Inlet temperature, K	1833	Heat capacity at constant pressure, J·kg^{-1}·K^{-1}	4500
The diameter of the particle, μm	10, 30, 50	Dynamic viscosity, Pa·s	0.0065
Particle density, kg·m^{-3}	3900	Surface heat loss, W·m^{-2}	15,000
Particle count released at a time	1000	Bottom heat loss, W·m^{-2}	1800
Particle precision order	5	Side heat loss, W·m^{-2}	4600
Molten steel density, kg·m^{-3}	7580	Channel heat loss, W·m^{-2}	2000

2.3. Mesh Independent Study

Since the research focus of this paper is the molten steel in the tundish, only grid-independent research is done on the tundish. The mesh division of the induction heating tundish is exhibited in Figure 3.

Figure 3. Mesh division diagram of induction heating tundish.

The grid division of the tundish is all free tetrahedral grids, and the positions of the channel connection, inlet, and outlet have been refined, and five layers of boundary layer grids have been set to improve the accuracy of settlement results. The results of grid-independent research on induction heating tundish are shown in Table 2, with a total number of 1.3 million grids.

Table 2. Grid-independent calculation results of induction heating tundish.

Number of Grids	400,000	600,000	1,000,000	1,300,000	1,800,000	2,000,000
Average velocity, m·s^{-1}	0.0194	0.0198	0.0203	0.0209	0.0209	0.0209

The tundish's average flow rate of molten steel is 0.0209 m/s, and the average flow velocity is kept with the number of grids increasing. Therefore, the optimal solution mesh number of the tundish three-dimensional (3D) model should be 1.3 million. Based on this, the number of tundish grids established in this study is 1,328,585.

2.4. Model Validation

Since the research mechanism of Vives et al. [26] is similar to the induction heating mechanism involved in this paper, the experimental results of Vives et al. are selected for verification. The comparison results are shown in Figure 4.

The magnetic field strength of the simulation results of this study and the results of Vives et al. [26] are both 0.04 T on the Z = −20 cm section of the left and middle channels. The intensity and direction distribution of the magnetic field are similar, and they all have the characteristics of a symmetrical magnetic field in the middle channel section. The magnetic field of the center channel is distributed symmetrically, and the magnetic field of the left channel is distributed eccentrically due to the proximity effect. It can be seen that the predicted results are in good agreement with the experimental data, which show the reliability of the simulation results of this model.

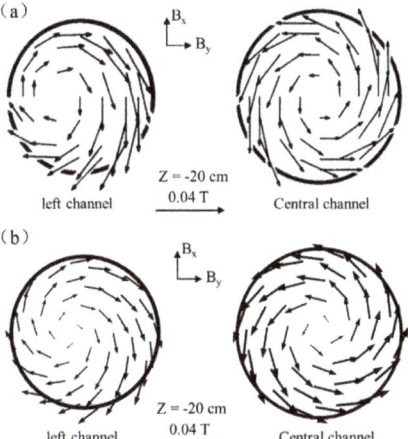

Figure 4. Comparison of magnetic field intensity distribution: (**a**) experimental results; (**b**) simulation results of this model.

3. Results and Discussion

3.1. Physical Simulation Results

Due to the symmetry of the T-shaped tundish, only the RTD curves and flow field change trends of outlet 1, outlet 2, and outlet 3 are analyzed. The RTD curves for the three outlets are shown in Figure 5.

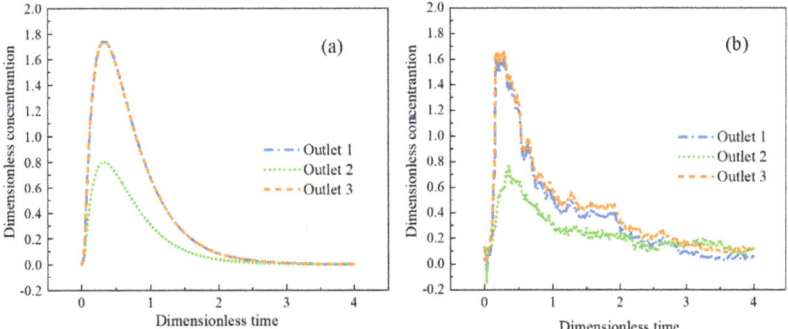

Figure 5. Comparison of simulated and physical-simulated RTD curves: (**a**) numerical simulation results; (**b**) physical simulation results.

It can be seen from Figure 5 that the RTD curve obtained by numerical simulation is in good agreement with the RTD curve obtained by physical simulation, which shows the reliability of the model in this paper. Outlet 1 and outlet 3 have faster response times than outlet 2. The consistency of stream 1 and stream 3 is high, and the concentration of none in outlet 2 is relatively low. The reason is that the molten steel passes through outlet 1 and outlet 3 after flowing out from the branch, then hits the inner wall of the tundish, the relatively flowing molten steel forms a vortex, and then part of the molten steel flows to the outlet 2. Due to various changing factors in the experimental process during actual detection, the measured curve is tortuous, but its trend is still consistent with the simulation results.

3.2. Mathematical Simulation Results

3.2.1. Simulation of the Temperature and Flow Fields

Compared with the traditional tundish, the induction heating tundish adopts an external heating source to keep molten steel's temperature constant during pouring. This technology has a great influence on the flow field, temperature field, and behavior characteristics of inclusions in the tundish.

The flow field and temperature field distribution in the molten steel are described through the horizontal section of the tundish at the height of the channel center. Figure 6 shows the schematic diagram of the horizontal section of the tundish at the height of the channel center.

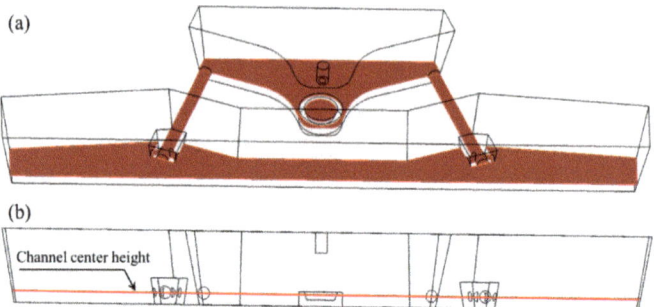

Figure 6. Schematic diagram of channel center height section: (**a**) upper view; (**b**) front view.

The flow and temperature field of molten steel on the center height section of the channel are shown in Figures 7 and 8.

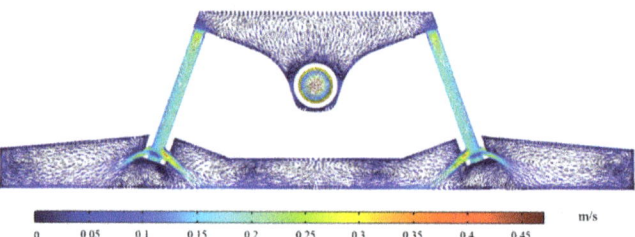

Figure 7. Flow field distribution on the center height section of tundish channel.

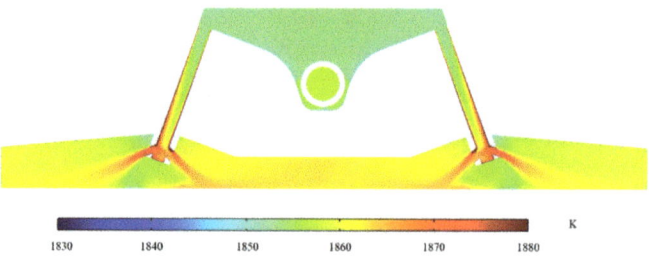

Figure 8. Temperature field distribution on the center height section of tundish channel.

As depicted in Figure 7, the molten steel travels from the two sub-channels of the channel to the pouring region after passing through the channel. According to the principle of the continuity equation, the flow velocity gradually decreases after flowing out of the sub-channels. Additionally, the molten steel forms a large vortex in the variable flow area of the pouring area (outlet 1 and outlet 6) and forms a small vortex at the middle nozzle position (outlet 2 to outlet 5). This phenomenon contributes to the extension of the molten steel's residence duration in the tundish, the decrease of the dead zone's volume percentage, and the complete mixing of the molten steel's constituent parts. The molten steel in the channel in Figure 8 is heated due to induction heating. After the molten steel flows through the left and right channels, the temperature rises from 1853.0 K and 1853.1 K to 1863.0 K and 1863.3 K, and the temperature increases by 10.0 K and 10.2 K, respectively. The average temperature of every outlet is shown in Table 3.

Table 3. The average temperature of each outlet.

Outlet's Number	1	2	3	4	5	6
Average temperature, K	1860.3	1857.1	1862.3	1862.5	1857.0	1860.5

Each outlet has a maximum temperature variation of 5.5 K. Because outlet 2 and outlet 5 are located between two sub-channels, the molten steel flowing from the sub-channels needs to go through a long path to reach these two water outlets. Therefore, the average temperature of the sections of outlet 2 and outlet 5 is relatively low. The analysis shown above reveals that the temperature distribution in the channel could be raised greatly using induction heating technology. In addition, the application of induction heating helps to improve the cooling of molten steel at the initial stage of pouring so that the molten steel can be poured with low superheat.

3.2.2. Simulation of Inclusion Removal from Tundish with and without Induction Heating

On the effectiveness of removing inclusions from the tundish with and without using induction heating technology, a simulated investigation was carried out, and Table 4 displays the outcomes. Inclusions particles have a diameter of 50 μm.

Table 4. Removal rate of inclusions with and without induction heating.

Local Area in the Tundish	Removal Rate without Induction Heating, %	Removal Rate with Induction Heating, %
The receiving chamber	97.500	96.400
Channel	2.100	3.000
The wall of discharging chamber	0.030	0.040
Liquid level in discharging chamber	0.000	0.000
Outlet	0.000	0.010

In the receiving chamber, where inclusion removal mostly occurs, the removal rate is 97.5% without induction heating and 96.4% with it. Compared with the condition without induction heating, the adsorption rate of inclusion particles in the channel increases under the condition of induction heating, which indicates that the induction heating technology is helpful to the removal of inclusion in the channel. The simulation calculation results of the movement trajectories of inclusion particles in the channel and discharging chamber illustrate that the inclusion particles are completely removed within 346 s without induction heating; the inclusion particles were completely removed within 302 s when induction heating was used; the time was shortened by 44 s and decreased by 12.72%. This phenomenon further shows that induction heating technology is helpful in promoting the movement and removal efficiency of inclusions.

3.2.3. Simulation of Inclusion Removal with Various Sizes Using Induction Heating

Only the motion distribution of inclusion particles in the channel region and pouring area is simulated so as to analyze the removal of inclusion particles there thoroughly. Based on the original model, the entrance of the channel is set as the entrance of inclusion particles. The inclusion particles with diameters of 10 μm, 30 μm, and 50 μm in the channel and discharging chamber were simulated, and the trajectory of inclusion particles under different times was intercepted, as shown in Figures 9–11.

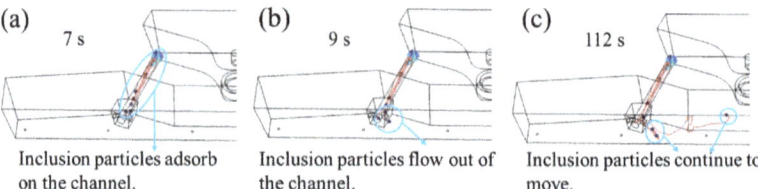

Figure 9. Trajectories and removal effects of 10 μm inclusion particles at (a) 7 s; (b) 9 s; (c) 112 s.

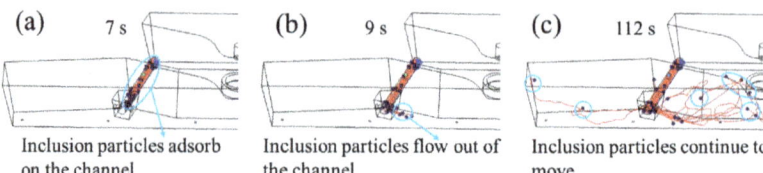

Figure 10. Trajectories and removal effects of 30 μm inclusion particles at (a) 7 s; (b) 9 s; (c) 112 s.

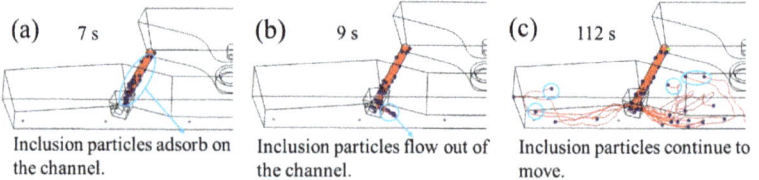

Figure 11. Trajectories and removal effects of 50 μm inclusion particles at (a) 7 s; (b) 9 s; (c) 112 s.

Among them, the blue particles represent the inclusion particles, and the curve represents the trajectory of the inclusion particles. Figures 9–11 show that the flow characteristics of inclusion particles with different diameters are similar, they are all adsorbed in a large amount in the channel, and only a small amount of inclusion particles flow to the discharging chamber.

The inclusion particles are released from 0 s, and the flow time of the inclusion particles in the channel is about 7 s. At 9 s, the inclusion particles first flow out from the sub-channel opening near the center, and some of the inclusion particles are adsorbed on the wall of the pouring area, as shown in the blue circle. The remaining particles continue to move, such as the particles in the circle at 112 s. Among them, most inclusion particles with a diameter of 10 μm were adsorbed at the channel, and the removal rate was 70.9%. A small part flowed to the discharging chamber and adhered to the wall of the discharging chamber. With a removal rate of 20.7%, the removal rate at the top surface of the discharging chamber was 4.2%, and no inclusion particles were detected at the outlet. Most inclusion particles with a diameter of 30 μm adhered to the channel, and the removal rate was 60.9%. Followed by the pouring area wall, the removal rate was 26.1%, 7.6% of the inclusion particles floated to the pouring area liquid surface, and no inclusion particles were detected at the nozzle. The inclusion particles with a diameter of 50 μm are mostly removed in the channel area, with a removal rate of 56.1%, followed by the wall of the pouring area, with a removal rate of 31.8%. A total of 7.9% of the inclusion particles floated to the top surface of the discharging chamber, and no inclusion particles were detected at the nozzle. According to the movement trajectories of inclusion particles, it can be seen that inclusion particles of 10 μm, 30 μm, and 50 μm were removed in 226 s, 344 s, and 368 s, respectively.

The removal effect and removal time of three kinds of inclusion particles with different diameters in each region are displayed in Figure 12. From Figure 12a, when the diameter of inclusion particles increases from 10 µm to 50 µm, the clearance rate of particles in the channel steadily declines from 70.9% to 56.1%. The clearance rate of particles on the wall of the discharging chamber increased from 20.7% to 31.8%, the clearance rate of particles at the top surface of the discharging chamber increased by 3.7%, and the removal rates of particles with three diameters at the nozzle were all 0%. The above analysis shows that the inclusion particles with smaller diameters are more likely to be adsorbed on the wall of the channel, and the inclusion particles with larger diameters are more likely to adhere to the wall of the pouring area or float up to the top surface of the discharging chamber.

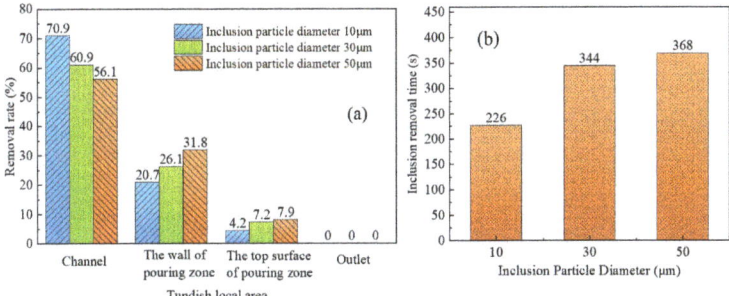

Figure 12. (**a**) Removal rate and (**b**) removal time of inclusion particles with different diameters.

Figure 12b shows that when the particle diameter continues to increase, the removal time of inclusion particles shows an overall upward trend. When the particle diameter is 10 µm, the particles are removed within 226 s, and when the particle diameter of inclusions is 50 µm, the removal time is extended to 368 s. The clearance time of particles in the pouring area is connected to their floating speed and particle concentration in molten steel. The expression of the floating velocity of inclusion particles is shown in Equation (8).

$$U_p = \frac{(\rho_l - \rho_p) \cdot g \cdot D^2}{18\mu} \tag{8}$$

where U_p is the floating speed of inclusion particles, m/s; ρ_l and ρ_p represent the density of steel and inclusions, respectively, kg/m^3; μ is the viscosity of molten steel, Pa·s; D means the diameter of inclusions, µm.

According to Equation (8), for the same kind of inclusions, the floating speed of the inclusion particles increases as the inclusions' diameter steadily rises, which is advantageous for the inclusion particles' removal. The simulation results under induction heating conditions show that the clearance rate of particles with small sizes in the channel area is obviously higher than that of particles with large sizes. When the total number of particles released at the entrance remains unchanged, the concentration gradient of particles with large diameters in the pouring area is higher than that with small diameters among the particles flowing into the pouring area. As a result, inclusion particles with larger diameters take longer to remove from the discharging chamber than particles with smaller diameters.

The distribution and removal of inclusion particles in tundish have always been a hot issue in tundish metallurgy. However, due to the sealing and danger of field operation, it is difficult to detect the distribution of inclusion particles in tundish with field data, which is why this paper does not verify the distribution of inclusion particles in various areas of tundish with factory data. Subsequently, the influence of different parameters on the removal of inclusion particles in tundish will be further studied, and comparative analysis will be carried out through the field data of inclusion particles as far as possible.

4. Conclusions

- The mathematical simulation results of the removal of inclusions with and without induction heating show that the induction heating technology contributes to promoting

the movement and removal of inclusions in the channel. Meanwhile, induction heating technology shortens inclusion particle removal time;
- Smaller inclusion particles are easier to be adsorbed in the channel, and larger inclusion particles are easier to be removed in the discharging chamber. The clearance rate of inclusion in the channel gradually decreases from 70.9% to 56.1%, with the diameter of inclusion particles increasing from 10 μm to 50 μm;
- The inclusion particles with large particle sizes are easier to be removed on the wall and liquid surface of the discharging chamber. Moreover, compared with small-size inclusions, the removal time of large-size inclusions is longer.

Author Contributions: Conceptualization, B.Y. and G.Z.; methodology, B.Y.; software, Q.J. and P.Z.; validation, P.Z., Z.F. and N.T.; formal analysis, B.Y. and P.Z.; investigation, B.Y.; resources, G.Z.; data curation, B.Y.; writing—original draft preparation, B.Y.; writing—review and editing, Q.J. and G.Z.; visualization, Q.J.; supervision, G.Z.; project administration, G.Z.; funding acquisition, G.Z. All authors have read and agreed to the published version of the manuscript.

Funding: This research was funded by the National Natural Science Foundation of China, grant number 52074140, and the Scientific Research Fund of the Yunnan Provincial Department of Education, grant number 2023J0130.

Institutional Review Board Statement: Not applicable.

Informed Consent Statement: Not applicable.

Data Availability Statement: Data are contained within the article.

Acknowledgments: This work was supported by the National Natural Science Foundation of China [grant number 52074140] and the Scientific Research Fund of the Yunnan Provincial Department of Education [grant number 2023J0130].

Conflicts of Interest: The authors declare no conflict of interest.

References

1. Sahai, Y. Tundish technology for casting clean steel: A review. *Metall. Mater. Trans. B* **2016**, *47*, 2095–2106. [CrossRef]
2. Jin, Y.; Dong, X.; Yang, F.; Cheng, C.; Li, Y.; Wang, W. Removal mechanism of microscale non-metallic inclusions in a tundish with multi-hole-double-baffles. *Metals* **2018**, *8*, 611. [CrossRef]
3. Yang, Y.; Jönsson, P.G.; Ersson, M.; Nakajima, K. Inclusion Behavior under a Swirl Flow in a Submerged Entry Nozzle and Mold. *Steel Res. Int.* **2015**, *86*, 341–360. [CrossRef]
4. Gildardo, S.D.; Rodolfo, D.M.; Jose de Jesus, B.S.; Hector Javier, V.H.; Angel, R.B.; Sergio, R.G. Numerical Modelling of Dissipation Phenomena in a New Ladle Shroud for Fluidynamic Control and Its Effect on Inclusions Removal in a Slab Tundish. *Steel Res. Int.* **2014**, *85*, 863–874.
5. Zhang, L.F.; Taniguchi, S.; Cai, K.K. Fluid flow and inclusion removal in continuous casting tundish. *Metall. Mater. Trans. B* **2000**, *31*, 253–266. [CrossRef]
6. Bao, Y.P.; Wang, M. Development trend of tundish metallurgical technology. *Contin. Cast.* **2021**, *33*, 2–11.
7. Wang, Z.M.; Li, Y.; Wang, X.Z.; Li, X.L.; Yue, Q.; Xiao, H. Channel-Type Induction Heating Tundish Technology for Continuous Casting: A Review. *Materials* **2023**, *16*, 493. [CrossRef]
8. Xing, F.; Zheng, S.G.; Zhu, M.Y. Motion and Removal of Inclusions in New Induction Heating Tundish. *Steel Res. Int.* **2018**, *89*, 1700542. [CrossRef]
9. Yuan, C.; Liu, Y.; Li, G.Q.; Zhou, Y.S.; Huang, A. Interaction between microporous magnesia castable and 38CrMoAl steel. *J. Iron Steel Res. Int.* **2023**, *30*, 516–524. [CrossRef]
10. Wang, P.; Xiao, H.; Chen, X.Q.; Li, X.S.; He, H.; Tang, H.Y.; Zhang, J.Q. Influence of Dual-Channel Induction Heating Coil Parameters on the Magnetic Field and Macroscopic Transport Behavior in T-Type Tundish. *Metall. Mater. Trans. B* **2021**, *52*, 3447–3467. [CrossRef]
11. Mabuchi, M.; Yoshii, Y.; Nozaki, T.; Kakiu, Y.; Kakihara, S.; Ueda, N. Investigation of the purification of molten steel by using tundish heater: Development on the controlling method of casting temperature in continuous casting V. *ISIJ Int.* **1984**, *70*, 118.
12. Wang, Q.; Qi, F.S.; Li, B.K.; Tsukihashi, F. Behavior of Non-metallic Inclusions in a Continuous Casting Tundish with Channel Type Induction Heating. *ISIJ Int.* **2014**, *54*, 2796–2805. [CrossRef]
13. Lei, H.; Yang, B.; Bi, Q.; Xiao, Y.Y.; Chen, S.F.; Ding, C.Y. Numerical Simulation of Collision-Coalescence and Removal of Inclusion in Tundish with Channel Type Induction Heating. *ISIJ Int.* **2019**, *59*, 1811–1819. [CrossRef]
14. Miki, Y.; Thomas, B.G. Modeling of inclusion removal in a tundish. *Metall. Mater. Trans. B* **1999**, *30*, 639–654. [CrossRef]
15. Xing, F.; Zheng, S.G.; Liu, Z.H.; Zhu, M.Y. Flow Field, Temperature Field, and Inclusion Removal in a New Induction Heating Tundish with Bent Channels. *Metals* **2019**, *9*, 561. [CrossRef]

16. Zhang, Q.; Xu, G.Y.; Iwai, K. Effect of an AC Magnetic-field on the Dead-zone Range of Inclusions in the Circular Channel of an Induction-heating Tundish. *ISIJ Int.* **2022**, *62*, 56–63. [CrossRef]
17. Yang, B.; Lei, H.; Bi, Q.; Jiang, J.M.; Zhang, H.W.; Zhao, Y.; Zhou, J.A. Electromagnetic Conditions in a Tundish with Channel Type Induction Heating. *Steel Res. Int.* **2018**, *89*, 1800145. [CrossRef]
18. Yang, B.; Deng, A.Y.; Li, Y.; Wang, E.G. Exploration of the Relationship between the Electromagnetic Field and the Hydrodynamic Phenomenon in a Channel Type Induction Heating Tundish Using a Validated Model. *ISIJ Int.* **2022**, *62*, 677–688. [CrossRef]
19. Yang, B.; Deng, A.Y.; Duan, P.F.; Kang, X.L.; Wang, E.G. "Power curve" key factor affecting metallurgical effects of an induction heating tundish. *J. Iron Steel Res. Int.* **2022**, *29*, 151–164. [CrossRef]
20. Yue, Q.; Zhang, C.B.; Pei, X.H. Magnetohydrodynamic flows and heat transfer in a twin-channel induction heating tundish. *Ironmak. Steelmak.* **2017**, *44*, 227–236. [CrossRef]
21. Chen, X.Q.; Xiao, H.; Wang, P.; Lan, P.; Tang, H.Y.; Zhang, J.Q. Effect of Channel Heights on the Flow Field, Temperature Field, and Inclusion Removal in a Channel-type Induction Heating Tundish. In *12th International Symposium on High-Temperature Metallurgical Processing*; Springer International Publishing: Cham, Switzerland, 2022; pp. 501–522.
22. Zhang, H.; Lei, H.; Ding, C.; Chen, S.; Niu, H.; Yang, B. Deep Insight into the Pinch Effect in a Tundish with Channel-Type Induction Heater. *Steel Res. Int.* **2022**, *93*, 2200181. [CrossRef]
23. Launder, B.E.; Spalding, D.B. *Numerical Prediction of Flow, Heat Transfer, Turbulence and Combustion*, 1st ed.; Pergamon: Oxford, UK, 1983; pp. 96–116.
24. Brenner, H.; Bielenberg, J.R. A continuum approach to phoretic motions: Thermophoresis. *Phys. A* **2005**, *355*, 251–273. [CrossRef]
25. Bielenberg, J.R.; Brenner, H. A hydrodynamic/Brownian motion model of thermal diffusion in liquids. *Phys. A* **2005**, *356*, 279–293. [CrossRef]
26. Vives, C.; Ricou, R. Magnetohydrodynamic flows in a channel-induction furnace. *Metall. Mater. Trans. B.* **1991**, *22*, 193–209. [CrossRef]

Disclaimer/Publisher's Note: The statements, opinions and data contained in all publications are solely those of the individual author(s) and contributor(s) and not of MDPI and/or the editor(s). MDPI and/or the editor(s) disclaim responsibility for any injury to people or property resulting from any ideas, methods, instructions or products referred to in the content.

Article

Prediction of Compressive Strength of Biomass–Humic Acid Limonite Pellets Using Artificial Neural Network Model

Haoli Yan, Xiaolei Zhou *, Lei Gao *, Haoyu Fang, Yunpeng Wang, Haohang Ji and Shangrui Liu

Faculty of Metallurgical and Energy Engineering, Kunming University of Science and Technology, Kunming 650093, China; ximengfunny@163.com (H.Y.); fanghaoyuya1@163.com (H.F.); wangypeng1997@163.com (Y.W.); jhhkust2022@163.com (H.J.); lsr190608@163.com (S.L.)
* Correspondence: zxl@kust.edu.cn (X.Z.); leigao@kust.edu.cn (L.G.)

Abstract: Due to the detrimental impact of steel industry emissions on the environment, countries worldwide prioritize green development. Replacing sintered iron ore with pellets holds promise for emission reduction and environmental protection. As high-grade iron ore resources decline, research on limonite pellet technology becomes crucial. However, pellets undergo rigorous mechanical actions during production and use. This study prepared a series of limonite pellet samples with varying ratios and measured their compressive strength. The influence of humic acid on the compressive strength of green and indurated pellets was explored. The results indicate that humic acid enhances the strength of green pellets but reduces that of indurated limonite pellets, which exhibit lower compressive strength compared to bentonite-based pellets. Furthermore, artificial neural networks (ANN) predicted the compressive strength of humic acid and bentonite-based pellets, establishing the relationship between input variables (binder content, pellet diameter, and weight) and output response (compressive strength). Integrating pellet technology and machine learning drives limonite pellet advancement, contributing to emission reduction and environmental preservation.

Keywords: neural network; limonite; pelletizing; compressive strength; organic binder

Citation: Yan, H.; Zhou, X.; Gao, L.; Fang, H.; Wang, Y.; Ji, H.; Liu, S. Prediction of Compressive Strength of Biomass–Humic Acid Limonite Pellets Using Artificial Neural Network Model. *Materials* **2023**, *16*, 5184. https://doi.org/10.3390/ma16145184

Academic Editor: Nicolas Sbirrazzuoli

Received: 25 June 2023
Revised: 18 July 2023
Accepted: 20 July 2023
Published: 24 July 2023

Copyright: © 2023 by the authors. Licensee MDPI, Basel, Switzerland. This article is an open access article distributed under the terms and conditions of the Creative Commons Attribution (CC BY) license (https://creativecommons.org/licenses/by/4.0/).

1. Introduction

Many countries worldwide are implementing green development initiatives to reduce carbon emissions [1,2]. The steel industry, as a heavy industry, generates a significant amount of exhaust gases, wastewater, and solid waste during the production process, leading to severe environmental pollution [3,4]. In comparison to sintered ore, pelletized ore has higher iron content, fewer harmful elements, and good reducibility, which results in increased production, reduced coke consumption, cost savings, optimized burden structure, and improved economic benefits in blast furnace smelting. Therefore, it is necessary to continuously increase the proportion of pelletized ore in the ironmaking process [5,6].

To date, extensive research has been conducted on the preparation of high-grade ore pellets [7]. However, with the rapid development of the steel industry, the output of rich ores has decreased, while the extraction of low-grade ores has increased [8–12]. The technology related to pellet production using low-grade ores is still not mature, necessitating further research on low-grade ore pellets. Limonite, characterized by low levels of harmful elements, is an example of low-grade iron ore. However, low-grade ores often contain numerous impurities, resulting in lower pellet strength when using limonite. Moreover, pelletized ores undergo multiple handling, transportation, stacking, and movement processes before and after entering the blast furnace, experiencing various harsh mechanical forces such as collisions, impacts, compression, and friction [13]. These mechanical forces can cause the breakage of some pellets, leading to the generation of fines and affecting furnace operation and production indicators. To address these challenges, the addition of binders to the pellets can fill the gaps and cracks between ore particles, bonding them together to increase the density and mechanical strength of the pellets [14].

Bentonite, as the main binder for pellets, cannot undergo thermal decomposition during the smelting process and mostly remains in the pellets, leading to a decrease in the iron grade of the pelletized ore [15]. Qiu et al. [16] utilized organic binders to prepare pelletized ore and found that, compared to pellets prepared using bentonite, the resulting pellets had a higher iron grade and lower impurity content [17]. Humic acid, as an organic binder and a biomass-derived substance, exhibits strong adsorption capacity on the surface of iron ore, thereby enhancing the strength of green pellets [18,19]. Currently, there is limited research on the preparation of limonite pellets using humic acid as a binder, necessitating further investigation into humic acid-based limonite pellets. However, most studies on pellet compressive strength have relied on single-factor analysis or orthogonal experiments to qualitatively describe the influencing factors of various parameters, without considering the interactions among them [20]. Due to the limited number of experiments, the predictability is relatively poor [21]. Artificial neural networks (ANNs) offer a promising solution for predicting material properties and have been widely used in this context [22]. ANNs are often referred to as "black-box" models because they can make accurate predictions based on training data without providing any physical explanation behind the phenomena [23,24].

Numerous studies have harnessed the power of Artificial Neural Networks (ANN) in diverse applications within the field of materials science and metallurgy. Chagas et al. [25] applied ANN to assess the sensitivity of variables in pellet bed formation, aiding in the generation of green pellets with reduced fuel and energy consumption and improved final quality. Dwarapudi et al. [26] developed an ANN model to predict the cold compressive strength (CCS) of pellets in a straight grate furnace. By considering variations in bentonite, alkalinity, FeO, and green pellet moisture, the model successfully predicted CCS with an error margin of less than 3%. Klippel et al. [27] introduced an early detection system for slope instability risks based on iron ore images, utilizing edge artificial intelligence. The field test results demonstrated an accuracy rate of 91% and a recall rate of 96%, highlighting the feasibility of employing deep learning for detecting iron ore types and preventing slope instability risks. Fan et al. [28] investigated the main factors affecting sintering quality, such as humidity, fuel ratio, sintering speed, and sintering drum strength. Utilizing backpropagation ANN, prediction models were constructed and applied in sintering pot experiments to optimize humidity and fuel ratio, ultimately improving sintering drum strength. Golmohammadi et al. [29] developed a Quantitative Structure–Property Relationship (QSPR) using Partial Least Squares (PLS) and ANN to predict the precipitation of trivalent iron during bioleaching. The neural network model exhibited reliable and accurate predictive capabilities during the bioleaching process. Li Guo et al. [30] proposed a novel method to estimate ore feed load directly from images of ore pellets using deep learning models. The introduction of a weakly supervised learning method and a two-stage model training algorithm allowed competitive model performance and real-time estimation of ore feed load in the grinding process optimization. Li et al. [31] developed an intelligent system named the Group Method of Data Handling (GMDH) for predicting iron ore prices. Compared to other techniques, the GMDH technique exhibited superior accuracy with a variance accounted for (VAF) value of 97.89%. Wang et al. [32] successfully implemented a hybrid ensemble model combining Extreme Learning Machine (ELM) with an improved AdaBoost. The RT algorithm is used to solve regression problems in sintering processes. This approach led to significant improvements in energy efficiency and sintering quality through the analysis of high-priority factors. Yachun Mao et al. [33] introduced a detection method for the magnetic properties of limonite using the improved Particle Swarm Optimization–Enhanced Extreme Learning Machine (IPSO–ELM) algorithm and spectroscopy. The IPSO–ELM predictive model exhibited excellent performance and generalization capability compared to ELM and PSO–ELM predictive models. Yanwei Yang et al. [34] successfully integrated laser-induced breakdown spectroscopy (LIBS) and machine learning for rapid and precise classification of iron ore, providing a novel method for iron ore selection in the metallurgical industry. Tunckaya et al. [35] utilized ANN, Multiple

Linear Regression (MLR), and auto-regressive integrated moving average (ARIMA) models to predict and track the flame temperature of a blast furnace. The computational results demonstrated satisfactory performance in the selected performance indicators, including regression coefficients and root mean square errors.

Previous studies on pellet compressive strength mostly relied on single-factor analysis or orthogonal experiments to qualitatively describe the influencing factors, resulting in limited predictability. Therefore, this study introduces an innovative approach by applying an Artificial Neural Network (ANN) for data analysis and optimization, offering a novel solution for the preparation and sintering process of limonite pellets. This contributes to enhanced process efficiency, reduced production costs, and mitigated environmental impact, providing significant practical application value.

2. Experiment

2.1. Experimental Material

The limonite used in this study was obtained from a factory in Yunnan Province, China. The limonite exhibits a rough surface, porous texture, and low levels of harmful elements. Its chemical composition is presented in Table 1, showing a high content of total iron (TFe) at 54.67 wt.%, along with 4.04 wt.% of SiO_2 and 3.47 wt.% of MnO. Notably, this limonite possesses strong adhesive properties and high water absorption capacity.

Table 1. Chemical composition of precious sand limonite (wt.%).

TFe	FeO	SiO_2	S	MnO	TiO_2	Pb	Zn	K_2O	Na_2O	Cu	V_2O_5	LOI
54.67	0.29	4.04	0.048	3.47	0.27	0.009	0.018	0.095	0.001	0.009	0.04	14.84

2.2. Experimental Procedure

The experimental procedure is illustrated in Figure 1:

(1) Pellet Manufacturing: 200 g of experimental material, limonite powder, is weighed. Binders in different proportions (0.4 wt.%, 0.8 wt.%, 1.2 wt.%, 1.6 wt.%, and 2.0 wt.%) are added, and the materials are thoroughly mixed using the multiple folding and stirring method. The balling machine is started, and approximately 2/3 of the mixed material is added to the machine. Water is slowly dripped onto the surface of the mixture to form pellets with a diameter of approximately 3 mm over a period of 3 min. Every 3 min, a mist of water is sprayed onto the surface of the mother pellets, and material is added to the wetted surface to allow the pellets to continuously roll and grow. The process is generally controlled within 12 min to achieve pellets of the desired size (9–16 mm). After stopping the addition of water and material, the pellets are allowed to continue rotating in the balling machine for approximately 2 min to achieve compaction. The pellets are removed using a small scoop. The pellets are separated using a vernier caliper, and the pellets with a diameter of 9–16 mm are considered qualified products, while the remaining pellets are deemed unqualified.

(2) Pellet Roasting: The manufactured pellets are transferred in batches into crucibles. The crucibles containing the pellets are placed into a muffle furnace for roasting. The roasting temperature is set as follows, as shown in Figure 2: Ramp up to 200 °C in 20 min and hold at 200 °C for 20 min; Ramp up to 700 °C in 20 min and hold at 700 °C for 20 min; ramp up to 1250 °C in 40 min and hold at 1250 °C for 25 min. Subsequently, the pellets are gradually cooled to room temperature and removed from the furnace.

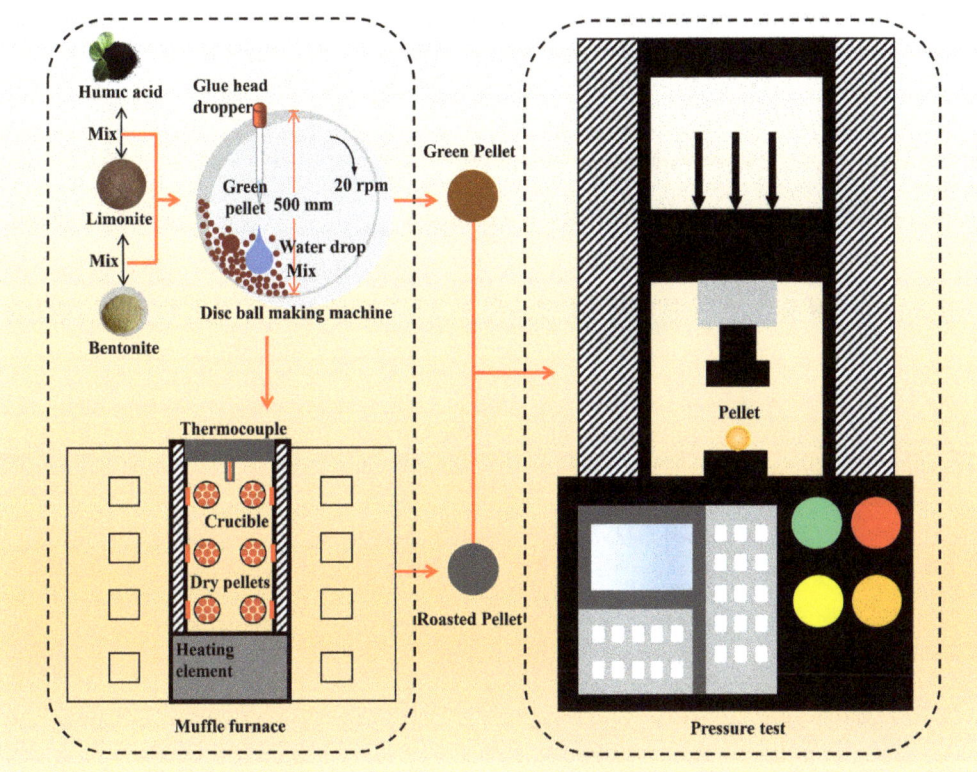

Figure 1. Experimental flow.

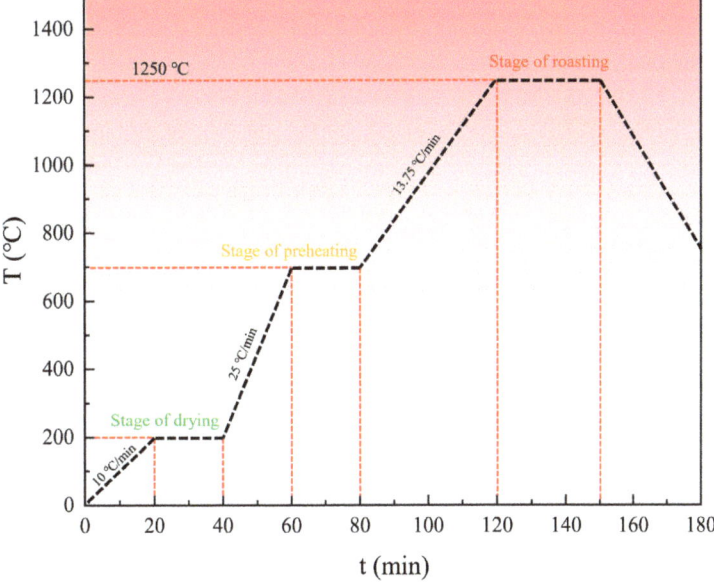

Figure 2. Pellet roasting temperature diagram.

(3) Measurement and Testing: The weight of both the raw pellets and the roasted pellets is measured using an electronic balance. The compressive strength of the raw pellets and the roasted pellets is measured using a YAW-100C testing machine. Pellets with a diameter of 9–16 mm are selected using a vernier caliper. Following the guidelines of YB/T 4848-2020, "Physical Test Methods for Roasted Pellets," the average compressive strength of the limonite raw pellets and the roasted pellets is recorded.

3. Results and Discussion

3.1. Effect of Humic Acid on Compressive Strength of Limonite Green Pellets

As shown in Figure 3A, the influence of humic acid content, pellet weight, and diameter on the compressive strength of limonite pellets is illustrated. The diameter of the pellets is represented by the size of the purple spheres, and the data for the four parameters have been normalized. The compressive strength of the pellets increases with the weight, and larger pellets are predominantly located at the top, while smaller pellets are mainly located at the bottom. The variation in compressive strength with increasing humic acid content is not significant and is primarily related to the diameter and weight. Figure 3B,C present the scatter plot and normal distribution plot illustrating the effect of different proportions of humic acid on the compressive strength of the pellets. The humic acid content has a positive correlation with the compressive strength of the pellets. The impact of different proportions of humic acid on the compressive strength of the pellets is depicted in Figure 3D; as the proportion of humic acid in the pellet increases from 0.4% to 2.0%, the average compressive strength of the pellets rises and reaches its maximum value of 18.7 N when the humic acid content reaches 2.0%.

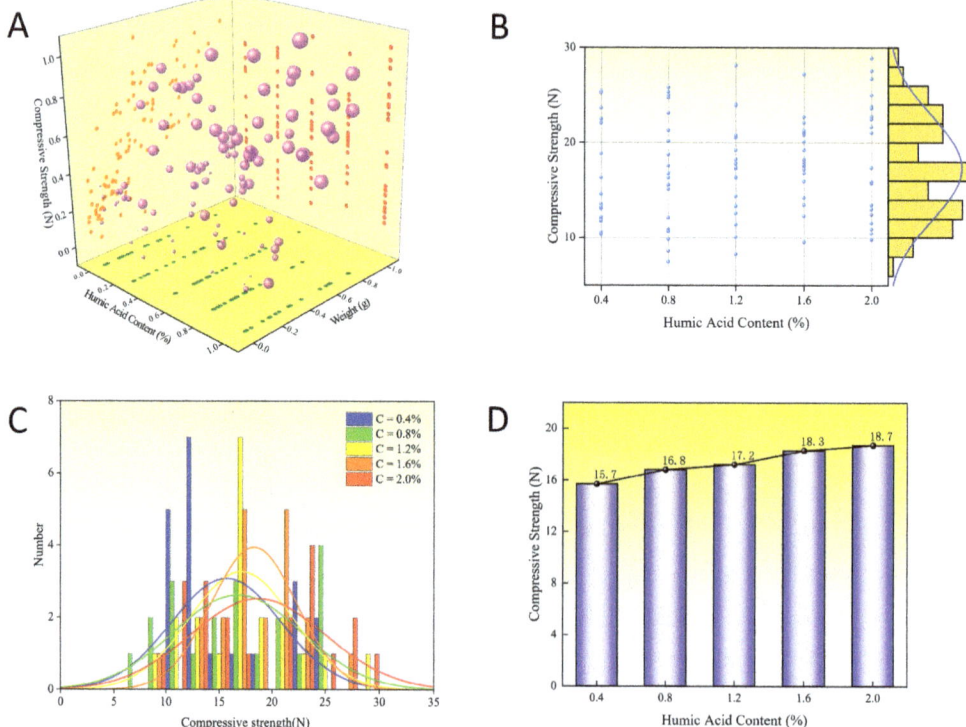

Figure 3. Effect of humic acid on compressive strength of limonite green pellets. (**A**) Stereo scatter data. (**B**) Planar scatter data. (**C**) Normal distribution of data. (**D**) The average of the scatter data.

When the mineral powder is wetted by water during the rolling process, it forms pellets of a certain size and imparts them with a certain strength through the combined action of capillary force, molecular attraction, and frictional force. The pellet size, moisture content, mechanical strength, and thermal stability of the pellets influence the subsequent roasting operation and are related to the yield and quality of the pelletized ore.

Dry mineral powders generally exhibit hydrophilic properties. As shown in Figure 4, under the molecular forces on the particle surface, water molecules are adsorbed onto the surface of the particles. Due to the action of molecular attraction, a thin film of water is formed outside the adsorbed water layer. The inner layer of the thin film water, which is closer to the particles, experiences stronger cohesive forces and is called bound water. It, together with the adsorbed water, is referred to as maximum dividable water, which enables the powder to be shaped but still lacks plasticity. The outer layer of the thin film water is closer to free water and can undergo plastic deformation under external forces. When the mineral powder is wetted by water, and the amount of water exceeds that of the thin film water, capillary water appears between the particles, initially in a contact state, connecting the particles. Further wetting leads to a honeycomb state, where the particles come closer together under the influence of water surface tension and external forces. Continued wetting results in the saturation state of capillary water, generating the strongest capillary forces between the particles.

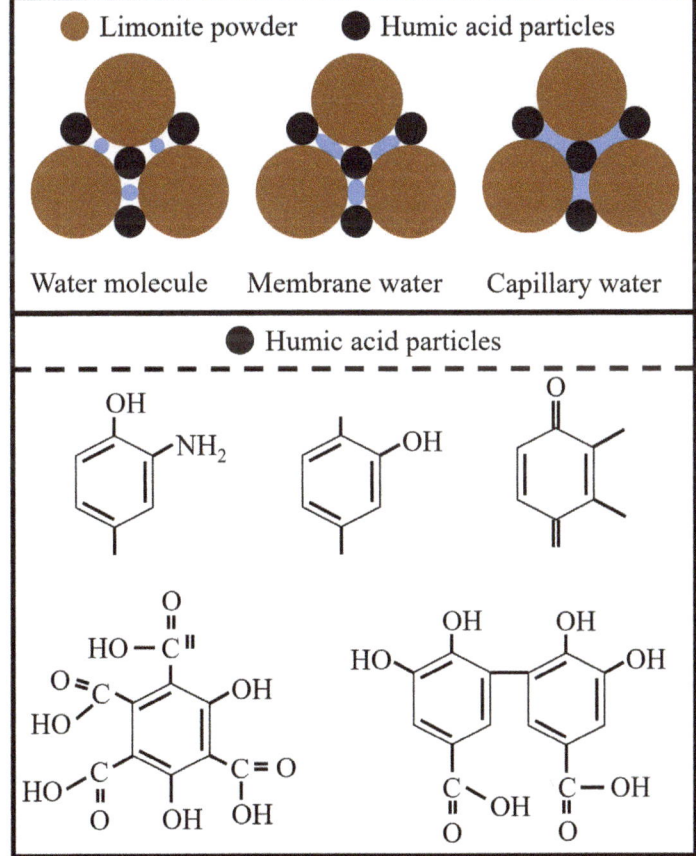

Figure 4. Pellet forming principle.

As an organic binder, humic acid not only improves the grade of metallic pellets but also accelerates reduction where possible. It exhibits high particle size at room temperature and high bonding strength after drying. The addition of a small amount of organic binder significantly enhances the compressive strength of the pellets. This is because humic acid contains a considerable amount of carboxylate ions, hydroxyl groups, and other oxygen-containing functional groups, indicating its strong hydrophilicity. Moreover, carboxyl groups can form complex or chelation reactions with metal ions and metal hydroxides, facilitating chemical adsorption between humic acid and the surface of iron ore particles, resulting in strong binding forces and improved pellet strength.

3.2. Effect of Humic Acid on Compressive Strength of Limonite Roasted Pellets

As shown in Figure 5A, the influence of humic acid content, roasted pellet weight, and diameter on the compressive strength of roasted limonite pellets is depicted. The size of the pellets is represented by the size of purple spheres, and the data for the four parameters have been normalized. The compressive strength of the pellets increases with increasing weight, with larger pellets mainly distributed towards the upper end and smaller pellets towards the lower end. With the increase in humic acid content, the compressive strength of the roasted pellets decreases. Figure 5B,C presents scatter plots and normal distribution plots illustrating the effect of different proportions of humic acid on the compressive strength of the roasted pellets. The humic acid content has a significant impact on the compressive strength of the roasted pellets, showing a negative correlation. The effect of different proportions of humic acid on the compressive strength of the roasted pellets is illustrated in Figure 5D; when the proportion of humic acid in the pellet ore increases from 0.4% to 2.0%, the average compressive strength of the roasted pellets decreases, reaching a minimum value of 417 N at a humic acid content of 2.0%.

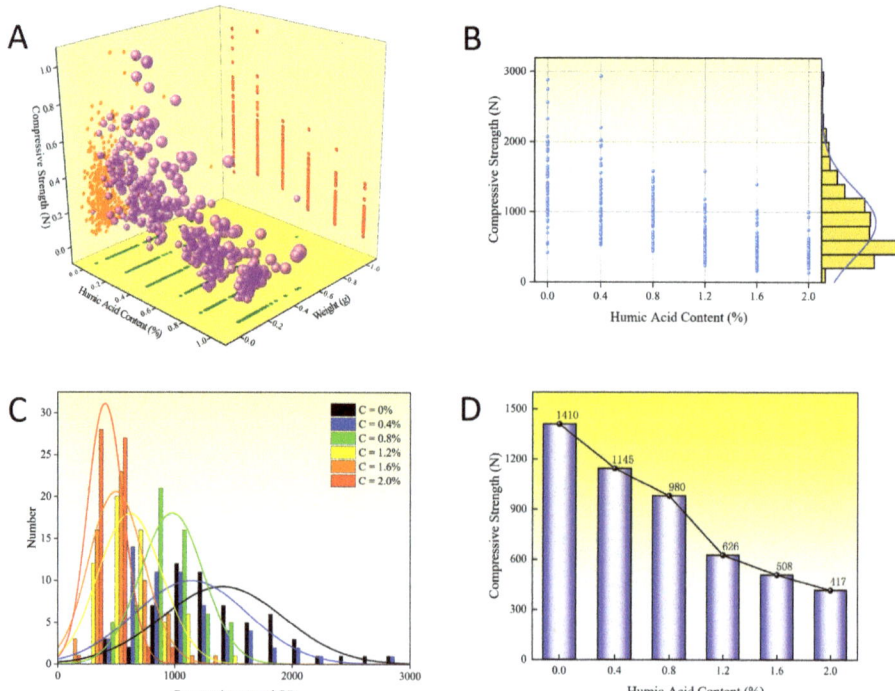

Figure 5. Effect of humic acid on compressive strength of limonite roasted pellets. (**A**) Stereo scatter data. (**B**) Planar scatter data. (**C**) Normal distribution of data. (**D**) The average of the scatter data.

The consolidation of pellet ore primarily relies on solid-state reactions. This includes the solid-state diffusion consolidation of individual-phase particles at high temperatures and the formation of compounds or solid solutions through solid diffusion in multicomponent systems. These processes generally occur below their melting temperatures without the generation of a liquid phase, enabling the consolidation of pellet ore with sufficient strength. It is important to note that the complete exclusion of a liquid phase is not necessary for pellet ore consolidation, although the presence of a liquid phase is minimal. From a consolidation principal perspective, pellet ore consolidation can occur without a liquid phase. Under microscopic observation, the liquid phase in pellet ore typically does not exceed 5%, although self-melting pellet ore may contain a higher amount of liquid phase.

As the roasting temperature increases, various physical and chemical reactions within the pellet ore are accelerated, leading to increased particle diffusion and contact area. The interparticle pores gradually become rounded and reduced. At high temperatures, two processes occur within the pellet: recrystallization and grain growth. These processes influence the microstructure of the pellet, including grain size and pore distribution, resulting in the formation of a dense sphere and improved strength of the final product. The compressive strength of the roasted pellet is significantly higher than that of the raw pellet.

When the content of humic acid in the pellet increases, the compressive strength of the roasted pellet decreases. For humic acid, excessively high solution viscosity has a negative impact on the compressive strength of the roasted pellet. A high concentration of humic acid leads to increased solution viscosity, inhibiting particle diffusion within the pellet and reducing the contact area. This hampers various reactions and also suppresses the recrystallization process within the pellet, thus impeding pellet ore consolidation. Additionally, as an organic binder, humic acid is prone to volatilization during the pellet roasting process, leaving behind voids within the pellet, further reducing its compressive strength.

3.3. Effect of Bentonite on Compressive Strength of Limonite Roasted Pellets

As shown in Figure 6A, the influence of bentonite content, roasting pellet weight, and diameter on the compressive strength of roasted limonite pellets is depicted. The size of the pellets is represented by the diameter of purple spheres, and the data for these four parameters have been normalized. The compressive strength of the roasted pellets increases with an increase in weight, with larger pellets predominantly distributed in the upper range and smaller pellets in the lower range. On the other hand, as the bentonite content increases, there is a downward trend observed in the compressive strength of the roasted pellets. Figure 6B,C illustrates the scatter plot and normal distribution plot, respectively, demonstrating the impact of different proportions of bentonite on the compressive strength of roasted pellets. The bentonite content indeed has an effect on the compressive strength of the roasted pellets, showing an overall negative correlation. Figure 6D presents the effects of different proportions of bentonite on the compressive strength of roasted pellets. It can be observed that as the bentonite content in the pellet ore increases from 0.4% to 2.0%, there is an overall decreasing trend in the average compressive strength of the roasted pellets. The minimum value of compressive strength is reached at a bentonite content of 1.6%, measuring 1085 N.

As shown in Figure 7, a comparative illustration of the influence of humic acid and bentonite on the compressive strength of roasted limonite pellets is presented. The compressive strength data is arranged in descending order on both sides, with a total of 360 data points. It can be observed that the compressive strength of the pellets with bentonite is generally higher than that of the pellets with humic acid, indicating that the compressive strength of the pellets produced using humic acid as a binder is inferior to those containing bentonite after roasting. In comparison to the bentonite pellets, the particles in the humic acid pellets are more dispersed and smaller in size. To form a connected crystalline structure, it is essential for the particles to come into contact with each other, which suggests that the particle contact in the humic acid pellets is not as close as in the bentonite pellets. Additionally, the generation of a liquid phase by low-melting substances facilitates solid-state diffusion and promotes crystal growth.

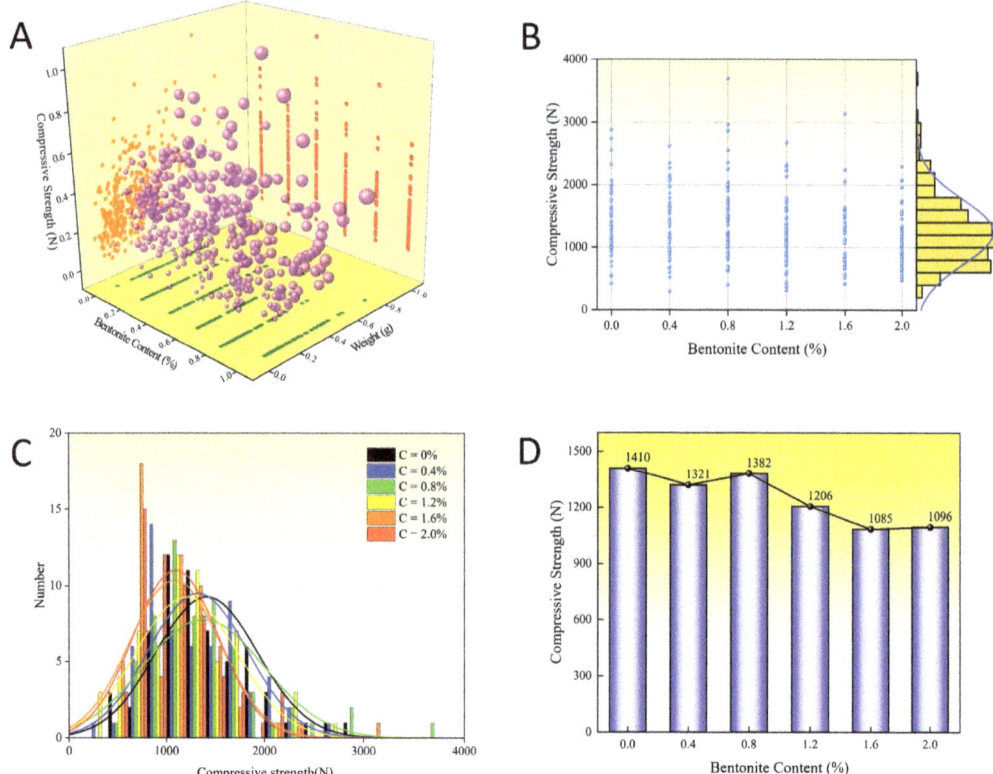

Figure 6. Effect of bentonite on compressive strength of limonite roasted pellets. (**A**) Stereo scatter data. (**B**) Planar scatter data. (**C**) Normal distribution of data. (**D**) The average of the scatter data.

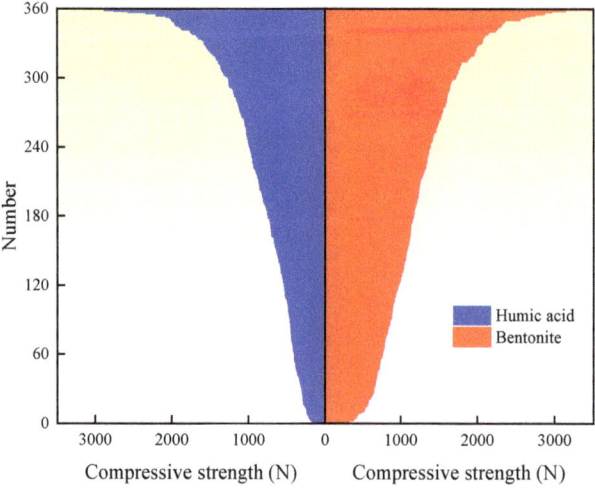

Figure 7. Comparison between humic acid pellets and bentonite pellets.

The difference in particle contact between these two types of pellets is primarily attributed to the small particle size and good dispersibility of bentonite, which fills the interstices between mineral particles in a colloidal state, thereby improving the particle size distribution of the pellet raw materials and reducing the pellet porosity. On the other hand, during the preheating process, humic acid undergoes combustion and decomposition, resulting in the formation of micro-pores in the pellets. Furthermore, the mechanisms of bentonite and humic acid differ. In the pellet formation process, the main binding forces between particles are interfacial forces and capillary forces. When the voids are completely filled with liquid, capillary forces play a major role in particle bonding. Bentonite fills the interstices between particles, reducing the capillary diameter within the green pellets and increasing capillary forces. It also enhances the molecular bonding forces between particles, resulting in closer particle arrangement. On the other hand, humic acid dissolves in water, exhibiting high viscosity and forming a network structure within the pellets to bind the particles together. However, due to the high viscosity, the liquid is difficult to be expelled from the capillaries, resulting in high moisture content within the green pellets and less compact particle arrangement compared to the bentonite pellets.

3.4. Predicting the Compressive Strength of Indurated Pellets Using an Artificial Neural Network Model

The Artificial Neural Network (ANN) is a computational model inspired by the biological neural system. It consists of a large number of interconnected artificial neurons (neuron models), resembling the connections between neurons in the human brain. Each artificial neuron receives multiple inputs, undergoes weighted processing, and generates output through an activation function. The training process of ANN resembles human learning. By iteratively adjusting connection weights, ANN can learn and extract features from input data, enabling tasks such as pattern recognition, data classification, and function approximation. It exhibits high parallel processing capability and adaptability, capable of handling complex nonlinear relationships and large-scale datasets.

In this study, the experimental investigation focused on the impact of binder content, pellet diameter, and pellet weight on compressive strength. The ANN model (Figure 8) design utilized binder content, pellet diameter, and pellet weight as input layers and compressive strength as the output layer. The experimental dataset was normalized and randomized using Excel's random function. Subsequently, the dataset was divided into two parts: a training dataset (top 80%) and a testing dataset (bottom 20%). The model's performance was evaluated using the correlation coefficient (R^2). The number of epochs is set to 1000.

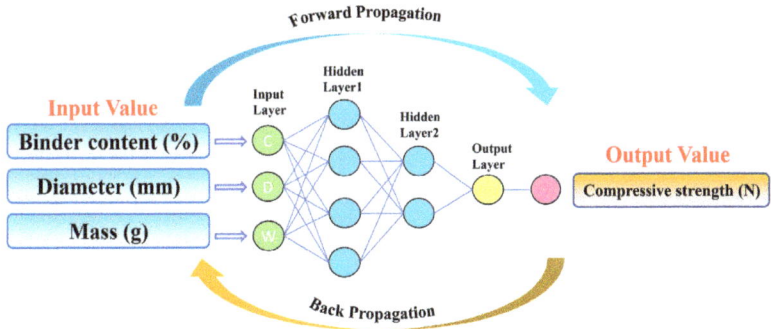

Figure 8. Artificial neural network model.

As shown in Figure 9, a scatter plot illustrates the correlation between the target values (experimental results) and the predicted values for the compressive strength of two types of indurated pellets, namely, humic acid-based and bentonite-based pellets. Linear regression analysis was performed on the data, yielding the linear regression equations and R-squared values, which were found to be 0.59095 and 0.30088 for humic acid-based and bentonite-based pellets, respectively.

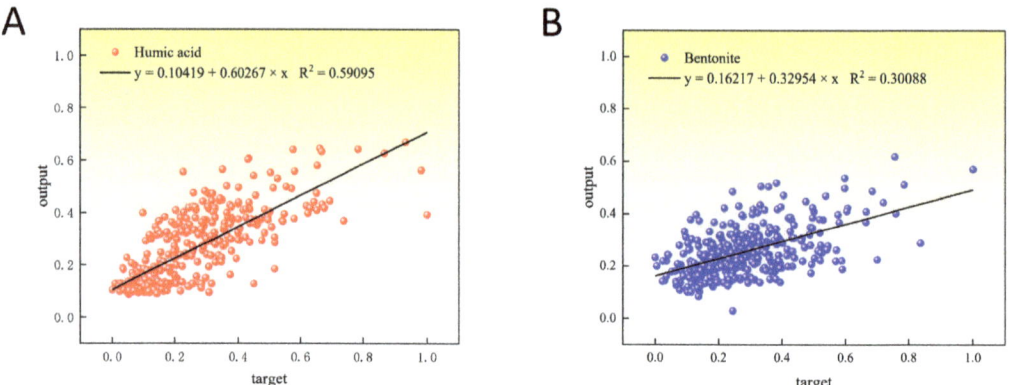

Figure 9. Data regression graph using artificial neural network. (**A**) Humic acid. (**B**) Bentonite.

Figure 10 depicts the scatter plot of target values (experimental results) versus predicted values for the compressive strength of the indurated pellets. Each target value corresponds to an output value, and all target values are arranged in ascending order. Smaller distances between target values and output values indicate higher predictive accuracy and smaller errors. The plot reveals a consistent trend among the 360 target values and their corresponding output values.

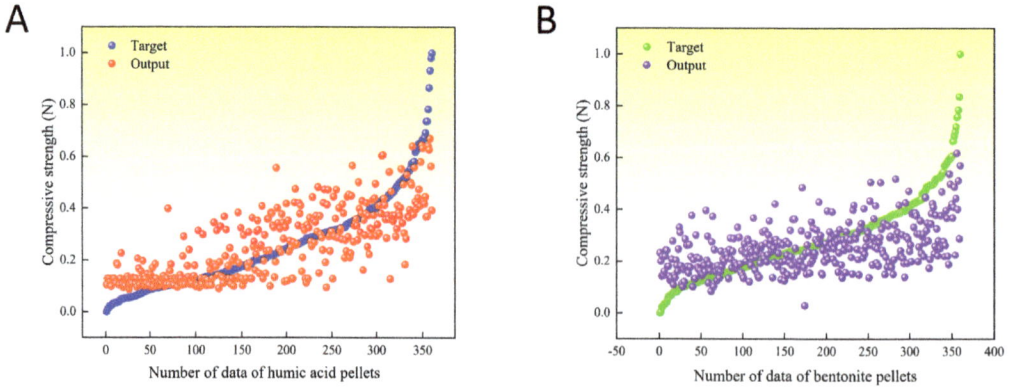

Figure 10. Scatter plot of target and output values. (**A**) Humic acid. (**B**) Bentonite.

Furthermore, Figure 11 displays the absolute errors (target value − output value) between the target values (experimental results) and predicted values for the compressive strength of the indurated pellets. The errors follow a normal distribution pattern centered around zero. These findings demonstrate that the artificial neural network model designed in this study provides highly accurate predictions for the compressive strength of humic acid-based indurated limonite pellets. Importantly, the model successfully establishes the relationship between input variables (binder content, pellet diameter, and pellet weight) and the output variable (compressive strength).

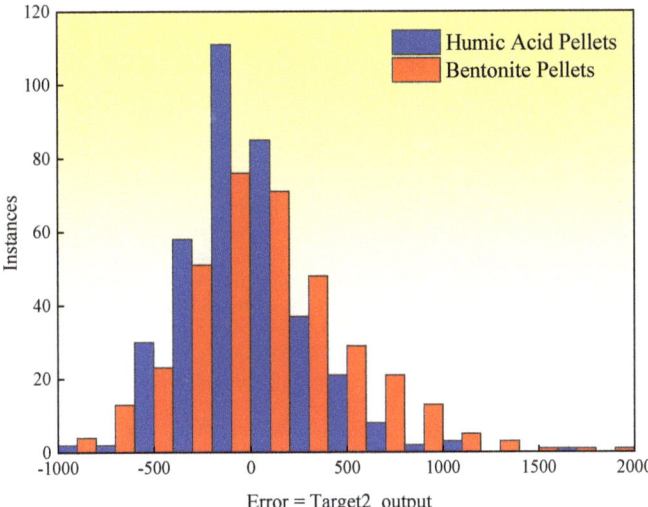

Figure 11. Error histogram.

4. Conclusions

Contributing to the reduction of industrial exhaust emissions and environmental protection, this study investigates the impact of humic acid on the compressive strength of limonite pellets, aiming to elucidate the physical properties of these pellets. The results demonstrate that as the content of humic acid increases, the compressive strength of raw limonite pellets shows an upward trend, with an increase from 15.7 N/pellet to 18.7 N/pellet. Conversely, the compressive strength of roasted limonite pellets exhibits a downward trend, decreasing from 1410 N/pellet to 417 N/pellet. Similarly, with an increase in bentonite content, the compressive strength of roasted limonite pellets shows a declining trend, decreasing from 1410 N/pellet to 1096 N/pellet. The lower strength of humic acid-roasted pellets compared to bentonite-roasted pellets can be attributed to the fewer contact points between particles within the pellets, higher porosity, lower probability of particle reactions, and fewer low-melting-point substances acting as binders within the pellets. The R^2 values for humic acid-roasted pellets and bentonite-roasted pellets are 0.59095 and 0.30088, respectively. The 360 experimental values align with the predicted values, and their absolute errors (target values–output values) exhibit a normal distribution centered around zero. These results indicate that the artificial neural network model designed in this study achieves high accuracy in predicting the compressive strength of humic acid-roasted limonite pellets.

Author Contributions: Conceptualization, X.Z.; Methodology, H.Y.; Software, H.Y.; Formal analysis, H.Y.; Investigation, S.L.; Resources, L.G., H.F., Y.W. and H.J.; Data curation, H.Y.; Writing—original draft, H.Y.; Visualization, L.G.; Supervision, X.Z. and L.G.; Funding acquisition, X.Z. All authors have read and agreed to the published version of the manuscript.

Funding: This research was funded by General Project of Yunnan Science and Technology Department (grant number KKS0202152010, 202101AT070083).

Institutional Review Board Statement: Not applicable.

Informed Consent Statement: Not applicable.

Data Availability Statement: Not applicable.

Acknowledgments: Yunnan Provincial Department of Science and Technology Fund Project for their support of this research on the mechanism of enhancing the performance of composite pellets made from limonite (Project No. KKS0202152010, 202101AT070083).

Conflicts of Interest: The authors declare no conflict of interest.

References

1. Han, Y.; Huo, Q. Research on Hebei Province's Iron and Steel Industry Transformation from the Perspective of Low-Carbon Economy. *IOP Conf. Ser. Earth Environ. Sci.* **2021**, *791*, 12187. [CrossRef]
2. Wang, R.; Jiang, L.; Wang, Y.; Roskilly, A. Energy saving technologies and mass-thermal network optimization for decarbonized iron and steel industry: A review. *J. Clean. Prod.* **2020**, *274*, 122997. [CrossRef]
3. Kolbe, N.; Cesário, F.; Ahrenhold, F.; Suer, J.; Oles, M. Carbon Utilization Combined with Carbon Direct Avoidance for Climate Neutrality in Steel Manufacturing. *Chem. Ing. Tech.* **2022**, *94*, 1548–1552. [CrossRef]
4. Bi, K.; Yu, B.; Guo, B.; Bi, Y. On the Basis of Computer Analysis of the Current Situation of Energy Conservation and Consumption Reduction in China's Iron and Steel Industry and Research Measures. *J. Physics Conf. Ser.* **2021**, *1992*, 022172. [CrossRef]
5. Gustafsson, G.; Häggblad, H.Å.; Jonsén, P.; Nishida, M. High-rate behaviour of iron ore pellet. *EPJ Web Conf.* **2015**, *94*, 05003. [CrossRef]
6. Fang, H.; Zhou, X.; Gao, L.; Yan, H.; Wang, Y.; Ji, H. Characteristic and kinetic study of the hot air-drying process of artificial limonite pellets. *J. Taiwan Inst. Chem. Eng.* **2023**, *147*, 104925. [CrossRef]
7. Prasad, J.; Venkatesh, A.S.; Sahoo, P.R.; Singh, S.; Kanouo, N.S. Geological controls on high-grade iron ores from kiriburu-meghahatuburu iron ore deposit, Singhbhum-Orissa Craton, Eastern India. *Minerals* **2017**, *7*, 197. [CrossRef]
8. Du, J.; Gao, L.; Yang, Y.; Guo, S.; Chen, J.; Omran, M.; Chen, G. Modeling and kinetics study of microwave heat drying of low-grade manganese ore. *Adv. Powder Technol.* **2020**, *31*, 2901–2911. [CrossRef]
9. Lin, S.; Li, K.; Yang, Y.; Gao, L.; Omran, M.; Guo, S.; Chen, J.; Chen, G. Microwave-assisted method investigation for the selective and enhanced leaching of manganese from low-grade pyrolusite using pyrite as the reducing agent. *Chem. Eng. Process. Process Intensif.* **2021**, *159*, 108209. [CrossRef]
10. Chen, G.; Ling, Y.; Li, Q.; Zheng, H.; Qi, J.; Li, K.; Chen, J.; Peng, J.; Gao, L.; Omran, M.; et al. Investigation on microwave carbothermal reduction behavior of low-grade pyrolusite. *J. Mater. Res. Technol.* **2020**, *9*, 7862–7869. [CrossRef]
11. Li, K.; Jiang, Q.; Chen, G.; Gao, L.; Peng, J.; Chen, Q.; Koppala, S.; Omran, M.; Chen, J. Kinetics characteristics and microwave reduction behavior of walnut shell-pyrolusite blends. *Bioresour. Technol.* **2021**, *319*, 124172. [CrossRef]
12. Lin, S.; Gao, L.; Yang, Y.; Chen, J.; Guo, S.; Omran, M.; Chen, G. Dielectric properties and high temperature thermochemical properties of the pyrolusite-pyrite mixture during reduction roasting. *J. Mater. Res. Technol.* **2020**, *9*, 13128–13136. [CrossRef]
13. Zhang, H.; Bai, K.; Liu, W.; Chen, Y.; Yuan, Y.; Zuo, H.; Wang, J. Effect of Magnetite Concentrate Particle Size on Pellet Oxidation Roasting Process and Compressive Strength. *ISIJ Int.* **2022**, *62*, 1792–1801. [CrossRef]
14. Jovanović, V.D.; Knežević, D.N.; Sekulić, Ž.T.; Kragović, M.M.; Stojanović, J.N.; Mihajlović, S.R.; Nišić, D.D.; Radulović, D.S.; Ivošević, B.B.; Petrov, M.M. Effects of bentonite binder dosage on the properties of green limestone pellets. *Hem. Ind.* **2017**, *71*, 135–144. [CrossRef]
15. Bhuiyan, I.U.; Mouzon, J.; Schröppel, B.; Kaech, A.; Dobryden, I.; Forsmo, S.P.; Hedlund, J. Microstructure of bentonite in iron ore green pellets. *Microsc. Microanal.* **2014**, *20*, 33–41. [CrossRef]
16. Qiu, G.; Jiang, T.; Huang, Z.; Zhu, D.; Fan, X. Characterization of preparing cold bonded pellets for direct reduction using an organic binder. *ISIJ Int.* **2003**, *43*, 20–25. [CrossRef]
17. Li, G.; Zhang, Y.; Zhang, X.; Meng, F.; Cao, P.; Yi, L. Effect of Humic Acid Binder on the Preparation of Oxidized Pellets from Vanadium-Bearing Titanomagnetite Concentrate. *Sustainability* **2023**, *15*, 6454. [CrossRef]
18. Zhao, H.X.; Zhou, F.S.; Bao, X.C.; Zhou, S.; Wei, Z.; Long, W.J.; Yi, Z. A review on the humic substances in pelletizing binders: Preparation, interaction mechanism, and process characteristics. *ISIJ Int.* **2023**, *63*, 205–215. [CrossRef]
19. Prasad, R.; Soren, S.; Kumaraswamidhas, L.A.; Pandey, C.; Pan, S.K. Experimental Investigation of Different Fineness and Firing Temperatures on Pellets Properties of Different Iron Ore fines from Indian Mines. *Materials* **2022**, *15*, 4220. [CrossRef] [PubMed]
20. Gao, Q.; Zhang, Y.; Jiang, X.; Zheng, H.; Shen, F. Prediction model of iron ore pellet ambient strength and sensitivity analysis on the influence factors. *Metals* **2018**, *8*, 593. [CrossRef]
21. Davraz, M.; Kilinçarslan, Ş.; Ceylan, H. Predicting the poisson ratio of lightweight concretes using artificial neural network. *Acta Phys. Pol. A* **2015**, *128*, B-184. [CrossRef]
22. Sitek, W.; Trzaska, J. Practical aspects of the design and use of the artificial neural networks in materials engineering. *Metals* **2021**, *11*, 1832. [CrossRef]
23. Yu, H.; Zheng, J.; Lin, Q. Strength prediction of seawater sea sand concrete based on artificial neural network in python. *Mater. Res. Express* **2022**, *9*, 035201. [CrossRef]
24. Kiyohara, S.; Tsubaki, M.; Mizoguchi, T. Prediction of ELNES and Quantification of Structural Properties Using Artificial Neural Network. *Microsc. Microanal.* **2020**, *26* (Suppl. 2), 2100–2101. [CrossRef]
25. Chagas, M.; Machado, M.L.P.; Souza, J.B.C.; Frigini, E.F.D.J. Use of an Artificial Neural Network in determination of iron ore pellet bed permeability. *REM-Int. Eng. J.* **2017**, *70*, 187–191. [CrossRef]

26. Dwarapudi, S.; Rao, S.M. Prediction of iron ore pellet strength using artificial neural network model. *ISIJ Int.* **2007**, *47*, 67–72. [CrossRef]
27. Klippel, E.; Bianchi, A.G.C.; Delabrida, S.; Silva, M.C.; Garrocho, C.T.B.; Moreira, V.d.S.; Oliveira, R.A.R. Deep learning approach at the edge to detect iron ore type. *Sensors* **2021**, *22*, 169. [CrossRef]
28. Fan, X.; Li, Y.; Chen, X. Prediction of iron ore sintering characters on the basis of regression analysis and artificial neural network. *Energy Procedia* **2012**, *16*, 769–776. [CrossRef]
29. Golmohammadi, H.; Rashidi, A.; Safdari, J.S. Prediction of ferric iron precipitation in bioleaching process using partial least squares and artificial neural network. *Chem. Ind. Chem. Eng. Q.* **2013**, *19*, 321–331. [CrossRef]
30. Guo, L.; Peng, Y.; Qin, R.; Liu, B. A new weakly supervised learning approach for real-time iron ore feed load estimation. *Expert Syst. Appl.* **2022**, *202*, 117469. [CrossRef]
31. Li, D.; Moghaddam, M.R.; Monjezi, M.; Armaghani, D.J.; Mehrdanesh, A. Development of a group method of data handling technique to forecast iron ore price. *Appl. Sci.* **2020**, *10*, 2364. [CrossRef]
32. Wang, S.-H.; Li, H.-F.; Zhang, Y.-J.; Zou, Z.-S. A hybrid ensemble model based on ELM and improved AdaBoost. RT algorithm for predicting the iron ore sintering characters. *Comput. Intell. Neurosci.* **2019**, *2019*, 4164296. [CrossRef]
33. Mao, Y.; Liu, C.; Xiao, D.; Wang, J.; Le, B.T. Study of the magnetic properties of haematite based on spectroscopy and the IPSO-ELM neural network. *J. Sens.* **2018**, *2018*, 6357905. [CrossRef]
34. Yang, Y.; Hao, X.; Zhang, L.; Ren, L. Application of scikit and keras libraries for the classification of iron ore data acquired by laser-induced breakdown spectroscopy (LIBS). *Sensors* **2020**, *20*, 1393. [CrossRef] [PubMed]
35. Tunckaya, Y.; Köklükaya, E. Comparative performance evaluation of blast furnace flame temperature prediction using artificial intelligence and statistical methods. *Turk. J. Electr. Eng. Comput. Sci.* **2016**, *24*, 1163–1175. [CrossRef]

Disclaimer/Publisher's Note: The statements, opinions and data contained in all publications are solely those of the individual author(s) and contributor(s) and not of MDPI and/or the editor(s). MDPI and/or the editor(s) disclaim responsibility for any injury to people or property resulting from any ideas, methods, instructions or products referred to in the content.

Article

Recovery of Zinc from Metallurgical Slag and Dust by Ammonium Acetate Using Response Surface Methodology

Xuemei Zheng [1,2], Jinjing Li [1], Aiyuan Ma [1,*] and Bingguo Liu [2,*]

[1] School of Chemistry and Materials Engineering, Liupanshui Normal University, Liupanshui 553004, China; zxm_lpssy19@163.com (X.Z.); 19985189154@163.com (J.L.)
[2] Key Laboratory of Unconventional Metallurgy, Faculty of Metallurgical and Energy Engineering, Kunming University of Science and Technology, Kunming 650093, China
* Correspondence: may_kmust11@163.com (A.M.); bingoliu@126.com (B.L.)

Abstract: Metallurgical slag and dust (MSD) are abundant Zn-containing secondary resources that can partially alleviate the shortage of zinc minerals, with hazardous characteristics and a high recycling value. In this work, the process conditions of recycling Zn from MSD materials leaching by ammonium acetate (NH_3-CH_3COONH_4-H_2O) were optimised using response surface methodology (RSM). The influences of liquid/solid ratio, stirring speed, leaching time, total ammonia concentration, and the interactions between these variables on the Zn effective extraction rate during the ammonium acetate leaching process were investigated. Additionally, the predicted regression equation between the Zn effective extraction rate and the four affecting factors was established, and the optimal process parameters were determined with a stirring speed of 345 r/min, leaching temperature of 25 °C, $[NH_3]/[NH_4]^+$ of 1:1, total ammonia concentration of 4.8 mol/L, liquid/solid ratio of 4.3:1, and leaching time of 46 min. The Zn effective extraction rates predicted by the proposed model and the measured values were 85.25% and 84.67%, respectively, with a relative error of 0.58% between the two values, indicating the accuracy and reliability of the proposed model. XRD and SEM-EDS analysis results showed that Zn_2SiO_4, ZnS, and $ZnFe_2O_4$ were among the main factors affecting the low extraction rate of zinc from metallurgical slag dust. This work established a new technology prototype for the effective and clean extraction of zinc resources, which can provide new routes to effectively utilise Zn-containing MSD materials and lay a foundation for developing other novel techniques for recycling Zn from Zn-containing secondary resources.

Keywords: metallurgical slag and dust; zinc recovery; ammonium acetate; response surface methodology

1. Introduction

Zinc (Zn) is an essential metal in modern life, and its main mineral source is sphalerite (ZnS). Zinc has a low melting point, good melt fluidity, and is easy to die-cast; hence, it is frequently utilised to produce precision castings [1]. Zinc is a negatively charged metal, and its electrode potential is more negative than iron (Fe), with the result that zinc can be corroded instead of iron through electrochemical action [2,3]. Therefore, zinc is also widely applied as battery anode materials and plated steel materials. With the increasing demand for galvanised materials in the battery industry, automobile industry, and construction industry, the mining volume of sphalerite has increased, and the ore grade has declined year by year [4]. The ever-increasing demand for galvanised materials and the scarcity of zinc resources have forced the development and utilisation of other zinc-containing hazardous waste resources, such as metallurgical slag and dust (MSD).

MSD materials source from the smelting process of steel, zinc (Zn), lead (Pb), and copper (Cu), etc., and their production, is rising sharply [5–7]. As hazardous materials, MSD materials are rich in heavy metals like Zn, Cu, and Fe, as well as toxic components such as As, Hg, Pb, Cr, and Cd [8,9]. Hence, simple burial and stockpile disposal for MSD materials are inadvisable. In addition, once the content of zinc and other elements in metallurgical slag exceeds the specified value, it will damage the refractory materials of the furnace cavity, further shortening the service life of smelting furnace, and cause the productivity of smelting furnace to decrease, as well as leading to operation difficulties [10,11]. Therefore, the research and development on the clean, efficient, and economical approaches for recycling MSD materials have good environmental and practical significance; moreover, the relevant metallurgical industry will gain considerable added value and economic benefits based on the developed approaches.

Metallurgical slag and dust have a high recycling value and contain diverse conventional smelting metals like Zn, Fe, and Pb, and precious metals like Ag, Au, and in [12,13]. However, MSD materials have diverse compositions, wherein Zn mainly presents as ZnO, frantzite, and zinc silicate; Fe mainly presents as Fe_3O_4 and frankite; and Ca mainly presents as $CaCO_3$. Further, Fe, Zn, and Ca also exist in the form of silicate [14–16]. Furthermore, the structures of MSD materials are complicated, with different metal oxides, chlorides, carbon-containing compounds, and gangues doped and wrapped together. The diverse compositions and complicated structures of MSD materials make the recycling process difficult [17,18]. At present, the widely applied method for zinc extraction is a hydrometallurgy leaching approach with sulfuric acid as the leaching agent, which consumes less energy than pyrometallurgy methods [19]. However, Zn-containing MSD contains zinc by-products with various impurities, including Fe (up to 14%), Ca (up to 19%), Cl (up to 12%), and F (up to 2%) [20,21]. Moreover, the high-content components like Fe, Cl, Ca, and gangues in the zinc-containing MSD materials, consume excessive acid and complicate the purification process become [22]. The application of zinc extraction by conventional acid methods is limited by the disadvantages of long purification time, large acid consumption, high energy consumption, low-quality electrolytic zinc, and low recovery rate [23–25]. Therefore, alkaline leaching is gradually proposed for zinc extraction. In the literature, applications of ammonium salts such as ammonium chloride, ammonium sulphate, and ammonium bicarbonate have been reported on the Zn extraction from single-phase zinc-containing minerals (e.g., ZnO, smithsonite, and hydrozincite), and the effects of zinc extraction are sound [26–28]. The essence of the alkaline leaching method is that ammonia compounds can form tetraammine zinc ion ($[Zn(NH_3)_4]^{2+}$) coordination compounds with zinc metal (Zn^{2+}) ions to prevent impurities containing Fe, Al, and Si from entering the leaching solution, thereby achieving the effective separation of Zn and impurities [29,30]. Additionally, Rao et al. [31] highlighted that the zinc effective extraction rate leaching using a mixed solution of NH_4Cl-NH_3-NTA was higher than that using single NH_4Cl-NH_3 solution leaching. The addition of nitrosotriacetic acid ($N(CH_2COOH)_3$, i.e., NTA) promotes the transformation of $[Zn(NH_3)_4]^{2+}$ and $[Zn(NTA)_2]^{4-}$ coordination compounds into more stable $[Zn(NTA)(NH_3)_2]^{-}$ coordination compounds, thereby effectively improving the zinc effective leaching rate [31]. Therefore, ammonium ion (NH_4^+) and carboxylate anion ($RCOO^-$) play crucial and complementary roles in the zinc leaching process. Compared with a single ammonia leaching method, the leaching solution mixed with NH_4^+ ion and $RCOO^-$ anion contributes to extracting zinc from MSD materials more efficiently. Therefore, the leaching solution mixed with NH_4^+ ions and $RCOO^-$ anions can be considered for introduction into the leaching process of zinc-containing metallurgical slag and dust with complex structures and diverse components.

In this work, response surface methodology (RSM) [32–34] was introduced into the process optimisation of recovering Zn from Zn-containing MSD materials by coordination leaching using NH_3-CH_3COONH_4-H_2O solution. The influences of liquid/solid ratio, stirring speed, leaching time, total ammonia concentration, and the interactions between them on the Zn effective extraction rate were explored using the central composite design (CCD) of RSM, and a mathematic model of the factors affecting the Zn effective extraction rate was established. Additionally, the fitting analysis, confidence analysis, and variance analysis of the regression equation of the proposed model; the linear correlation between the experimental and predicted values of the Zn effective extraction rate; and the normal probability characteristics of residuals for the Zn effective extraction rate were systematically investigated to confirm the credibility and accuracy of the proposed model. Moreover, the accuracy of the optimisation parameters obtained by RSM was verified through comparing the predicted value and the average value of the Zn effective extraction rate, as determined by three parallel experiments.

2. Materials and Methods

2.1. Chemical Composition of MSD Materials

The metallurgical slag and dust (MSD) materials studied in this work were drawn from a local enterprise located in Yunnan province (Qujing, China), which is mainly engaged in the recovery and utilization of zinc secondary resources. The MSD sample is a mixture of various MSD materials. After completely drying at 85 °C until no further mass loss was observed, the composition analysis of the MSD sample was determined by the ICP method using Agilent 5110 (OES, Agilent Technologies, Santa Clara, CA, USA), and the analytical results are displayed in Table 1. As determined in Table 1, the MSD sample has a complex composition: the contents of Zn, Fe, Ca, and Cl were high, gangue, and scattered In were observed. Therefore, the MSD sample exhibits a high recycling value.

Table 1. Chemical compositions of the MSD sample.

Compositions Mass (w%)	Zn 24.74	Fe 21.66	C 9.14	Si 2.66	Cl 2.94	S 1.39
Compositions Mass (w%)	Mg 1.14	Bi 0.97	Pb 1.13	In (g/t) 354		

2.2. Particle Size Distribution of MSD Materials

Figure 1 shows the particle size distribution of the MSD samples. Table 2 presents the particle size values at the particle size level of D10, D50, D90, and D98 of the MSD samples, together with the volume average particle size, area average particle size, and corresponding surface area to volume ratio for the MSD sample. Table 2 shows that 90% of MSD samples have a particle size of 24.85 μm.

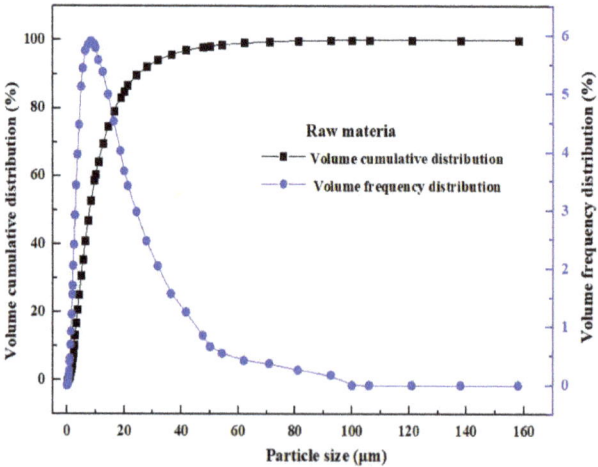

Figure 1. Particle size distribution of the MSD sample.

Table 2. Particle size distribution parameters of the MSD sample.

D_{10} (μm)	D_{50} (μm)	D_{90} (μm)	D_{98} (μm)	Volume Average Particle Size (μm)	Area Average Particle Size (μm)	Surface Area to Volume Ratio (m²/cm³)
2.47	7.89	24.85	47.91	10.78	5.59	6.42

2.3. Experimental Design of Response Surfaces and Leaching Experimental Methods

2.3.1. Experimental Design of Response Surfaces

In this study, the stirring speed (X_1, r/min), leaching time (X_2, min), total ammonia concentration (X_3, mol/L), and liquid/solid ratio (X_4, mL/g) were selected as the variables in the leaching experiments. The Zn effective extraction rate by the coordination leaching process using NH_3-CH_3COONH_4-H_2O solution was defined as the response value (Y, %). The leaching experimental conditions were provided with the molar ratio of $[NH_3]/[NH_4]^+$ as 1:1, and the leaching temperature as 25 °C.

Table 3 summarises the codes of the centre combination design (CCD) optimisation based on RSM, with four factors and three levels. In the centre combination design (CCD), each factor has 2^4 sufficient factorials, including eight factorial points, eight axial points, and six repeated centre points. The full-factor centre was optimised over 30 experiments to identify the optimised values of the dependent variable and the independent variables. The number (i.e., 30) of experiments was calculated using the following formula:

$$N = 2^n + 2n + n_c = 2^4 + 2 \times 4 + 6 = 30 \tag{1}$$

where N denotes the number of the needed experiments, n presents the number of factors, and n_c indicates the number of repeated centre points.

Table 3. Response surface method factors' level coding.

Factors	Levels		
	−1	0	1
Stirring speed (X_1, r/min)	250	300	350
Leaching time (X_2, min)	20	35	50
Total ammonia concentration (X_3, mol/L)	3	4	5
Liquid/solid ratio (X_4, mL/g)	3	4	5

During response surface optimisation, the model accuracy was verified before data analysis. The accuracy of the proposed model was investigated using Design Expert (STAT-EASE, Stat Ease Inc., Minneapolis, MN, USA). The Zn effective extraction rate (Y, %) was the dependent variable, and the stirring speed (X_1, r/min), leaching time (X_2, min), total ammonia concentration (X_3, mol/L), and liquid/solid ratio (X_4, mL/g) were selected as the variables.

2.3.2. Leaching Experimental Methods

Before the XRD and SEM analysis, the MSD samples were dried at 85 °C until no further mass loss was observed. Then, 20.00 g of the dried MSD material was sampled and mixed with the freshly prepared ammonium acetate (NH_3-CH_3COONH_4-H_2O) leaching agent in a 300 mL conical flask with stirring. The detailed experimental parameters for leaching zinc from the MSD sample were as follows: the leaching time was set from 5 min to 65 min; the stirring speed was controlled between 200 r/min and 400 r/min; the total ammonia concentration was set between 2 mol/L and 6 mol/L; the liquid/solid ratio was adjustable from 2 mL/g to 6 mL/g, and the leaching temperature was 25 °C. This leaching process was performed in a thermostatic water bath. After leaching, solids and liquids were separated, and the Zn concentration in the leaching solution was measured by EDTA titration. The Zn effective extraction rate (η_{Zn}, %) was determined by Equation (2):

$$\eta_{Zn} = \frac{C_{Zn} \times V}{m \times \omega_{Zn}} \times 100\% \qquad (2)$$

In Equation (1), C_{Zn} indicates the Zn concentration in the leaching solution, g/L; V denotes the volume of the leaching solution, L; m presents the MSD sample mass, g; and w_{Zn} denotes the mass percent of zinc in the MSD sample, 24.74%.

After obtaining the Zn effective extraction rates under different leaching conditions using the above calculation, response surface methodology (RSM) was introduced into the process optimisation of recycling Zn from Zn-containing MSD materials by coordination leaching using NH_3-CH_3COONH_4-H_2O solution.

2.4. Leaching Reaction of Zinc Extraction Process

For the leaching system of ZnO-NH_3-CH_3COONH_4-H_2O, the dissolved zinc oxide (ZnO) can combine with ammonium ions (NH_4^+) and ammonia (NH_3) to form soluble $[Zn(NH_3)_i]^{2+}$ complexes, as shown in Equations (3) and (4). In addition, zinc ions can combine with carboxylate anion ($RCOO^-$) to form stability complexes, as shown in Equation (5).

$$ZnO + iNH_4^+ = [Zn(NH_3)_i]^{2+} + H_2O + (i-2)H^+ \qquad (3)$$

$$ZnO + iNH_3 + H_2O = Zn(NH_3)_i^{2+} + 2OH^- \qquad (4)$$

$$2RCOO^- + ZnO \overset{(H_3O^+)}{\leftrightarrow} (RCOO)_2Zn + H_2O \qquad (5)$$

3. Results and Discussion

3.1. Results of the Optimization of Response Surfaces

To reduce the system errors during the NH_3-CH_3COONH_4-H_2O leaching process, the sequence of experiments was determined randomly by Design Expert (STAT-EASE, USA), and the central composite experimental design and corresponding zinc effective leaching rates are illustrated in Table 4.

Table 4. Centre composite design plan and Zn effective leaching rate results.

Number	X_1 (r/min)	X_2 (min)	X_3 (mol/L)	X_4 (L/S)	Y (%)
1	300	35	4	6	81.88
2	300	35	6	4	82.75
3	300	35	4	4	79.19
4	350	20	5	3	78.08
5	300	35	4	4	79.19
6	300	35	4	4	79.19
7	350	20	5	5	83.33
8	250	20	5	5	81.03
9	350	50	3	5	76.30
10	300	35	4	4	79.19
11	350	50	5	3	78.51
12	350	20	3	5	73.82
13	250	50	5	5	82.98
14	250	20	3	5	72.69
15	250	20	5	3	77.83
16	200	35	4	4	77.62
17	250	50	3	5	75.27
18	250	50	5	3	78.01

Table 4. Cont.

Number	X_1 (r/min)	X_2 (min)	X_3 (mol/L)	X_4 (L/S)	Y (%)
19	350	50	5	5	84.50
20	350	50	3	3	53.21
21	300	65	4	4	78.28
22	300	35	4	2	48.67
23	350	20	3	3	52.09
24	250	50	3	3	52.59
25	300	35	4	4	79.19
26	300	5	4	4	73.50
27	400	35	4	4	78.36
28	250	20	3	3	51.91
29	300	35	4	4	79.19
30	300	35	2	4	48.39

3.2. Model Verification and Statistical Analysis

The secondary polynomial regression equation of the Zn effective extraction rate by coordination leaching from the MSD sample using NH_3-CH_3COONH_4-H_2O solution was obtained using the least-squares method, as follows:

$$Y = 79.19 + 0.38X_1 + 0.84X_2 + 8.55X_3 + 7.25X_4 - 0.012X_1X_2 + 0.10X_1X_3 + 0.28X_1X_4 - 0.20X_2X_3 + 0.36X_2X_4 - 4.30X_3X_4 - 0.16X_1^2 - 0.69X_2^2 - 3.27X_3^2 - 3.34X_4^2 \quad (6)$$

The model accuracy can be further demonstrated using a variance analysis to identify the significances of all factors in the polynomial equation, as well as to judge the effectiveness of the model. Tables 5–7 indicate the fitting analysis, confidence analysis, and variance analysis for the regression equation of the proposed model, respectively.

Table 5. Model fitting parameters for the designed experiments.

	The Sequential Model Sum of Squares					
Source	Sum of Squares	df	Mean Square	F Value	p-Value Probd > F	
Mean vs. total	160,900	1	1.609×10^5			
Linear vs. mean	3036.32	4	759.08	22.16	<0.0001	
2FI vs. linear	300.53	6	50.09	1.71	0.1726	
Quadratic vs. 2FI	536.69	4	134.17	104.59	<0.0001	Suggested
Cubic vs. quadratic	15.57	8	1.95	3.71	0.0505	Aliased
Residual	3.67	7	0.52			
Total	1.647×10^5	30	5491.61			
	Lack of fit tests					
Source	Sum of squares	df	Mean square			
Linear	856.46	20	42.82			
2FI	555.93	14	39.71			
Quadratic	19.24	10	1.92		Suggested	
Cubic	3.67	2	1.84		Aliased	
Pure error	0.00	5	0.00			
	Model summary statistics					
Source	Std. Dev.	R-squared	Adjusted R-squared	Predicted R-squared	PRESS	
Linear	5.85	0.7800	0.7448	0.6832	1233.10	
2FI	5.41	0.8572	0.7820	0.7741	879.23	
Quadratic	1.13	0.9951	0.9904	0.9715	110.84	Suggested
Cubic	0.72	0.9991	0.9961	0.8641	528.85	Aliased

Table 6. Credibility analysis for Zn effective leaching rate.

Std. Dev.	1.13	R-Squared	0.9951
Mean	73.22	Adj R-Squared	0.9904
C.V.%	1.55	Pred R-Squared	0.9715
PRESS	110.84	Adeq Precision	44.8820

Table 7. Variance analysis for response surface quadratic model.

Source	Sum of Squares	df	Mean Square	F Value	p-Value (Probd > F)	Significance
Model	3873.53	14	276.68	215.67	<0.0001	significant
X_1	3.38	1	3.38	2.64	0.1252	
X_2	16.92	1	16.92	13.19	0.0025	significant
X_3	1752.92	1	1752.92	1366.39	<0.0001	significant
X_4	1263.10	1	1263.10	984.57	<0.0001	significant
$X_1 X_2$	2.26×10^{-3}	1	2.26×10^{-3}	1.76×10^{-3}	0.9671	
$X_1 X_3$	0.16	1	0.16	0.13	0.7273	
$X_1 X_4$	1.23	1	1.23	0.96	0.3437	
$X_2 X_3$	0.61	1	0.61	0.48	0.5002	
$X_2 X_4$	2.08	1	2.08	1.62	0.2222	
$X_3 X_4$	296.44	1	296.44	231.07	<0.0001	significant
X_1^2	0.72	1	0.72	0.56	0.4655	
X_2^2	12.94	1	12.94	10.09	0.0063	significant
X_3^2	292.75	1	292.75	228.20	<0.0001	significant
X_4^2	306.12	1	306.12	238.61	<0.0001	significant
Residual	19.24	15	1.28			
Lack of Fit	19.24	10	1.92			
Pure Error	0	5	0			
Cor Total	3892.78	29				

Table 5 displays the numerical analysis of the experimental response surface. During the CCD of the response surface, the value of (Prob > F) of the high-accuracy regression model should be lower than 0.05, in order to guarantee an effective simulation. Moreover, the value of (Prob > F) is required to exceed 0.05, which denotes a high fitting degree of the regression equation. As depicted in Table 4, the quadratic model had a Prob > F value of below 0.0001 and an anomalistic term of 0.0674, demonstrating that the designed model had a significant fitting effect. Hence, the quadratic model was used as the fitting model for the centre combination design (CCD).

The applicability and accuracy of the designed model are indicated by its correlation coefficient (R^2). Table 6 presents the credibility analysis results of the zinc effective leaching rate. As presented in Table 6, the R^2 value of the quadratic model was 0.9951, demonstrating that the model presented a good fit effect with the experimental data. Generally, the difference between the predicted R^2 and calibrated R^2 values should be below 0.2. In this case, the R^2_{Pred} value was 0.9715 and the R^2_{adj} value 0.9902, suggesting that the proposed model can accurately predict the experimental data. In addition, Adeq precision can reflect the signal-to-noise ratio, and Adeq precision > 4 indicates a reasonable signal-to-noise ratio. In this case, Adeq precision was determined at 44.882 (Table 6), indicating a high signal-to-noise ratio. Furthermore, according to the MYERS theory, the correlation coefficient of a model should be higher than 0.8, where the specific value indicates good fitting performance. In this study, $R^2 = 0.9951$, $R^2_{adj} = 0.9902$, and $R^2_{Pred} = 0.9715$ demonstrated a good fitting performance of the proposed model. Therefore, it can be surmised from the above analysis that the proposed model is applicable to the examined case.

Table 7 depicts the variance analysis results of the response surface quadratic model. As illustrated in Table 7, the F value of the proposed model was 215.67. The probability that the signal-to-noise ratio is exposed to error was 0.01% (Prob > F < 0.0001), indicating a high accuracy and good fitting performance of the proposed regression model. The Prob > F value of a variable less than 0.05 indicates that the variable significantly affects the response value. Among all the listed affecting factors, X_2, X_3, X_4, X_3X_4, X_2^2, X_3^2, and X_4^2 had significant effects on the Zn effective extraction rate using NH_3-CH_3COONH_4-H_2O leaching. The variance analysis demonstrated that the proposed model showed a good fit to the experimental data, and the model can accurately predict the Zn effective extraction rate achieved by coordination leaching from MSD using NH_3-CH_3COONH_4-H_2O.

Figure 2 displays the linear correlation between the predicted and experimental values of the Zn effective extraction rate. It can be concluded from Figure 2 that the experimental values were highly consistent with the predicted values. The experimental values were uniformly scattered on both sides of the predicted values, demonstrating that the quadratic model was suitable to describe the correlation of experimental factors and the Zn effective extraction rate. In other words, the proposed model accurately reflected correlations between different parameters.

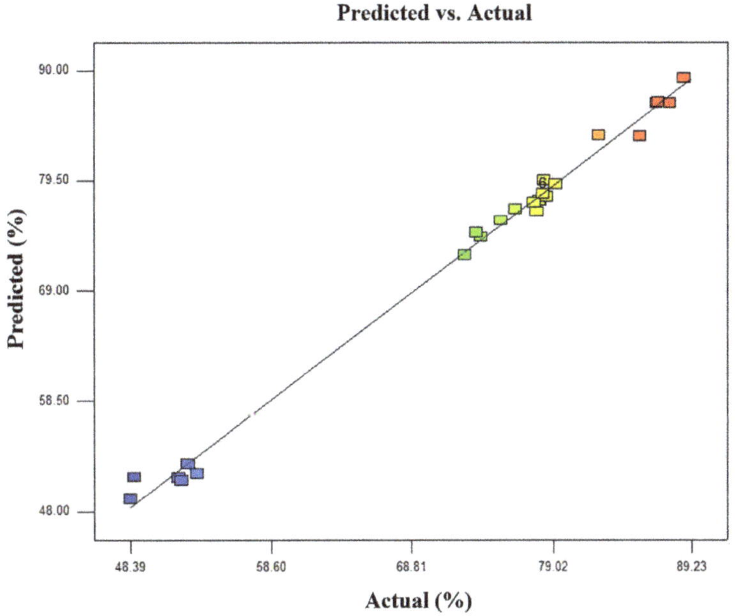

Figure 2. Linear correlation between the experimental and predicted values of Zn effective leaching rate.

Figure 3 shows the normal probability plot of the residuals for the Zn effective extraction rate. Herein, the division of normal probability on the Y-axis reflects a normal distribution of residuals. As displayed in Figure 3, the residuals of the Zn effective extraction rate are distributed along a straight line, demonstrating a normal distribution of the experimental residuals. The residuals on the X-axis reflect the differences between the experimental responses and the predicted values by the model. The residuals are concentrated in the middle region in an S-shaped curve, suggesting a good accuracy of the proposed model.

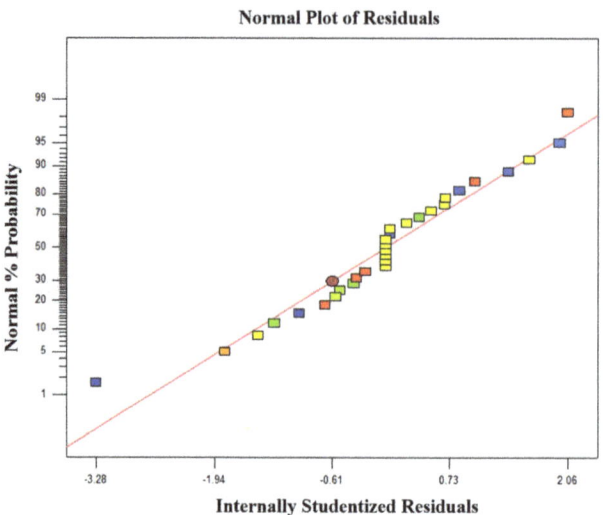

Figure 3. Normal probability plot of studentized Zn effective extraction rate.

3.3. Analysis of the Response Surface Model

Based on the above regression analysis and variance analysis, the effects of different factors on the Zn effective extraction rate were investigated, by establishing 3D response surfaces of the regression model based on statistical calculations of the regression coefficients. Based on the optimised model, the response surfaces of the influences of stirring speed, leaching time, total ammonia concentration, and liquid/solid ratio, as well as interactions between these factors on the Zn effective extraction rate, were obtained, and the results are plotted in Figure 4.

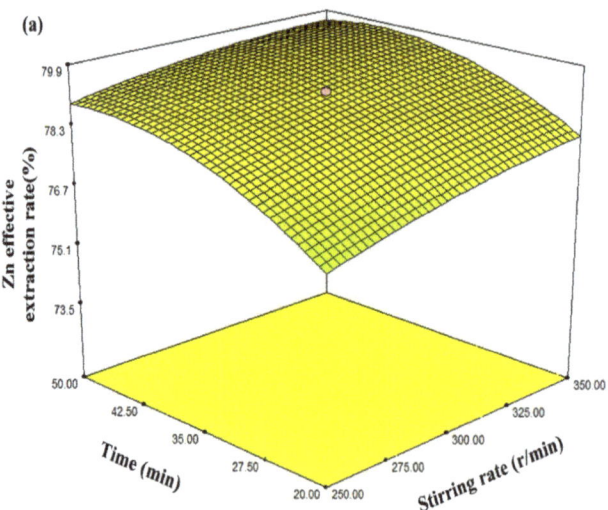

Figure 4. *Cont.*

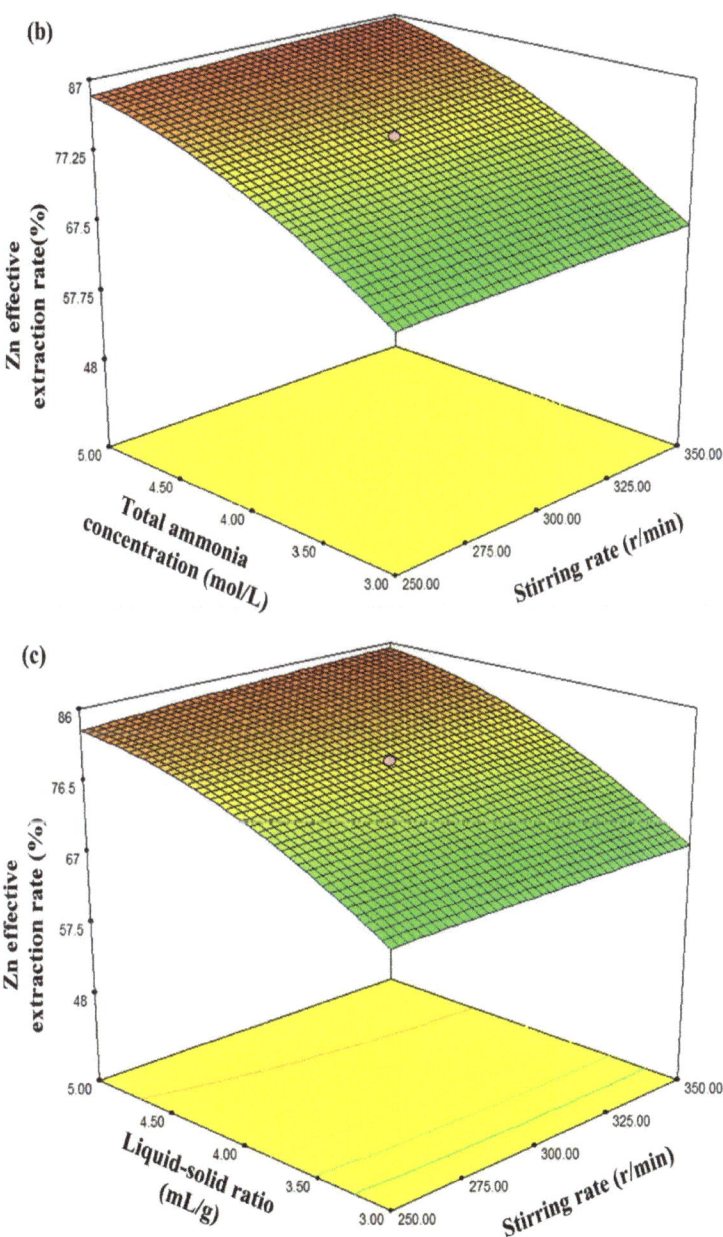

Figure 4. *Cont.*

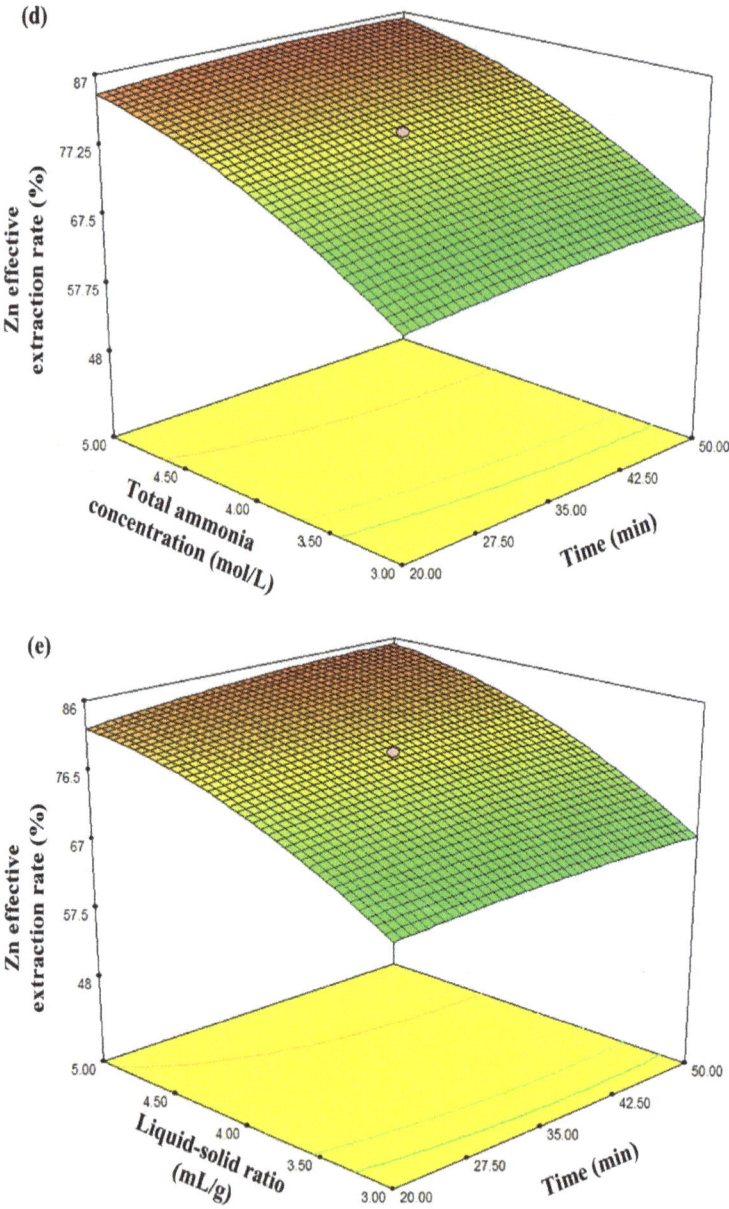

Figure 4. Cont.

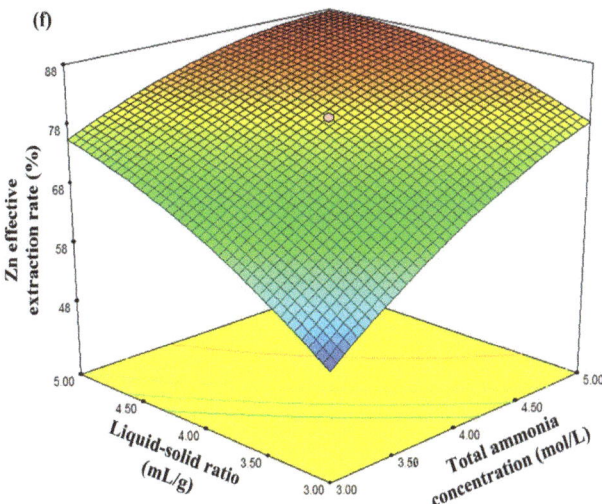

Figure 4. Response surface plots for stirring speed vs. leaching time vs. total ammonia concentration vs. liquid/solid ratio, (**a**) leaching time vs. stirring speed; (**b**) total ammonia concentration vs. stirring speed; (**c**) liquid/solid ratio vs. stirring speed; (**d**) total ammonia concentration vs. leaching time; (**e**) liquid/solid ratio vs. leaching time; (**f**) liquid/solid ratio vs. total ammonia concentration.

As illustrated in Figure 4a–c, leaching time, total ammonia concentration, and liquid/solid ratio presented greater effects on the Zn effective extraction rate than stirring speed. It can be observed in Figure 4d,e that the influences of total ammonia concentration and liquid/solid ratio showed greater effects than leaching time. Moreover, it is depicted in Figure 4f that the interactions between the total ammonia concentration and liquid/solid ratio exerted the most significant effects on the Zn effective extraction rate, and this conclusion is consistent with the above variance analysis. Furthermore, as shown in Figure 4f, simultaneously increasing total ammonia concentration and liquid/solid ratio can significantly improve the Zn effective extraction rate, before reaching a plateau value. The influences of total ammonia concentration and liquid/solid ratio are different, wherein increasing the total ammonia concentration increases the concentration of coordination agents, while the increase in L/S ratio can enhance the rapid dissolution of zinc oxide, and the solubility of $[Zn(NH_3)]^{2+}$ and $(RCOO)_2Zn$ in solution can be increased accordingly, further promoting the dissolution of zinc in MSD samples. As a result, the ion diffusion resistance decreases, and the leaching behaviour of Zn ion is strengthened, further enhancing the Zn effective extraction rate. However, once the values of the total ammonia concentration and liquid/solid ratio exceeded critical levels, the further increase in the two factors values had negligible effects on the Zn effective extraction rate. Since the materials in the container remain constant, the interfacial areas of Zn-containing minerals and NH_3-CH_3COONH_4-H_2O leaching agents are constant. Therefore, under the conditions of the sufficiently large quantity of the leaching agent and maintaining the effective mass transfer and maximal reactions between Zn-containing minerals, the Zn effective extraction rate can be maximised by leaching from the MSD sample using NH_3-CH_3COONH_4-H_2O leaching agent.

3.4. Condition Optimisation and Verification

Based on the predictions of the response surface methodology (RSM), the leaching time, total ammonia concentration, liquid/solid ratio, ammonia/ammonium ratio, stirring speed, and leaching temperature were optimised, and the results were experimentally verified.

Table 8 lists the predicted values and experimental values of the Zn effective extraction rate achieved by coordination leaching from the MSD materials by $NH_3\text{-}CH_3COONH_4\text{-}H_2O$ solution, as well as the optimised conditions and model verification results. Three leaching experiments based on the optimised process parameters were conducted to confirm the accuracy of the optimization parameters obtained by RSM. The average value of the Zn effective extraction rate in three parallel experiments was 84.67%; in contrast, the predicted value was 85.25%, demonstrating that RSM provides a reliable method to optimise the recovery process of Zn from Zn-containing MSD materials by coordination leaching using $NH_3\text{-}CH_3COONH_4\text{-}H_2O$ solution.

Table 8. The optimised process parameters determined by the regression model.

Total Ammonia Concentration (mol/L)	$[NH_3]/[NH_4]^+$ Mole Ratio	Temperature (°C)	Time (min)
4.78	1:01	25	46.20
Liquid/solid ratio (mL/g)	Stirring speed (r/min)	Zn effective leaching rate (%)	
		Predicted value	Experimental value
4.29	344.78	85.25	84.67

3.5. Characterization Analysis

3.5.1. XRD Analysis

To determine the metal ions and impurities existing in the MSD sample and the leaching residues, the MSD samples and the leaching residues were characterized by a rotating target multi-functional X-ray diffractometer (TTRA III, Rigaku, Japan). The operation power of the X-ray generator was 18 kW, CuKα irradiation (λ = 1.54056 Å) was applied, and the voltage and the current were 40 kV and 200 mA, respectively. Under filtering using a graphite monochromator with a high reflection efficiency, scanning was conducted with a scanning rate of 4°/min from 10°–90°. The XRD pattern of the MSD sample is illustrated in Figure 5a. The XRD pattern revealed that Zn presented as ZnO, $Zn_5(OH)_8Cl_2\cdot H_2O$, ZnS, $ZnFe_2O_4$, and Zn_2SiO_4, while Fe mainly presented as Fe_3O_4 and Fe_2O_3. The diversity of the zinc phase and the presence of iron suggest that the process of extracting zinc from the MSD sample is very difficult. Figure 5b shows an XRD comparison of leaching residues under optimized conditions. As shown in Figure 5, the diffraction peaks of ZnO and $Zn_5(OH)_8Cl_2H_2O$ disappear, and the main residual zinc phases in the leaching slag are zinc ferrite ($ZnFe_2O_4$), zinc silicate ($ZnSiO_4$), and zinc sulfide (ZnS).

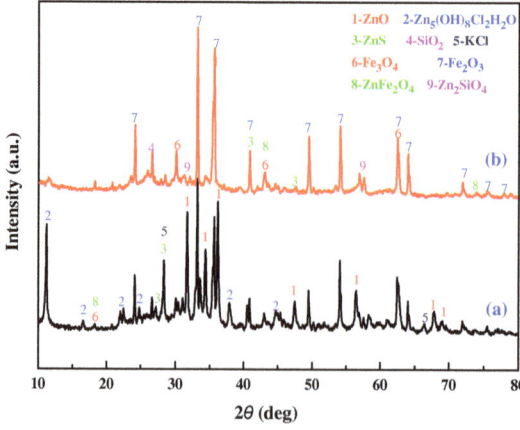

Figure 5. XRD pattern of the MSD sample (a) and leaching residue (b).

3.5.2. SEM-EDS Analysis

To further investigate the particle distribution, morphology, and composition of the MSD sample and the leaching residue, the microstructures of the MSD sample and zinc leaching residue were determined using the SEM apparatus (XL30ESEM-TMP, Philips, The Netherlands), allocated with an EDS detector (EDS-Genesis, EDAX, Mahwah, NJ, USA), and the SEM-ESD spectra are displayed as shown in Figures 6 and 7, respectively. Figure 6 shows that the grey flocculent amorphous structures in area A contained Zn, metals (Fe, Pb, and Al), and gangue components (Si, Ca, and Mg); the phase of area B was mainly quartz (SiO_2). Moreover, metal inclusions and gangue components were observed. Figure 7 shows that the structural topography and particle element distribution of the residue, and the presence of C, O, Mg, Al, Si, S, Pb, Cl, Ca, Fe, P, and Zn, are relevant to the investigated system. There are three main morphologies in this sample: Point (A), dispersed flocculent structure particles; Point (B), bright grey massive structural particles; and Point (C), dark grey massive structural particles. The three structural particles are inlaid and tightly bound to each other.

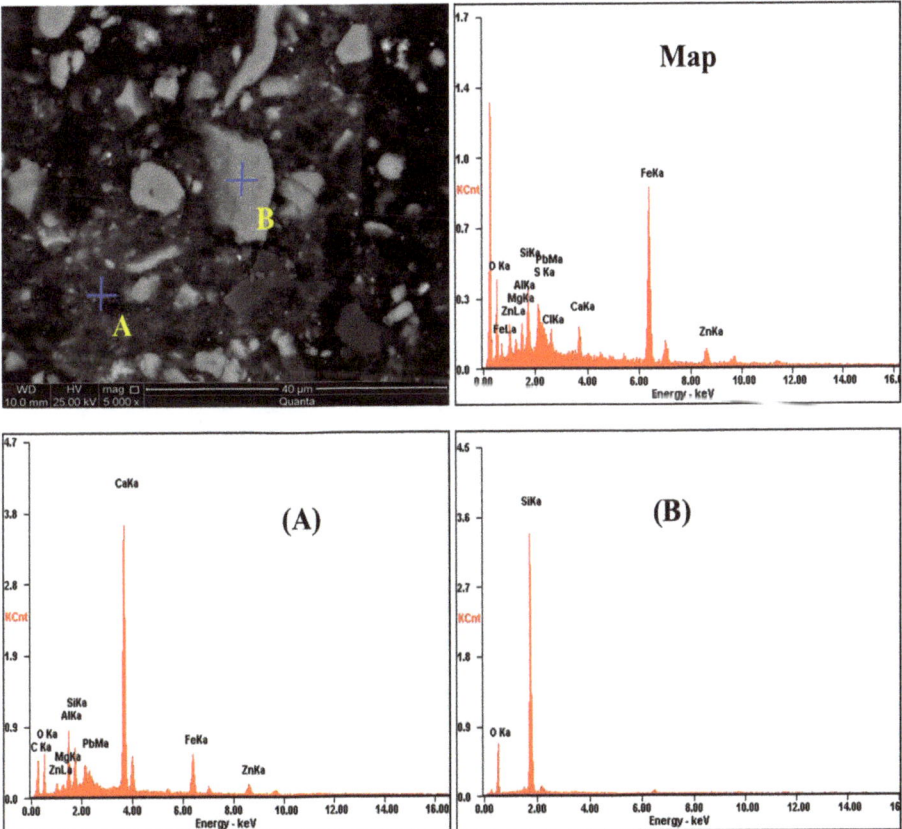

Figure 6. SEM-ESD spectra of the MSD sample.

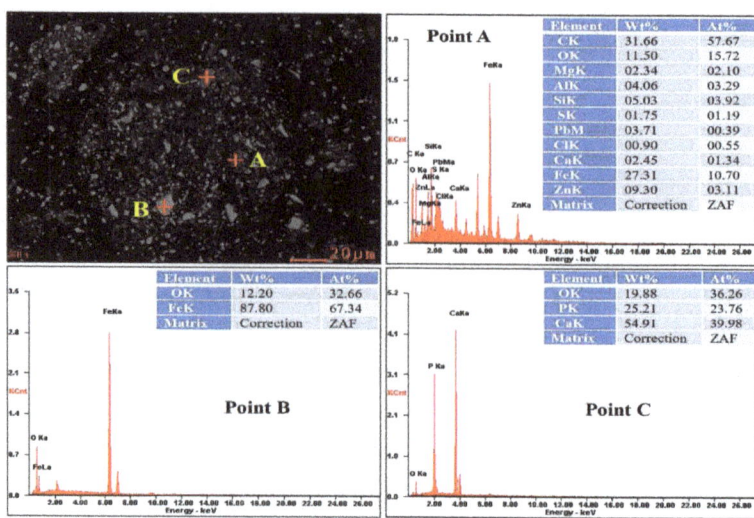

Figure 7. SEM-ESD pattern of MSD leaching residue.

The X-ray EDS maps of Cl, Al, Mg, Ca, O, Fe, Si, and Zn are analysed as shown in Figure 8. The EDS analysis of Point (A) shows that the dispersed flocculent structure particles are mainly composed of C, O, Mg, Al, Si, S, Pb, Cl, Ca, Fe, P, and Zn. The EDS analysis of Point (B) shows that the bright grey massive structural particles are mainly iron oxides. The EDS analysis of Point (C) shows that dark grey massive structural particles are mainly gangue minerals.

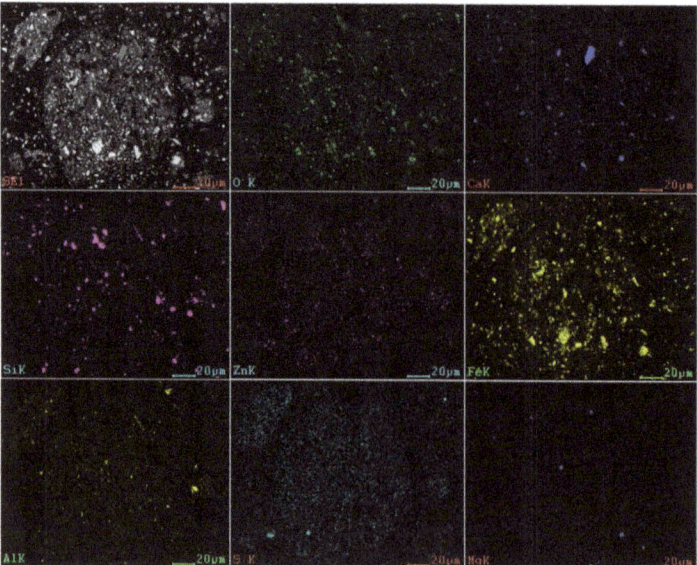

Figure 8. SEM-EDS map scanning pattern of MSD leaching residue.

The SEM-EDS surface-scanning pattern shows that the residue comprises a metallic minerals phase in addition to a gangue minerals phase. Furthermore, the surface-scanning pattern makes it clear that Zn_2SiO_4, ZnS, and $ZnFe_2O_4$ do not leach in the NH_3-CH_3COONH_4-H_2O system.

4. Conclusions

In this study, Zn-containing metallurgical slag and dust (MSD) materials were utilised as the research object, and response surface methodology (RSM) was introduced into the process optimisation of recovering Zn from Zn-containing MSD materials by coordination leaching using NH_3-CH_3COONH_4-H_2O solution. The main findings were as follows:

(1) At a constant leaching temperature and ammonia/ammonium ratio, the influences of liquid/solid ratio, stirring speed, leaching time, total ammonia concentration, and interactions between them on the Zn effective extraction rate were investigated using the central composite design. The mathematic model of the factors affecting the Zn effective extraction rate was established as:

$$Y = 79.19 + 0.38X_1 + 0.84X_2 + 8.55X_3 + 7.25X_4 - 0.012X_1X_2 + 0.10X_1X_3 + 0.28X_1X_4 - 0.20X_2X_3 + 0.36X_2X_4 - 4.30X_3X_4 - 0.16X_1^2 - 0.69X_2^2 - 3.27X_3^2 - 3.34X_4^2$$

(2) The optimised parameters for the leaching experiments were obtained: the leaching temperature was 25 °C, the total ammonia concentration was 4.8 mol/L, the leaching time was 46 min, the liquid/solid ratio was 4.3:1, $[NH_3]/[NH_4]^+$ was 1:1, the stirring speed was 345 r/min, the measured Zn effective extraction rate was 84.67%, and the Zn predicted effective extraction rate was 85.25%. The experimental and predicted values of the Zn effective extraction rate were similar, indicating the reliability of the proposed model and the suitability of the optimised process parameters.

(3) The XRD and SEM-EDS analysis results showed that Zn_2SiO_4, ZnS, and $ZnFe_2O_4$ were among the main factors affecting the low extraction rate of zinc from metallurgical slag dust under the CH_3COONH_4 system.

Author Contributions: X.Z.: Writing—Original Draft, Conceptualization, Methodology, Investigation, Validation; J.L.: Visualization, Resources; A.M.: Investigation, Validation, Funding Acquisition; B.L.: Supervision, Writing—Review and Editing. All authors have read and agreed to the published version of the manuscript.

Funding: This work was supported by the scientific research and cultivation project of Liupanshui Normal University (LPSSY2023KJYBPY06), the Discipline team of Liupanshui Normal University (LPSSY2023XKTD07), the Guizhou Provincial First-class professional (GZSylzy202103), and the key cultivation disciplines of Liupanshui Normal University (LPSSYZDXK202001).

Institutional Review Board Statement: Not applicable.

Informed Consent Statement: Not applicable.

Data Availability Statement: The data presented in this study are available on request from the corresponding author. The data are not publicly available due to technical or time limitations.

Acknowledgments: The authors acknowledge the financial support used to carry out this work.

Conflicts of Interest: All authors declare that they have no conflict of interest.

References

1. Hua, X.; Yang, Q.; Zhang, D.; Meng, F.; Chen, C.; You, Z.; Zhang, J.; Lv, S.; Meng, J. Microstructures and mechanical properties of a newly developed high-pressure die casting Mg-Zn-RE alloy. *J. Mater. Sci. Technol.* **2020**, *53*, 174–184. [CrossRef]
2. Wu, H.; Zhang, L.; Liu, C.; Mai, Y.; Zhang, Y.; Jie, X. Deposition of Zn-G/Al composite coating with excellent cathodic protection on low-carbon steel by low-pressure cold spraying. *J. Alloys Compd.* **2020**, *821*, 153483. [CrossRef]
3. Dong, L.; Yang, W.; Yang, W.; Tian, H.; Huang, Y.; Wang, X.; Xu, C.; Wang, C.; Kang, F.; Wang, G. Flexible and conductive scaffold-stabilized zinc metal anodes for ultralong-life zinc-ion batteries and zinc-ion hybrid capacitors. *Chem. Eng. J.* **2020**, *384*, 123355. [CrossRef]
4. Abkhoshk, E.; Jorjani, E.; Al-Harahsheh, M.; Rashchi, F.; Naazeri, M. Review of the hydrometallurgical processing of non-sulfide zinc ores. *Hydrometallurgy* **2014**, *149*, 153–167. [CrossRef]

5. Matsukevich, I.; Kulinich, N.; Romanovski, V. Direct reduced iron and zinc recovery from electric arc furnace dust. *J. Chem. Technol. Biotechnol.* **2022**, *97*, 3453–3458. [CrossRef]
6. Zhu, X.-L.; Xu, C.-Y.; Tang, J.; Hua, Y.-X.; Zhang, Q.-B.; Liu, H.; Wang, X.; Huang, M.-T. Selective recovery of zinc from zinc oxide dust using choline chloride based deep eutectic solvents. *Trans. Nonferrous Met. Soc. China* **2019**, *29*, 2222–2228. [CrossRef]
7. Ma, A.; Zheng, X.; Li, S.; Wang, Y.; Zhu, S. Zinc recovery from metallurgical slag and dust by coordination leaching in NH_3–CH_3COONH_4–H_2O system. *R. Soc. Open Sci.* **2018**, *5*, 180660. [CrossRef]
8. Trinkel, V.; Mallow, O.; Aschenbrenner, P.; Rechberger, H.; Fellner, J. Characterization of Blast Furnace Sludge with Respect to Heavy Metal Distribution. *Ind. Eng. Chem. Res.* **2016**, *55*, 5590–5597. [CrossRef]
9. Trinkel, V.; Mallow, O.; Thaler, C.; Schenk, J.; Rechberger, H.; Fellner, J. Behavior of Chromium, Nickel, Lead, Zinc, Cadmium, and Mercury in the Blast Furnace—A Critical Review of Literature Data and Plant Investigations. *Ind. Eng. Chem. Res.* **2015**, *54*, 11759–11771. [CrossRef]
10. Lanzerstorfer, C.; Bamberger-Strassmayr, B.; Pilz, K. Recycling of Blast Furnace Dust in the Iron Ore Sintering Process: Investigation of Coke Breeze Substitution and the Influence on Off-gas Emissions. *ISIJ Int.* **2015**, *55*, 758–764. [CrossRef]
11. Ma, N. Recycling of basic oxygen furnace steelmaking dust by in-process separation of zinc from the dust. *J. Clean. Prod.* **2016**, *112*, 4497–4504. [CrossRef]
12. Chang, J.; Zhang, E.-D.; Zhang, L.-B.; Peng, J.-H.; Zhou, J.-W.; Srinivasakannan, C.; Yang, C.-J. A comparison of ultrasound-augmented and conventional leaching of silver from sintering dust using acidic thiourea. *Ultrason. Sonochem.* **2017**, *34*, 222–231. [CrossRef]
13. Li, H.; Ma, A.; Srinivasakannan, C.; Zhang, L.; Li, S.; Yin, S. Investigation on the recovery of gold and silver from cyanide tailings using chlorination roasting process. *J. Alloys Compd.* **2018**, *763*, 241–249. [CrossRef]
14. Cantarino, M.V.; Filho, C.d.C.; Mansur, M.B. Selective removal of zinc from basic oxygen furnace sludges. *Hydrometallurgy* **2012**, *111*, 124–128. [CrossRef]
15. Miki, T.; Chairaksa-Fujimoto, R.; Maruyama, K.; Nagasaka, T. Hydrometallurgical extraction of zinc from CaO treated EAF dust in ammonium chloride solution. *J. Hazard. Mater.* **2016**, *302*, 90–96. [CrossRef]
16. Kukurugya, F.; Vindt, T.; Havlík, T. Behavior of zinc, iron and calcium from electric arc furnace (EAF) dust in hydrometallurgical processing in sulfuric acid solutions: Thermodynamic and kinetic aspects. *Hydrometallurgy* **2015**, *154*, 20–32. [CrossRef]
17. Das, B.; Prakash, S.; Reddy, P.S.R.; Misra, V.N. An overview of utilization of slag and sludge from steel industries. *Resour. Conserv. Recycl.* **2007**, *50*, 40–57. [CrossRef]
18. Tsakiridis, P.; Papadimitriou, G.; Tsivilis, S.; Koroneos, C. Utilization of steel slag for Portland cement clinker production. *J. Hazard. Mater.* **2008**, *152*, 805–811. [CrossRef]
19. Safari, V.; Arzpeyma, G.; Rashchi, F.; Mostoufi, N. A shrinking particle—Shrinking core model for leaching of a zinc ore containing silica. *Int. J. Miner. Process.* **2009**, *93*, 79–83. [CrossRef]
20. Dutra, A.; Paiva, P.; Tavares, L. Alkaline leaching of zinc from electric arc furnace steel dust. *Miner. Eng.* **2006**, *19*, 478–485. [CrossRef]
21. Brunelli, K.; Dabalà, M. Ultrasound effects on zinc recovery from EAF dust by sulfuric acid leaching. *Int. J. Miner. Met. Mater.* **2015**, *22*, 353–362. [CrossRef]
22. Xu, H.; Wei, C.; Li, C.; Fan, G.; Deng, Z.; Zhou, X.; Qiu, S. Leaching of a complex sulfidic, silicate-containing zinc ore in sulfuric acid solution under oxygen pressure. *Sep. Purif. Technol.* **2012**, *85*, 206–212. [CrossRef]
23. Halli, P.; Agarwal, V.; Partinen, J.; Lundström, M. Recovery of Pb and Zn from a citrate leach liquor of a roasted EAF dust using precipitation and solvent extraction. *Sep. Purif. Technol.* **2019**, *236*, 116264. [CrossRef]
24. Steer, J.M.; Griffiths, A.J. Investigation of carboxylic acids and non-aqueous solvents for the selective leaching of zinc from blast furnace dust slurry. *Hydrometallurgy* **2013**, *140*, 34–41. [CrossRef]
25. Sethurajan, M.; Huguenot, D.; Jain, R.; Lens, P.N.L.; Horn, H.A.; Figueiredo, L.H.A.; van Hullebusch, E.D. Leaching and selective zinc recovery from acidic leachates of zinc metallurgical leach residues. *J. Hazard. Mater.* **2017**, *324*, 71–82. [CrossRef]
26. Ma, A.; Zhang, L.; Peng, J.; Zheng, X.; Li, S.; Yang, K.; Chen, W. Extraction of zinc from blast furnace dust in ammonia leaching system. *Green Process. Synth.* **2016**, *5*, 23–30. [CrossRef]
27. Wang, R.-X.; Tang, M.-T.; Yang, S.-H.; Zhagn, W.-H.; Tang, C.-B.; He, J.; Yang, J.-G. Leaching kinetics of low grade zinc oxide ore in NH_3–NH_4Cl–H_2O system. *J. Cent. South Univ. Technol.* **2008**, *15*, 679–683. [CrossRef]
28. Liu, Z.; Li, Q.; Cao, Z.; Yang, T. Dissolution behavior of willemite in the $(NH_4)_2SO_4$–NH_3–H_2O system. *Hydrometallurgy* **2012**, *125*, 50–54. [CrossRef]
29. Ma, A.; Zheng, X.; Zhang, L.; Peng, J.; Li, Z.; Li, S.; Li, S. Clean recycling of zinc from blast furnace dust with ammonium acetate as complexing agents. *Sep. Sci. Technol.* **2018**, *53*, 1327–1341. [CrossRef]
30. Yang, S.; Zhao, D.; Jie, Y.; Tang, C.; He, J.; Chen, Y. Hydrometallurgical Process for Zinc Recovery from C.Z.O. Generated by the Steelmaking Industry with Ammonia–Ammonium Chloride Solution. *Metals* **2019**, *9*, 83. [CrossRef]
31. Rao, S.; Yang, T.; Zhang, D.; Liu, W.; Chen, L.; Hao, Z.; Xiao, Q.; Wen, J. Leaching of low grade zinc oxide ores in NH_4Cl–NH_3 solutions with nitrilotriacetic acid as complexing agents. *Hydrometallurgy* **2015**, *158*, 101–106. [CrossRef]
32. Li, Z.; Yu, D.; Liu, X.; Wang, Y. The Fate of Heavy Metals and Risk Assessment of Heavy Metal in Pyrolysis Coupling with Acid Washing Treatment for Sewage Sludge. *Toxics* **2023**, *11*, 447. [CrossRef] [PubMed]

33. Wang, S.; Xu, C.; Lei, Z.; Li, J.; Lu, J.; Xiang, Q.; Chen, X.; Hua, Y.; Li, Y. Recycling of zinc oxide dust using ChCl-urea deep eutectic solvent with nitrilotriacetic acid as complexing agents. *Miner. Eng.* **2022**, *175*, 107295. [CrossRef]
34. Javanshir, S.; Qashoqchi, S.S.; Mofrad, Z.H. Silver production from spent zinc–silver oxide batteries via leaching–cementation technique. *Sep. Sci. Technol.* **2021**, *56*, 1956–1964. [CrossRef]

Disclaimer/Publisher's Note: The statements, opinions and data contained in all publications are solely those of the individual author(s) and contributor(s) and not of MDPI and/or the editor(s). MDPI and/or the editor(s) disclaim responsibility for any injury to people or property resulting from any ideas, methods, instructions or products referred to in the content.

Article

Preparation of Spherical Ultrafine Silver Particles Using Y-Type Microjet Reactor

Xiaoxi Wan [1,2,3], Jun Li [1,2,3], Na Li [1,2,3], Jingxi Zhang [1,2,3], Yongwan Gu [4], Guo Chen [1,*] and Shaohua Ju [1,2,3,*]

1. Faculty of Metallurgical and Energy Engineering, Kunming University of Science and Technology, Kunming 650093, China
2. Key Laboratory of Unconventional Metallurgy, Ministry of Education, Kunming 650093, China
3. National Local Joint Laboratory of Engineering Application of Microwave Energy and Equipment Technology, Kunming 650093, China
4. Kunming Institute of Precious Metals, Kunming 650106, China
* Correspondence: guochen@kust.edu.cn (G.C.); shaohuaju@kust.edu.cn (S.J.)

Abstract: Herein, micron-sized silver particles were prepared using the chemical reduction method by employing a Y-type microjet reactor, silver nitrate as the precursor, ascorbic acid as the reducing agent, and gelatin as the dispersion at room temperature (23 °C ± 2°C). Using a microjet reactor, the two reaction solutions collide and combine outside the reactor, thereby avoiding microchannel obstruction issues and facilitating a quicker and more convenient synthesis process. This study examined the effect of the jet flow rate and dispersion addition on the morphology and size of silver powder particles. Based on the results of this study, spherical and dendritic silver particles with a rough surface can be prepared by adjusting the flow rate of the reaction solution and gelatin concentration. The microjet flow rate of 75 mL/min and the injected gelatin amount of 1% of the silver nitrate mass produced spherical ultrafine silver particles with a size of 4.84 μm and a tap density of 5.22 g/cm^3.

Keywords: Y-type microjet reactor; spherical ultrafine silver particles; wet chemical reduction; dendritic particle

Citation: Wan, X.; Li, J.; Li, N.; Zhang, J.; Gu, Y.; Chen, G.; Ju, S. Preparation of Spherical Ultrafine Silver Particles Using Y-Type Microjet Reactor. *Materials* **2023**, *16*, 2217. https://doi.org/10.3390/ma16062217

Academic Editor: Alexey N. Pestryakov

Received: 10 February 2023
Revised: 28 February 2023
Accepted: 5 March 2023
Published: 10 March 2023

Copyright: © 2023 by the authors. Licensee MDPI, Basel, Switzerland. This article is an open access article distributed under the terms and conditions of the Creative Commons Attribution (CC BY) license (https://creativecommons.org/licenses/by/4.0/).

1. Introduction

Silver is widely used in various industrial fields due to its good electrical conductivity, thermal conductivity, and ductility. As a functional material, silver at the micronano level shows a structure between crystalline and amorphous states, exhibits a modified surface molecular arrangement and crystal structure, and features enhanced surface activity [1]. The ultrafine silver powder shows spherical (or quasispherical), flake-like, dendritic, and microcrystalline morphologies [2–4]. Additionally, the micronano silver powder exhibits excellent performance in the fields of sound, light, electricity, magnetism, heat, and catalysis due to its small particle size, large specific surface area, high surface activity, and good catalytic activity. Furthermore, silver powder has antibacterial and sterilization capabilities because of its adsorption capacities and excellent optical properties due to its surface plasmon resonance. These characteristics have expanded the application of silver powder to fields such as electronics, the chemical industry, medicine, aerospace, the military industry, and metallurgy [5,6].

Various methods can be used to prepare silver powders, including physical, chemical, and biological methods [7]. Common physical preparation methods include the high-energy ball-milling method [8], spray thermal decomposition method [9], plasma evaporation condensation method [10], liquid phase reduction method [11,12], microemulsion method [13,14], liquid–solid phase reduction method [15], and microbial reduction [16]. The chemical method is widely used in large-scale industrial production because of its low equipment requirements and energy consumption.

Microchemical reactors mainly provide controllable and high-throughput chemical synthesis methods with good stability, low energy consumption, small reaction volume, and uniform reaction conditions [17–19]. Therefore, they provide a new process strategy for materials science, chemical synthesis, biomedical diagnosis, and drug screening [20–22]. Fouling (i.e., unnecessary deposition on the surface) often occurs and causes local constriction in microstructure equipment, which changes the flow rate and increases the pressure drop or even completely blocks the microchannel. Hence, this is the biggest obstacle to the effective operation of microstructure equipment [23]. Rathi [24] optimized the continuous synthesis of crosslinked chitosan sodium tripolyphosphate (CS-TPP) nanoparticles using a microreactor and compared it with a batch-stirred reactor. Lim [25] described a straightforward and adaptable coaxial turbulent jet mixer that not only synthesized various nanoparticles (NPs) at high throughput but also maintained the benefits of homogeneity, reproducibility, and tunability that could typically be attained only in specialized microscale mixing equipment. Using various conditions, Sebastian [26] obtained complex metal nanomaterials, such as Pt–Pd heterostructures, Ag–Pdcore–shell NPs, and Au–Pd dumbbell structures and achieved fine control of material size and morphology using the homogeneous microfluidic reactor. Baber [27] investigated $AgNO_3$ reduction by $NaBH4$ in an impinging jet reactor (IJR) to prepare silver NPs. Under certain conditions, the size of the silver NPs could be controlled at 4.3 ± 1 nm and 4.7 ± 1.3 nm. Sahoo [28] reported that the small size and uniformity (5.2 ± 0.9 nm) of silver NPs can be controlled using a free impinging stream reactor at room temperature. The unique IJR characteristics are effective mixing and the lack of channel walls to avoid fouling.

The Y-type microjet reactor makes the reaction solution converge outside the reactor; therefore, the solutions are uniformly mixed and reacted. The process is safe, efficient, and controllable, as required by modern chemical technologies, while effectively avoiding precipitation blockage problems in the microchannel. In this study, the Y-type microjet reactor was used to make two reactant solutions that are uniformly mixed and reacted at room temperature (23 °C $\pm$ 2 °C). Additionally, the microjet method's effects on micron silver powder's morphology, particle size, and dispersion performance were investigated by controlling the gelatin amount in the system and the jet flow rate. It is hoped that silver powder's morphology and particle size can be controlled within a certain range.

2. Materials and Methods

2.1. Materials

Gelatin (industrial gelatin) was purchased from Shanghai Maclean Biochemical Technology Co., Ltd. (Shanghai, China) Silver nitrate ($AgNO_3$) was purchased from Tongbai Hongxin New Material Co., Ltd. (Henan, China) Ascorbic acid ($C_6H_8O_6$) was purchased from Zhengzhou Tuoyang Industrial Co., Ltd. (Zhengzhou, China) Sodium chloride (NaCl), nitric acid (HNO_3), and absolute ethanol (C_2H_6O) were purchased from Chengdu Kelong Chemical Co., Ltd. (Chengdu, China) Water used in this study was deionized. All the reagents were of analytical purity and were used without additional purification.

2.2. Experimental Methods

The experimental device for preparing silver powder particles, the Y-type microjet reactor, is shown in Figure 1. The Y-type microjet reactor is made of 3D-printed photosensitive resin. The channel's inner diameter is 1.0 mm, the distance between the two outlets is d = 10 mm, and the jet's intersection angle is 45°.

Ascorbic acid and silver nitrate undergo the following chemical reduction reaction:

$$2AgNO_3 + C_6H_8O_6 = 2Ag\downarrow + C_6H_6O_6 + 2HNO_3 \tag{1}$$

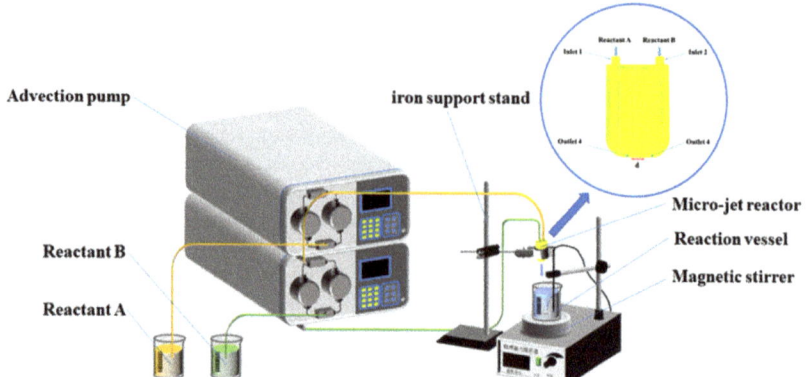

Figure 1. Schematic flow diagram of the reaction device.

The synthesis procedure and other experimental conditions for the silver particles' preparation in this study are shown in Figure 2. Solutions (A and B) were prepared as follows: A certain amount of $AgNO_3$ and $C_6H_8O_6$ were dissolved in deionized water, and an appropriate amount of HNO_3 was added to adjust the pH value of the $C_6H_8O_6$ solution. Further, gelatin (1.0–3.0% of the $AgNO_3$ mass) was added to the $C_6H_8O_6$ solution as a dispersant.

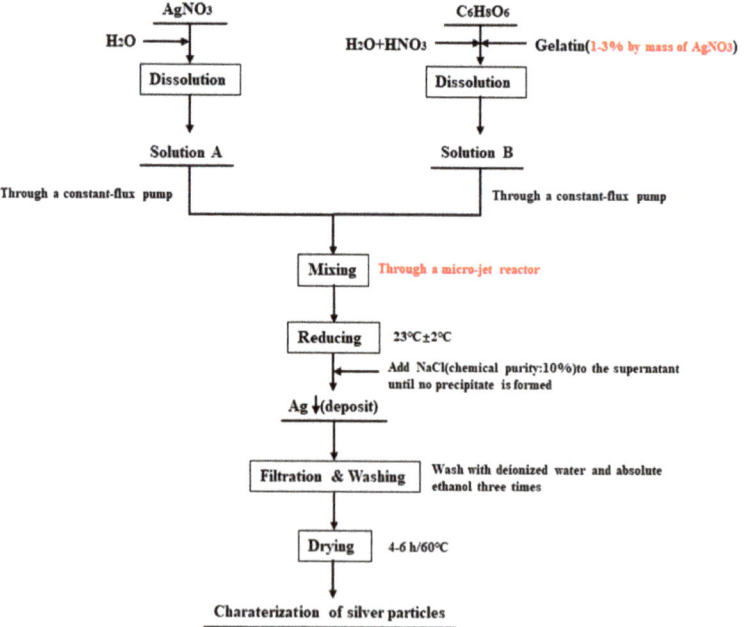

Figure 2. Experimental flow chart of silver particle preparation.

The prepared solutions A and B were delivered to the two inlets of the microjet mixing reactor by advection pumps, thereby providing appropriate flow rates and producing the desired jets at the two outlets. When the two jets collide, the solvents mix and subsequently react. The mixed solutions were poured vertically into a lower beaker filled with 100 mL of deionized water and rotated at 200 rpm. The spraying of the two solutions was arranged to ensure that the total volume of the final solution was 200 mL and the solution in the beaker was stirred for 30 s. Following the reaction, the solution stratified and precipitated after standing. The supernatant was removed and mixed with a 10% NaCl solution without white flocculent precipitation, which revealed that the silver nitrate had been totally reduced. The reaction was conducted at room temperature (23 °C ± 2 °C); all concentrations stated are those of the inflow before reagent mixing. The layered solution was filtered, washed, and dried before yielding the silver powders.

2.3. Characterization Testing

The morphology of silver powders was investigated using Nova Nano SEM450 field emission scanning electron microscopy (SEM, American FEI Company, Hillsboro, OR, USA). The physical phases of the silver powders were characterized by X-ray diffraction (XRD, Xpert powder, PANalytical, Amsterdam, The Netherlands). The particle size distribution of the silver powder was determined using a laser particle size meter (Rise-2002, Jinan Runzhi Technology Co., Ltd., Jinan, China). The specific surface area of silver powders was tested with BET-specific surface area measurement (DX 400, Beijing Jingwei Gaobo Science and Technology Co., Ltd., Beijing, China).

3. Results and Discussion
3.1. Effect of Preparation Method on the Morphology of Silver Powders

The silver powder prepared by the conventional method, and the microjet reactor was characterized by XRD, and the results are shown in Figure 3.

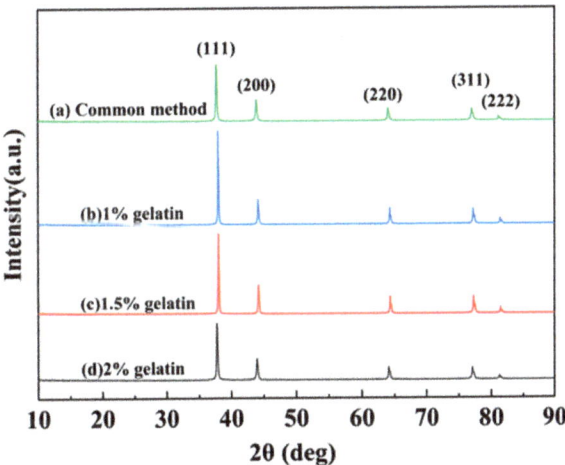

Figure 3. XRD patterns of silver powders were obtained using different preparation methods. (a) Conventional method; (b) microinjection method by adding 1% gelatin; (c) microinjection method by adding 1.5% gelatin; and (d) microinjection method by adding 2% gelatin.

As seen in Figure 3, the spectral lines have characteristic peaks at 38.1°, 44.23°, 64.37°, 77.36°, and 81.46°, which correspond to the (111), (200), (220), (311), and (222) crystal planes of cubic crystalline silver, respectively. Furthermore, these spectral lines are consistent with the monolithic silver standard pattern (JCPDS 04-0783). There were no other diffraction peaks in the spectrum, and the diffraction peaks of the curve were quite sharp. This indicates that the silver powder products obtained by the two experimental methods were highly crystalline and comprised monolithic silver.

The gelatin addition of 1% was chosen in the conventional approach to configure the $AgNO_3$ solution and the $C_6H_8O_6$ solution under the same other conditions. The $AgNO_3$ solution was added into a beaker containing $C_6H_8O_6$ mixed solution. The reaction solution was rinsed three times with deionized water and anhydrous ethanol before being dried at 60 °C for 4–6 h to obtain the silver powder product.

The SEM images of the silver powder prepared using the conventional and microjet methods are shown in Figure 4.

Figure 4. SEM of silver particles prepared by the conventional drop-in method. (**a**) Conventional method and (**b**) microjet method.

Figure 4a demonstrates that the silver powder particles prepared using the conventional method were not agglomerated and were monodispersed and polyhedral in shape. The silver powder particle size was about 2–5 μm, and the surface was smoother than the silver particles prepared using the Y-type microjet reactor. Figure 4b shows that the silver particles prepared using the Y-type microjet reactor aggregated from smaller particles into large spherical particles. Moreover, the silver powder morphology was mostly spherical with a rough surface.

3.2. Effect of Dispersant Dosage on the Morphology under Microjet Conditions

The microjet flow rate effect on the silver powder morphology was investigated at different contents of gelatin addition.

The flow rates of the microjet reactor were set at 50, 75, and 100 mL/min. A sufficient amount of gelatin was adsorbed on the silver particles' surface, successfully preventing the particles from adhering together. The silver powder was produced by mixing the reactant solutions (A and B) at a certain gelatin dispersion amount (1%, 1.5%, and 2%). Other conditions were consistent with the description of the experimental procedure presented in Section 2.2. Prior research has demonstrated that the amount of mixing between the two reactant solutions as well as the morphology and size of the micron and nanoparticles formed as a result of chemical reduction were all considerably influenced by the flow rate of the microjet reactor [27].

The high-magnification SEM images of the silver powder prepared under different jet flow rates by adding 1% gelatin are shown in Figure 5. Furthermore, Figure 5a demonstrates that when the solution flow rate was 50 mL/min, the formed silver powder particles contained both a sizable number of symmetrical dendritic particles as well as near-spherical particles with rough surfaces. When the flow rate was 75 mL/min, the formed silver powder particles were spherical particles with a rough surface. When the solution flow rate was 100 mL/min, the prepared silver powder particles were spherical particles with a rough surface, similar to that observed in the case of the particles in Figure 5b.

The high-magnification SEM images of the silver powder prepared using the Y-type microjet reactor by adding 1.5% gelatin while keeping other experimental conditions constant is shown in Figure 6.

Figure 5. SEM images of the silver powder prepared under different flow rates by adding 1% gelatin. (**a**) 50 mL/min; (**b**) 75 mL/min; and (**c**) 100 mL/min.

Figure 6. SEM images and particle size distribution of silver powders prepared under different flow rates by adding 1.5% gelatin. (**a**) 50 mL/min; (**b**) 75 mL/min; and (**c**) 100 mL/min.

Figure 6 shows that the morphology of the silver powder changed considerably with an increase in the solution flow rate. It can be seen from Figure 6a that when the solution flow rate was 50 mL/min, the prepared silver powder particles were mostly spherical with a few dendritic particles. However, when the solution flow rate was 75 mL/min, the formed silver powder particles were mostly dendritic with a certain thickness and fewer sphere-like particles (Figure 6b). It can be observed from Figure 6c that when the solution flow rate was 100 mL/min, the formed silver powder particles were almost dendritic with very few spherical particles. However, compared with Figure 5a (the 1.5% gelatin addition), the branches of dendritic particles formed in Figure 6b were wider and thicker.

Figure 7 shows the high-magnification SEM images of the silver powder fabricated with the microjet method under the same experimental circumstances except that here, 2% gelatin is added.

Figure 7. SEM images and particle size distribution of silver powders prepared under different flow rates by adding 2% gelatin. (**a**) 50 mL/min; (**b**) 75 mL/min; and (**c**) 100 mL/min.

Under this condition, the change in silver powder morphology was not obvious with an increase in the solution flow rate. When the solution flow rate was varied between 50, 75, and 100 mL/min, the morphological changes in the synthesized silver powder particles were not obvious; moreover, all of them were irregular dendrites with irregular particle surfaces. The dendritic particles formed by adding 2% gelatin had wider branches and rougher surfaces than those prepared with less gelatin addition (Figures 5 and 6).

3.3. Effect of Dispersant Dosage on the Particle Size of Silver Powder under Microjet Conditions

The effect of the microjet flow on the particle size of silver powder was investigated by adding different amounts of gelatin. The microjet reactor's flow rates were set at 50, 75, and 100 mL/min. Similarly, the silver powder was prepared by mixing the reactant solutions (A and B) with three different gelatin amounts (1, 1.5, and 2%) using the Y-type microjet reactor, and other conditions were consistent with the description of the experimental procedure in Section 2.2.

Figure 8d displays the low magnification SEM pictures, distribution map, and diameter D_{50} distribution map of the silver powder created at different jet flow rates with a 1% gelatin addition. When the gelatin addition was 1%, it can be seen that the particle size distribution of the silver powder did not change much with the modification of the jet flow rate. As shown in Figure 8d, the jet flow rates of 50, 75, and 100 mL/min correspond to the diameters D_{50} of the silver powder of 4.88, 4.84, and 5.35 μm. This demonstrates that when the amount of gelatin added is 1%, jet flow has little effect on the particle size of silver powder.

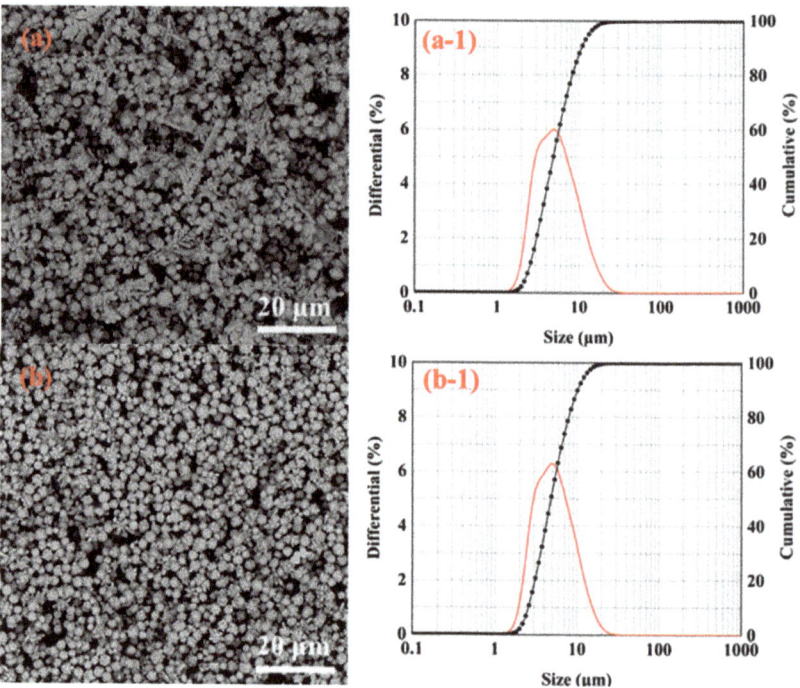

Figure 8. *Cont.*

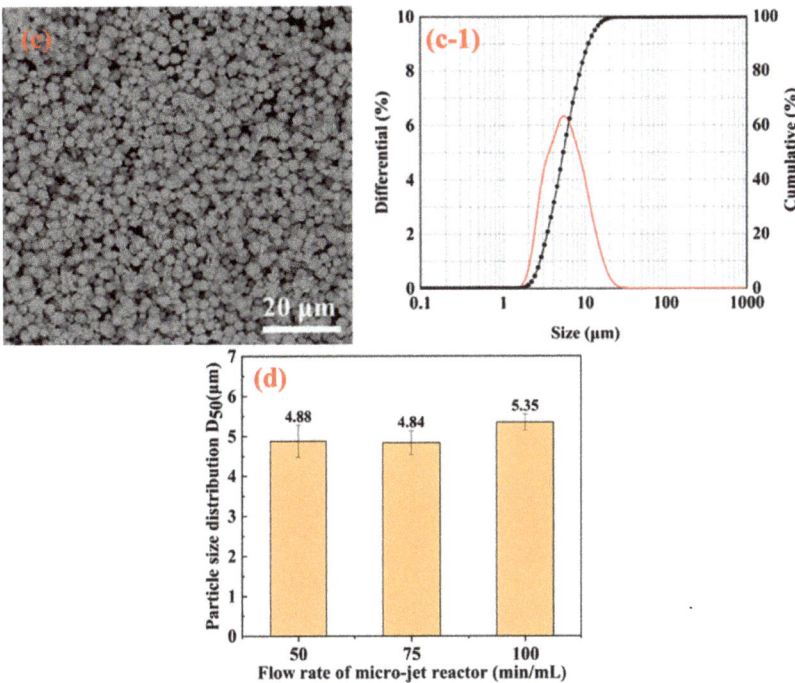

Figure 8. SEM images, particle size distributions, and D$_{50}$ distributions of the silver powder prepared by adding 1% gelatin under different jet flow rates. (**a**) 50 mL/min; (**b**) 75 mL/min; (**c**) 100 mL/min; and (**d**) D$_{50}$ distribution of silver powder.

The low-magnification SEM images, particle size distributions, and D$_{50}$ diameter distributions of the silver powder prepared under different jet flow rates by adding.5% gelatin are shown in Figure 9. From Figure 9d, it can be seen that the particle size of silver powder particles gradually increases with an increase in the microjet flow rate. Figure 9a shows that when the jet flow rate was low (50 mL/min), the silver powder particles were dispersed and had smaller particle sizes compared to Figure 8b,c.

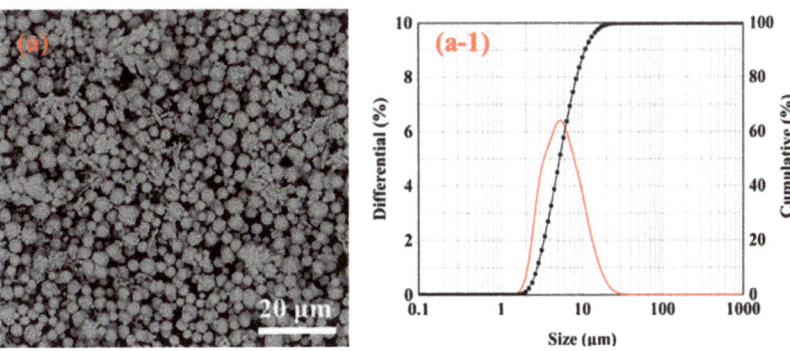

Figure 9. Cont.

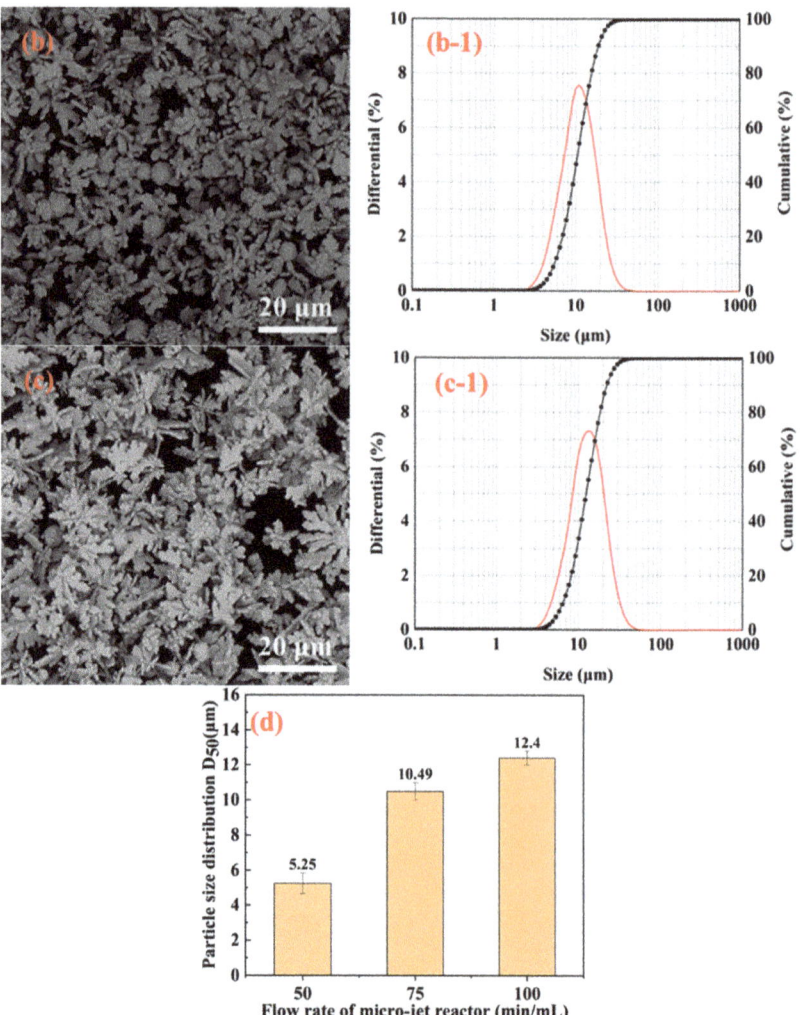

Figure 9. SEM images, particle size distributions, and the D_{50} distributions of silver powders were prepared by adding 1.5% gelatin and under different jet flow rates. (**a**) 50 mL/min; (**b**) 75 mL/min; (**c**) 100 mL/min; and (**d**) D_{50} distribution of silver powders.

Therefore, when the gelatin addition was 1.5%, the sphericity of the synthesized silver powder particles decreased. The particle size increased considerably when the microjet flow rate was higher. When the jet flow rate was too high, irregular and flaky silver powder particles with large particle sizes and low dispersion were formed.

The low-magnification SEM images, particle size distributions, and D_{50} diameter distributions of the silver powder prepared under different jet flow rates by adding 2% gelatin are shown in Figure 10. It is obvious from the figure that when too much gelatin was added (2%), the silver powder particles had poor dispersion and large particle sizes, and the maximum D_{50} diameter size of particles reached 17.75 μm.

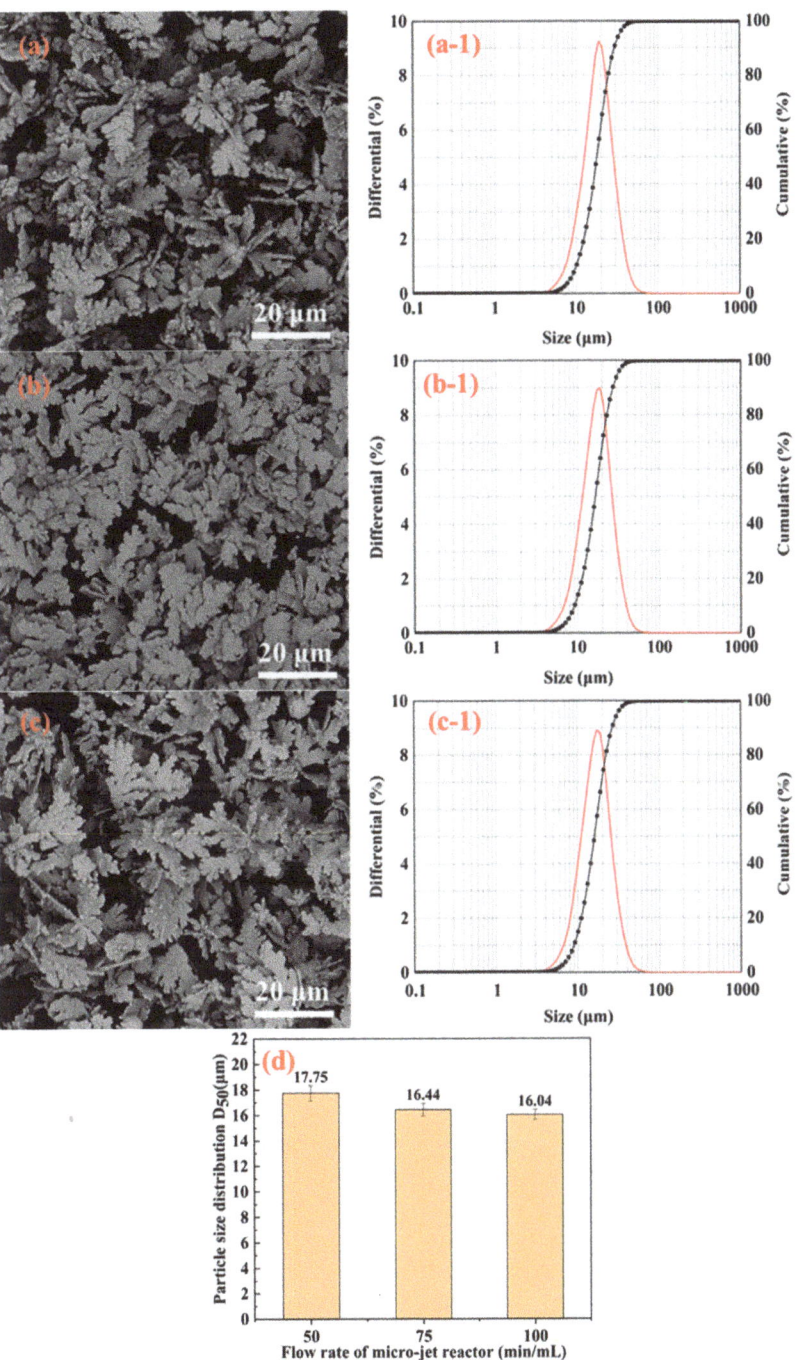

Figure 10. SEM images, particle size distribution, and D_{50} distribution of silver powders prepared under different jet flow rates by adding 2% gelatin. (**a**) 50 mL/min; (**b**) 75 mL/min; (**c**) 100 mL/min; and (**d**) D_{50} distribution of silver powder.

However, excessive gelatin will make the reaction liquid viscous, thereby decreasing the contact area of ascorbic acid and silver nitrate and slowing down the reaction rate. Simultaneously, the elemental silver formation also slows down because of the thicker gelatin film diffusion. This results in particle nucleation and growth from the two phases that cannot be effectively separated, and the resulting silver powder would have larger particle sizes [29]. Moreover, the amount of gelatin used is too large and inconvenient for washing and filtration later. The microreactor has a high jet flow rate when using a Y-type microjet reactor to prepare silver powder particles. When impinging and mixing, faster formation of silver crystal nuclei occurs but not fast enough to consume all silver ions. Hence, there will still be regular crystal growth, forming a large number of dendritic silver crystals. Therefore, the added dispersant gelatin amount should not exceed 1% of the $AgNO_3$ mass, and the jet flow rate should not be too high when preparing silver powder particles by impinging the jet method.

3.4. Effect of Dispersant Addition on Silver Powder Parameters

A potential synthesis mechanism of silver particles by conventional and microjet methods is shown in Figure 11. The morphology of the silver powder produced by various preparation techniques was quite diverse, as can be seen from the image. Ag^+ in the solution was gradually converted to silver atoms by adding the reducing agent ascorbic acid. The silver atoms developed into polyhedral, spherical, and dendritic particles due to the addition of gelatin and the mixing method of reaction solutions.

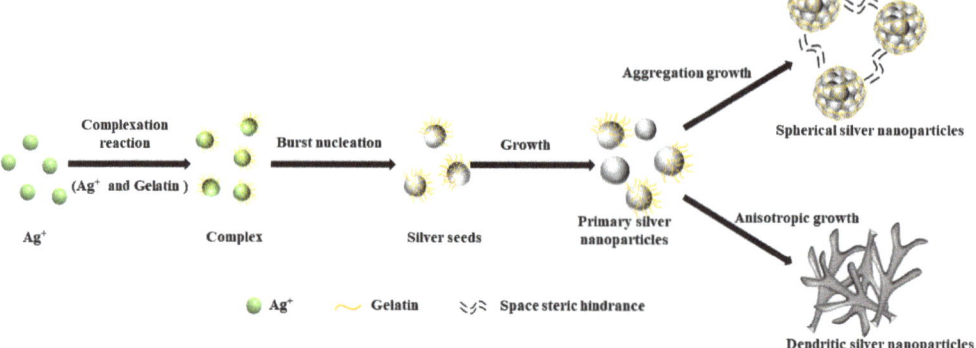

Figure 11. A schematic diagram of the growth process of spherical and dendritic micron silver particles.

Figure 12 demonstrates the connection between the added gelatin amount and the D_{90}/D_{10} ratio, tap density, and specific surface area of the manufactured silver powder particles.

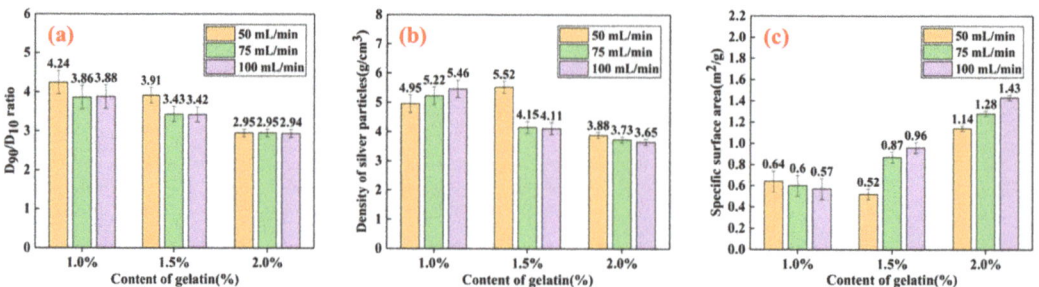

Figure 12. Properties of the silver powders prepared using different gelatin additions and jet flow rates. (**a**) D_{10}/D_{90}; (**b**) density of silver particles; and (**c**) specific surface area.

A larger D_{90}/D_{10} ratio indicates a wide particle size distribution and low dispersion [30]. The D_{90}/D_{10} ratio of the six samples was low, and the greatest value achieved was 4.24, as shown in Figure 12a. The D_{90}/D_{10} ratio essentially exhibited a decreasing trend when the amount of gelatin was added. When the gelatin amount injected was fixed, a slightly higher jet flow rate could result in a smaller D_{90}/D_{10} ratio.

As can be observed from Figure 12b, the maximum tap density (5.46 g/cm^3) was obtained when 1% gelatin was added. When the tap density of silver powders was high, the crystallinity was enhanced, the buildup between the silver powder particles in the natural state showed enhanced density, and the void ratio was small. When it is used for silver paste and other applications, the conductive film obtained after sintering the slurry has fewer and smaller voids, the resistance of series in the circuit is small, and the electrode conductivity is excellent [31]. The tap density of the silver powder particles fell as the gelatin amount increased. When 1% gelatin was introduced, the tap density increased as the jet flow rate increased. When 2% and 3% of gelatin were applied, the tap density decreased as the jet flow rate rose.

The morphology of the silver particles changed from spherical to dendritic as more gelatin was added, and the specific surface area gradually expanded, as illustrated in Figure 8c. As the particular surface area increased, the surface activity of silver powder increased. The maximum specific surface area of the produced silver particles was 1.43 m^2/g when the gelatin content was 2%. When 1% gelatin was added, the specific surface area of the silver powder particles gradually reduced with an increasing flow rate, although the difference was not considerable. Alternatively, the specific surface area of the silver powder particles gradually increased at 2% and 3% of added gelatin.

4. Conclusions

(1) The outcomes of this study demonstrate that under specific circumstances and within a specific range, the Y-type microjet reactor may be utilized to regulate silver particles' morphology and particle size.

(2) By changing the experimental conditions, spherical and dendritic silver particles can be obtained using a Y-type microjet reactor. When the microjet flow rate was 75 mL/min, and the gelatin content was 1% of the AgNO$_3$ mass, the ultrafine spherical silver powder with a particle size of 4.84 µm and a tap density of 5.22 g/cm^3 could be synthesized using the microreactor at room temperature.

(3) Compared to conventional stirred reactors, the Y-type microjet reactor can quickly and efficiently mix reactant solutions, and the process is controllable. This reactor, unlike other microchemical reactors, does not have synthetic product deposition or channel blockage problems. The controlled synthesis of silver nanoparticles offers a potential future application for the microjet reactor.

Author Contributions: X.W.: writing—original draft preparation, conceptualization, and investigation; N.L.: visualization; J.Z.: resources; Y.G.: supervision, writing—review, and editing; J.L.: formal analysis, supervision, and writing—review; G.C.: visualization and funding acquisition; S.J.: investigation, validation, and funding acquisition. All authors have read and agreed to the published version of the manuscript.

Funding: This work was supported by the National Scientific Foundation of China (No. 51964032).

Institutional Review Board Statement: Not applicable.

Informed Consent Statement: Not applicable.

Data Availability Statement: The data presented in this study are available in this article.

Conflicts of Interest: The authors declare no conflict of interest.

References

1. Kim, S.W.; Kwon, S.N.; Na, S.I. Stretchable and electrically conductive polyurethane-silver/graphene composite fibers prepared by wet-spinning process. *Compos. B Eng.* **2019**, *167*, 573–581. [CrossRef]
2. Kim, T.H.; Kim, H.; Jang, H.J.; Lee, N.; Nam, K.H.; Chung, D.; Lee, S. Improvement of the thermal stability of dendritic silver-coated copper microparticles by surface modification based on molecular self-assembly. *Nano Converg.* **2021**, *8*, 15. [CrossRef] [PubMed]
3. Liu, X.; Wu, S.; Chen, B.; Ma, Y.; Huang, Y.; Tang, S.; Liu, W. Tuning the electrical resistivity of conductive silver paste prepared by blending multi-morphologies and micro-nanometers silver powder. *J. Mater. Sci. Mater. Electron.* **2021**, *32*, 13777–13786. [CrossRef]
4. Gulyaev, A.I.; Besednov, K.L.; Petrova, A.P.; Antyufeeva, N.V. The Influence of Silver Powder Filler on the Curing Process and Structure of Conducting Adhesives. *Polym. Sci. Ser. D* **2022**, *15*, 389–393. [CrossRef]
5. Yaseen, B.; Gangwar, C.; Kumar, I.; Sarkar, J.; Naik, R.M. Detailed Kinetic and Mechanistic Study for the Preparation of Silver Nanoparticles by a Chemical Reduction Method in the Presence of a Neuroleptic Agent (Gabapentin) at an Alkaline pH and its Characterization. *ACS Omega* **2022**, *7*, 5739–5750. [CrossRef]
6. Hu, J.Q.; Chen, Q.; Xie, Z.X.; Han, G.B.; Wang, R.H.; Ren, B.; Zhang, Y.; Yang, Z.L.; Tian, Z.Q. A Simple and Effective Route for the Synthesis of Crystalline Silver Nanorods and Nanowires. *Adv. Funct. Mater.* **2004**, *14*, 183–189. [CrossRef]
7. Sinha, A.; Sharma, B.P. Preparation of silver powder through glycerol process. *Bull. Mater. Sci.* **2005**, *28*, 213–217. [CrossRef]
8. Khayati, G.R.; Janghorban, K. The nanostructure evolution of Ag powder synthesized by high energy ball milling. *Adv. Powder Technol.* **2012**, *23*, 393–397. [CrossRef]
9. Yang, S.Y.; Kim, S.G. Characterization of silver and silver/nickel composite particles prepared by spray pyrolysis. *Powder Technol.* **2004**, *146*, 185–192. [CrossRef]
10. Lee, S.H.; Oh, S.M.; Park, D.W. Preparation of silver nanopowder by thermal plasma. *Mater. Sci. Eng. C* **2007**, *27*, 1286–1290. [CrossRef]
11. Tian, Q.H.; Deng, D.; Li, Y.; Guo, X.Y. Preparation of ultrafine silver powders with controllable size and morphology. *Trans. Nonferrous Met. Soc. China* **2018**, *28*, 524–533. [CrossRef]
12. Nersisyan, H.H.; Lee, J.H.; Son, H.T.; Won, C.W.; Maeng, D.Y. A new and effective chemical reduction method for preparation of nanosized silver powder and colloid dispersion. *Mater. Res. Bull.* **2003**, *38*, 949–956. [CrossRef]
13. Salabat, A.; Saydi, H. Microemulsion route to fabrication of silver and platinum-polymer nanocomposites. *Polym. Compos.* **2014**, *35*, 2023–2028. [CrossRef]
14. Pedroza-Toscano, M.A.; Rabelero-Velasco, M.; Díaz de León, R.; Saade, H.; López, R.G.; Mendizábal, E.; Puig, J.E. Preparation of Silver Nanostructures from Bicontinuous Microemulsions. *J. Nanomater.* **2012**, *2012*, 975106. [CrossRef]
15. Cui, G.H.; Qi, S.; Wang, X.; Tian, G.; Sun, G.; Liu, W.; Yan, X.; Wu, D.; Wu, Z.; Zhang, L. Interfacial Growth of Controllable Morphology of Silver Patterns on Plastic Substrates. *J. Phys. Chem. B* **2012**, *116*, 12349–12356. [CrossRef]
16. Aziz, N.; Faraz, M.; Pandey, R.; Shakir, M.; Fatma, T.; Varma, A.; Barman, I.; Prasad, R. Facile Algae-Derived Route to Biogenic Silver Nanoparticles: Synthesis, Antibacterial, and Photocatalytic Properties. *Langmuir* **2015**, *31*, 11605–11612. [CrossRef]
17. Lim, J.-M.; Bertrand, N.; Valencia, P.M.; Rhee, M.; Langer, R.; Jon, S.; Farokhzad, O.C.; Karnik, R. Parallel microfluidic synthesis of size-tunable polymeric nanoparticles using 3D flow focusing towards in vivo study. *Nanomedicine* **2014**, *10*, 401–409. [CrossRef]
18. Ying, Y.; Chen, G.; Zhao, Y.; Li, S.; Yuan, Q. A high throughput methodology for continuous preparation of monodispersed nanocrystals in microfluidic reactors. *Chem. Eng. J.* **2008**, *135*, 209–215. [CrossRef]
19. Liu, K.; Zhu, Z.; Wang, X.; Goncalves, D.; Zhang, B.; Hierlemann, A.; Hunziker, P. Microfluidics-based single-step preparation of injection-ready polymeric nanosystems for medical imaging and drug delivery. *Nanoscale* **2015**, *7*, 16983–16993. [CrossRef]
20. Elvira, K.S.; Casadevall i Solvas, X.; Wootton, R.C.; deMello, A.J. The past, present and potential for microfluidic reactor technology in chemical synthesis. *Nat. Chem.* **2013**, *5*, 905–915. [CrossRef]
21. Ortiz de Solorzano, I.; Uson, L.; Larrea, A.; Miana, M.; Sebastian, V.; Arruebo, M. Continuous synthesis of drug-loaded nanoparticles using microchannel emulsification and numerical modeling: Effect of passive mixing. *Int. J. Nanomed.* **2016**, *11*, 3397–3416. [CrossRef]
22. Feng, H.; Zheng, T.; Li, M.; Wu, J.; Ji, H.; Zhang, J.; Zhao, W.; Guo, J. Droplet-based microfluidics systems in biomedical applications. *Electrophoresis* **2019**, *40*, 1580–1590. [CrossRef] [PubMed]
23. Schoenitz, M.; Grundemann, L.; Augustin, W.; Scholl, S. Fouling in microstructured devices: A review. *Chem. Commun.* **2015**, *51*, 8213–8228. [CrossRef] [PubMed]
24. Rathi, N.; Gaikar, V.G. Optimization of Continuous Synthesis of Cross-Linked Chitosan Nanoparticles Using Microreactors. *Chem. Eng. Technol.* **2017**, *40*, 506–513. [CrossRef]
25. Lim, J.-M.; Swami, A.; Gilson, L.M.; Chopra, S.; Choi, S.; Wu, J.; Langer, R.; Karnik, R.; Farokhzad, O.C. Ultra-High Throughput Synthesis of Nanoparticles with Homogeneous Size Distribution Using a Coaxial Turbulent Jet Mixer. *ACS Nano* **2014**, *8*, 6056–6065. [CrossRef] [PubMed]
26. Sebastian, V.; Jensen, K.F. Nanoengineering a library of metallic nanostructures using a single microfluidic reactor. *Nanoscale* **2016**, *8*, 15288–15295. [CrossRef]
27. Baber, R.; Mazzei, L.; Thanh, N.T.K.; Gavriilidis, A. Synthesis of silver nanoparticles using a microfluidic impinging jet reactor. *J. Flow Chem.* **2016**, *6*, 268–278. [CrossRef]

28. Sahoo, K.; Kumar, S. Green synthesis of sub 10 nm silver nanoparticles in gram scale using free impinging jet reactor. *Chem. Eng. Process. Process Intensif.* **2021**, *165*, 108439. [CrossRef]
29. Chen, Z.W.; Gan, G.Y.; Yan, J.K.; Liu, J. Preparation of Spherical Ultra-fine Silver Powder Using Gelatin as Dispersant. *Rare Met. Mater. Eng.* **2011**, *40*, 741–744.
30. Sannohe, K.; Ma, T.; Hayase, S. Synthesis of monodispersed silver particles: Synthetic techniques to control shapes, particle size distribution and lightness of silver particles. *Adv. Powder Technol.* **2019**, *30*, 3088–3098. [CrossRef]
31. Liu, Z.; Qi, X.; Wang, H. Synthesis and characterization of spherical and mono-disperse micro-silver powder used for silicon solar cell electronic paste. *Adv. Powder Technol.* **2012**, *23*, 250–255. [CrossRef]

Disclaimer/Publisher's Note: The statements, opinions and data contained in all publications are solely those of the individual author(s) and contributor(s) and not of MDPI and/or the editor(s). MDPI and/or the editor(s) disclaim responsibility for any injury to people or property resulting from any ideas, methods, instructions or products referred to in the content.

Article

Preparation of Micro-Size Spherical Silver Particles and Their Application in Conductive Silver Paste

Na Li [1,2,3], Jun Li [1,2,3], Xiaoxi Wan [1,2,3], Yifan Niu [1,2,3], Yongwan Gu [4], Guo Chen [1,*] and Shaohua Ju [1,2,3,*]

1. Faculty of Metallurgical and Energy Engineering, Kunming University of Science and Technology, Kunming 650093, China
2. Key Laboratory of Unconventional Metallurgy, Ministry of Education, Kunming 650093, China
3. National Local Joint Laboratory of Engineering Application of Microwave Energy and Equipment Technology, Kunming 650093, China
4. Kunming Institute of Precious Metals, Kunming 650106, China
* Correspondence: guochen@kust.edu.cn (G.C.); shaohuaju@kust.edu.cn (S.J.)

Abstract: In this paper, micro-size spherical silver particles were prepared by using a wet-chemical reduction method. The silver particles were characterized by scanning electron microscopy (SEM), X-ray diffraction (XRD) and a laser particle-size analyzer. The results indicate that different types and the content of surfactants can be used to prevent the accumulation, and control the morphology and particle size distribution, of silver particles. Moreover, the morphology of silver particles was changed from polyhedral to spherical when the pH was raised from 1 to 3. Under the optimal synthesis conditions (0.1 mol/L silver nitrate, 0.06 mol/L ascorbic acid, gelatin (5% by weight of silver nitrate), pH = 1), the micro-size spherical silver particles with diameter of 5–8 μm were obtained. In addition, the resistivity of conductive silver paste that prepared with the as-synthesized spherical silver particles was discussed in detail and the average resistivity of the conductive silver paste was 3.57×10^{-5} Ω·cm after sintering at 140 °C for 30 min.

Keywords: spherical silver particle; micro-size; monodispersed; resistivity

Citation: Li, N.; Li, J.; Wan, X.; Niu, Y.; Gu, Y.; Chen, G.; Ju, S. Preparation of Micro-Size Spherical Silver Particles and Their Application in Conductive Silver Paste. *Materials* 2023, 16, 1733. https://doi.org/10.3390/ma16041733

Academic Editor: Daeho Lee

Received: 19 December 2022
Revised: 2 February 2023
Accepted: 17 February 2023
Published: 20 February 2023

Copyright: © 2023 by the authors. Licensee MDPI, Basel, Switzerland. This article is an open access article distributed under the terms and conditions of the Creative Commons Attribution (CC BY) license (https://creativecommons.org/licenses/by/4.0/).

1. Introduction

With the rapid development of the electronic components industry, conductive silver paste is widely used in liquid crystal display (LCD), SMD light-emitting diodes (LED), integrated circuit (IC) chips, printed circuit board assemblies (PCBA), ceramic capacitors, membrane switches, smart cards, RF identification and other electronic components because of its excellent electrical conductivity, adhesion, solderability, bending resistance and simple process, precise wiring and easy cost control [1–5]. Metallic silver powder is the main component of conductive silver paste. Its conductive properties mainly rely on silver powder. The content of silver powder in the paste directly affects the conductive properties. Generally, micron-size silver powder has a lower surface resistance, and when applied to silver paste, it has higher vibrancy density and lower cost than nano silver powder [6].

Many methods have been developed for the preparation of silver particles, such as the direct current arc plasma method, spray thermal decomposition method, thermal decomposition method, electrolysis method, ultrasonic chemical method, precipitation conversion method, microemulsion method, mechanical ball grinding method and so on [7–10]. The wet-chemical reduction method has become the primary method for preparing silver particles because of its advantages such as ease of operation, low cost and energy-saving [11]. By optimizing the synthesis conditions, such as controlling the strength of solution, chemical properties of stabilizer, stirring method, stirring speed, reaction temperature, solution pH value and other process parameters, the morphological characteristics and related properties of particles can be adjusted [12–17].

Qin et al. [18] synthesized quasi-spherical silver nanoparticles with ascorbic acid as reductant and citrate as stabilizer in 30 °C water bath. Spherical silver particles with sizes of 30–72 nm were successfully obtained by changing the pH value of the reaction system. An et al. [19] has proposed a new wet chemistry method for the large-scale preparation of ultra-fine homogeneous spherical silver powder with ascorbic acid as reductant and citrate as stabilizer. It was found that the reaction temperature has an important effect on the particle size. The mean diameter of the particles decreased from 3.5 μm to 1.6 μm as the reaction temperature increased from 8 to 15 °C. Xie et al. [20] used PVP as dispersant, ascorbic acid as reductant and $AgNO_3$ as silver source to prepare highly dispersed silver micro/nanoparticles. It was found that $PVP/AgNO_3$ mass ratio could significantly alter the particle shape and size, while pH and temperature were the main factors influencing the particle size. A possible mechanism for the preparation of highly dispersed silver nanoparticles for shape-controlled synthesis is that PVP is applied to the surface of the particles to make them grow evenly on each surface. At 40 °C, $PVP/AgNO_3$ mass ratio of 0.6 and pH of 7, the mean particle size of silver powder is 0.2 μm. Tan et al. [21] studied the effect of silver ion concentration on the particle size and morphology of ultrafine silver powder and found that the particle size of silver powder increased with the increasing of silver ion concentration. Still, there was no significant change in the morphology of silver powder as the concentration of silver ions increased. This is because the increase of silver ion concentration accelerates the reduction reaction rate, and there is enough silver ion in the system to grow crystal nuclei. Hence, the silver powder particle size is large.

Generally speaking, the electrical conductivity of silver particles can be adjusted by controlling the size, shape, particle size distribution and crystallinity [22]. The regular particle shape and uniform particle size distribution are conducive to the closer filling of silver powder particles in conductive silver paste, so as to form more conductive pathways and improve electrical conductivity. Therefore, studying the preparation and application of micron silver particles is of great significance. In this paper, we present a method for preparing micro-size spherical silver particles. Silver nitrate was used as the silver source and ascorbic acid was used as reducing agent. The influence of different dispersants, the solution pH and the stirring speed on the shape and size of the spherical silver particles were investigated. The effects of sintering temperature and sintering time on the conductivity of the silver paste are discussed.

2. Materials and Methods

2.1. Materials

All materials used in this study are shown in Table 1, and ultrapure water was used in all of the experimental processes.

Table 1. Materials used in this study.

Materials	Chemical Formula	Specification
Silver nitrate	$AgNO_3$	≥99.8%
Ascorbic acid	$C_6H_8O_6$	≥99.7%
Nitric acid	HNO_3	98%
Absolute ethanol	C_2H_6O	95%
Gelatin	—	Industrial-grade
Stearic acid	$C_{18}H_{36}O_2$	Analytical Reagent
succinic acid	$C_4H_6O_4$	Analytical Reagent
glutaric acid	$C_5H_8O_4$	Analytical Reagent

2.2. Preparation of Silver Particles

Figure 1 shows the experimental device for the synthesis of spherical silver particles. The two reactant solutions were prepared as follows: 0.1 mol/L silver nitrate solution was prepared by dissolving 1.7 g silver nitrate in 100 mL ultrapure water, and the pH was adjusted by nitric acid, then stirred to obtain solution 1. The 0.06 mol/L ascorbic acid solution was prepared by dissolved 1.056 g ascorbic acid in 100 mL ultrapure water, then 0.085 g of surfactant (5% by weight of silver nitrate) was added to the solution and stirred evenly to obtain reduction solution 2.

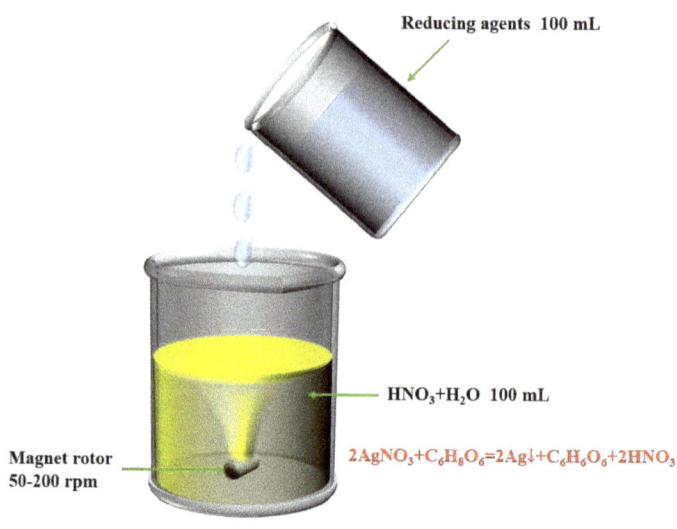

Figure 1. Experimental device for silver particle synthesis.

Figure 2 shows the synthetic process and conditions for preparation of monodispersed silver particles. Solution 2 was quickly added to solution 1 and the mixed solution was stirred at 25 °C (±3 °C) for 30 min. After the stirring was stopped, the solution was precipitated layer by layer. The reacted solution was filtered, the precipitate was washed with ultrapure water and then the precipitate was dried to obtain silver powder.

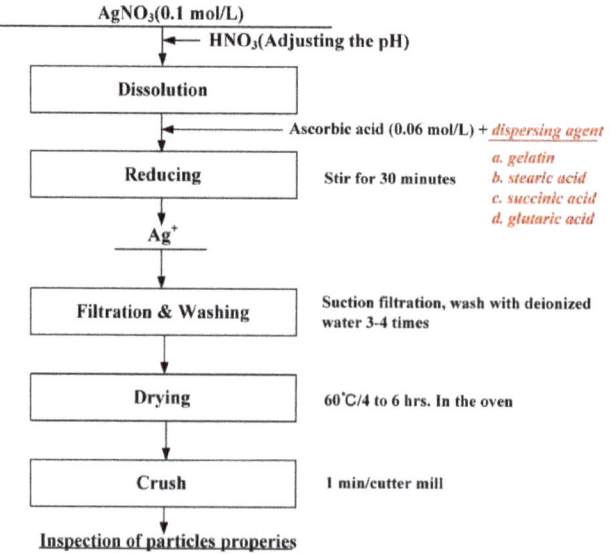

Figure 2. Synthetic flow chart for preparation of monodispersed silver particles.

2.3. Characterization

The surface morphologies of the silver particles were detected using a scanning electron microscope (SEM, JSM-6610A, Tokyo, Japan) operating at an acceleration voltage of 20 kV. The crystal structure of the silver particles was described by X-ray diffraction (XRD, XPert Powder, the Netherlands) with a scanning speed of 8° min^{-1} from 10 to 90°. The particle size distribution of silver powders was detected by a laser particle sizer (Rise-2002, Jinan). The amount of organic residue on the silver particle surface was measured by thermogravimetric analysis (TG, Mettler TGA2, Bern, Switzerland) and temperature rate of 30 to 800 °C using a flow rate of gas is 20 °C/min.

The conductive silver paste was prepared by mixing 0.8 g synthesized silver sample and 0.2 g organic carrier evenly in the agate. Subsequently, the prepared paste was uniformly printed on a polyethylene terephthalate (PET) substrate with screen-printing mesh, to form different shapes of conductive films (100 mm × 10 mm, 1500 mm × 0.3 mm and 1500 mm × 0.4 mm). Then the conductive film was sintered at different temperatures (100–180 °C) and times (5–45 min), and the resistivity of the sintered film was measured by a four-probe instrument.

The calculation formula of the resistivity is:

$$\rho = RS/L \tag{1}$$

where ρ (Ω·cm), R (Ω), S (cm^2) and L (cm) represent the resistivity, surface resistance, cross-sectional area and length of the cured silver paste.

3. Results and Discussion

3.1. Characterization of the Synthetic Silver Particles

The XRD spectrum of the particles prepared with different dispersants is exhibited in Figure 3. The diffraction peaks have five characteristics, corresponding to the crystal faces of (111), (200), (220), (311) and (222) of face-centered cubic silver, which are consistent with the pattern of standard crystal silver card (JCPDS#04-0783). It means that the prepared sample is metallic silver. It shows that the use of different dispersants does not affect the composition of silver powder. The intensity of the peaks reflects the high crystallinity of the silver nanoparticles, and it can be seen from Figure 3 that the silver powder prepared with gelatin as dispersant has the highest crystallinity, so gelatin was used as the dispersant in this experiment.

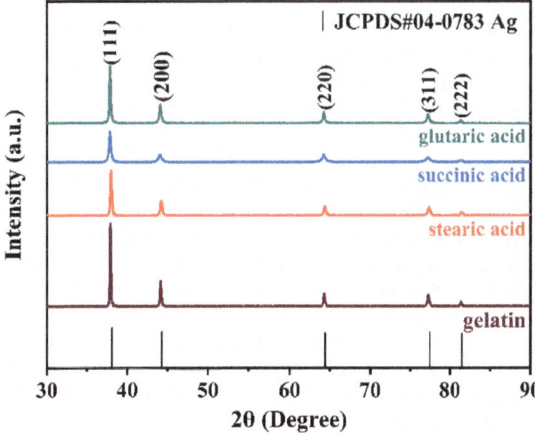

Figure 3. XRD of silver powder prepared with different dispersants.

To study the thermal stability, thermogravimetric analysis was carried out. The thermal analysis of the silver powder (TGA curve) is shown in Figure 4. It shows that the curve drops sharply between 130 °C and 250 °C, which may be due to the volatilization of solvent in the silver powder. The weight loss of the whole sample was only 0.73%, indicating that there was little organic material in the silver nanoparticles.

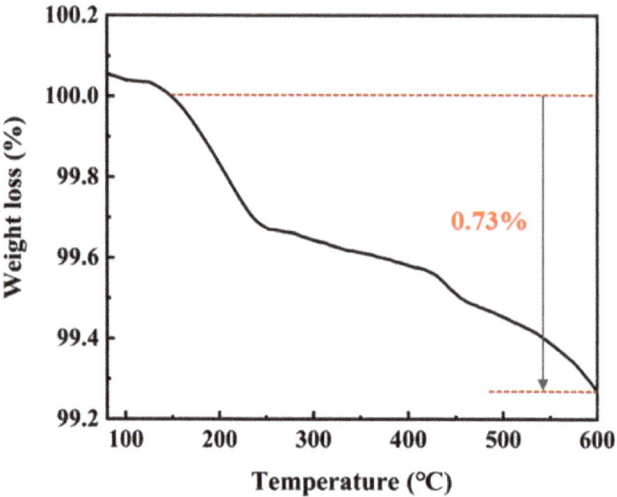

Figure 4. TGA thermogram of silver powder.

3.2. Effects of Experimental Condition on Morphology and Particle Size of Silver Particles

3.2.1. Effect of Dispersant

Using the reaction solution (1) and (2), when 4 g of nitric acid were added, the silver nitrate was stirred at 100 rpm in the beaker and amounts of different dispersants were 5% by weight of silver nitrate (other conditions correspond to the description of the test procedure in Section 2.2). The reaction was performed by adding the solution 2-dropwise to the solution 1 in 2 min. Silver powders made by different dispersants were characterized by SEM, as shown in Figure 5.

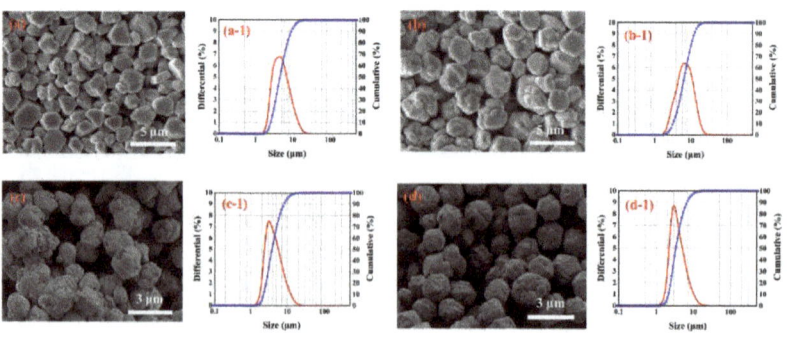

Figure 5. SEM images of silver particles prepared with different dispersants: (**a**) gelatin; (**b**) stearic acid; (**c**) succinic acid; (**d**) glutaric acid. Size distribution of silver particles prepared with different dispersants: (a-1) gelatin; (b-1) stearic acid; (c-1) succinic acid.

As can be seen from Figure 5, different dispersants have a great influence on the shape of silver powder, and the particle size distribution of silver powder formed by four dispersants is relatively concentrated. As illustrated in Figure 5a–d, when gelatin is used as dispersing agent, silver powder is a high-crystal spheroid with smooth crystal surface and good dispersion. The mean particle size is 5 to 8 µm, but when stearic acid, succinic acid and glutaric acid are used as dispersants, the results show that they are spherical in shape and appear to be highly agglomerating. It is obvious that the silver powder prepared with gelatin as a dispersant has better monodispersity than other silver powders, and the particle surface smoothness is higher. Sannohe et al. [23] also found that the polymer compound structure has an important role for silver with monodisperse particles due to the adsorption of polymer compounds on the surface of the growing silver particles.

The relationship between the D_{50} of the silver powder and the dispersant is given in Figure 6. From the results of Figures 5 and 6, it can be seen that the trend of particle size distribution of silver powder is consistent with the SEM results. Therefore, the dispersant is a key factor affecting the appearance and particle size of silver powder, and the selection of suitable dispersant is very important for the synthesis of micron-size spherical silver powder.

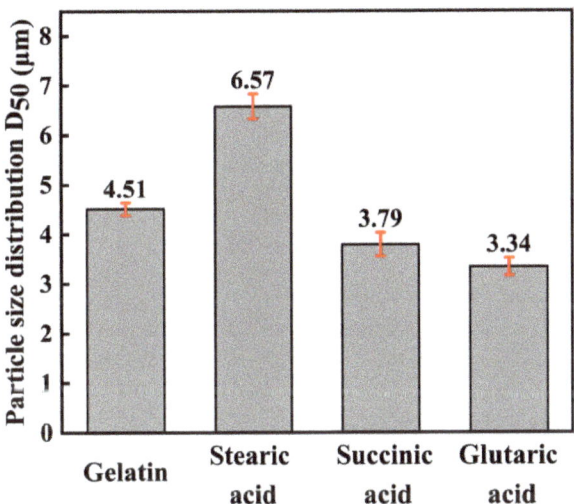

Figure 6. Variation of the average size of silver particles (D_{50}) with different dispersants.

3.2.2. Effect of pH Values

According to the results in Section 3.2.1, the influence of different pH values on the shapes and particle size distribution of the spherical silver powders were studied. The pH of the silver nitrate solution was adjusted to 1.0, 2.0 and 3.0 by adding different amounts of nitric acid to solution 1. The reaction was performed in a dropwise manner with the addition of solution 2 in solution 1 for 2 min, and the other conditions correspond to the description of the test procedure in Section 2.2. The SEM image and particle size distribution of the spherical silver powder are shown in Figure 7.

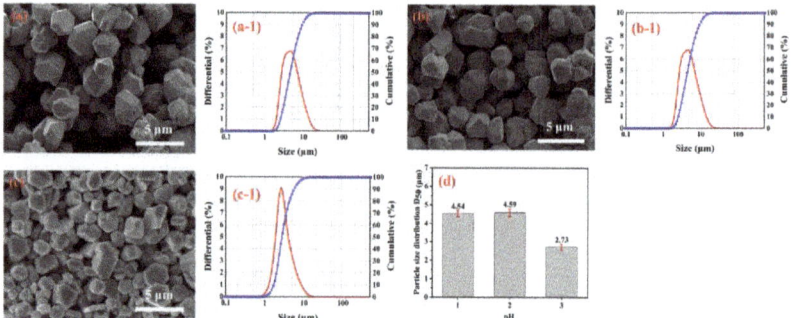

Figure 7. SEM images of silver particles prepared with different pH values: (**a**) pH = 1.0; (**b**) pH = 2.0; (**c**) pH = 3.0; (**d**) Variation of the average size of silver particles (D_{50}) with pH value. Size distribution of silver particles prepared with different pH values: (a-1) pH = 1.0; (b-1) pH = 2.0; (c-1) pH = 3.0.

From Figure 7, with the increase of pH from 1.0 to 3.0, the shape of silver particles is changed from polyhedral to partially spherical, and the mean diameter of silver particles is reduced. A possible reason is that the growth of silver powder is affected by the solution's pH value, which directly affects the ionization degree of ascorbic acid in it. This results in different driving forces for the reduction of silver powder.

The half-reaction of ascorbic acid is $C_6H_6O_6 + 2H^+ + 2e = C_6H_8O_6$. The reduction electrode potential of $C_6H_8O_6$ can be calculated by the following equation: $E = E^0 - 0.059$ pH [24]. Therefore, the reducing ability of ascorbic acid can be adjusted by changing the pH value, and the reducing ability of ascorbic acid increases with the increase of pH value. When the pH value is low, the reduced ability of $C_6H_8O_6$ is relatively low, and only a tiny amount of silver nuclei is generated in the solution. The silver nuclei formed grow in a specific crystal direction, leading to the formation of polyhedron structure [25–27]. With the increase of solution pH value, the reduced ability of $C_6H_8O_6$ solution increases, and more silver nuclei will be formed. The increase in the number of silver nuclei will limit the growth of crystal nuclei, leading to the reduction of silver particle size.

3.2.3. Effect of Stirring Speed

The effect of stirring speed on particle shapes and particle size distribution of spherical silver powders was investigated. The pH of solution 1 was adjusted to 1 and solution 2 was protected with 5% gelatin by weight of silver nitrate. Solution 2 was added dropwise to solution 1 for reaction, and the speed of the reaction solution was adjusted to 50 rpm, 100 rpm, and 200 rpm, the other conditions correspond to the description of the test procedure in Section 2.2.

The SEM images and the particle size distribution of the spherical silver powder at different stirring speeds were presented in Figure 8. From the SEM images a to c, it can be seen that the stirring speed significantly influences the shape of the silver powder. As illustrated in Figure 8, when the stirring speed was 50 rpm, the silver powder was polyhedral and spherical, and its surface was smooth. When the speed of stirring was increased to 100 rpm, there were obvious large and small particles, and the distribution of silver particles was uneven. This may be caused by the agglomerative growth of small particles. In addition, when the stirring speed was increased to 200 rpm, the smaller particles gradually disappeared, and the particle size distribution of the silver particles became more uniform. In this case, the particle size distribution of the silver particles became wider, and the dispersibility became lower. This is probably due to the partial destruction of the interaction between the dispersing agent and the silver particles, which results in the formation of new aggregates at high stirring speed.

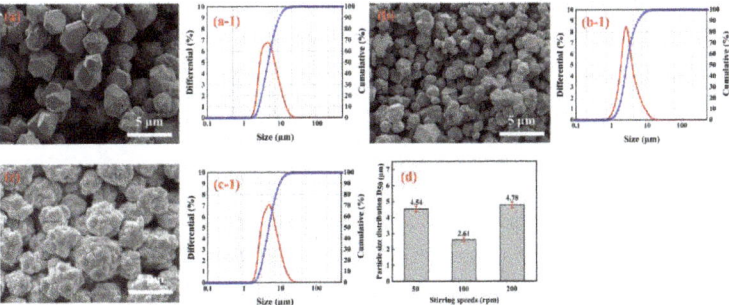

Figure 8. SEM images of silver particles prepared with different stirring speeds: (**a**) rpm = 50; (**b**) rpm = 100; (**c**) rpm = 200; (**d**) Variation of the average size of silver particles (D_{50}) with stirring speeds. Size distribution of silver particles prepared with different stirring speeds: (a-1) rpm = 50; (b-1) rpm = 100; (c-1) rpm = 200.

3.3. Possible Formation Mechanism of Spherical Silver Particles

This study used nitric acid as a pH regulator and gelatin as a stabilizer and dispersant. The results show that the type of dispersant, solution pH and stirring speed play an essential role in the formation and growth of silver particles, which may affect the morphology or distribution of silver particles. Figure 9 shows the mechanism of the growth process of spherical silver particles. Firstly, some Ag^+ nucleates on the surface of polyhedral spherical silver particles grow and penetrate polyhedral, spherical silver particles under the action of diffusion. However, the rest of the Ag^+ nucleates uniformly and contacts the first nucleated silver particles to form aggregates. By aggregate growth, the primary particles will further grow into secondary particles [28]. Due to the adsorption of gelatin on the surface of silver particles, the long chain structure of the gelatin can provide a good space constraint, thereby preventing the aggregation of silver particles and the formation of secondary particles [29]. Consequently, the silver particles are well dispersed, and it is possible to control particle size.

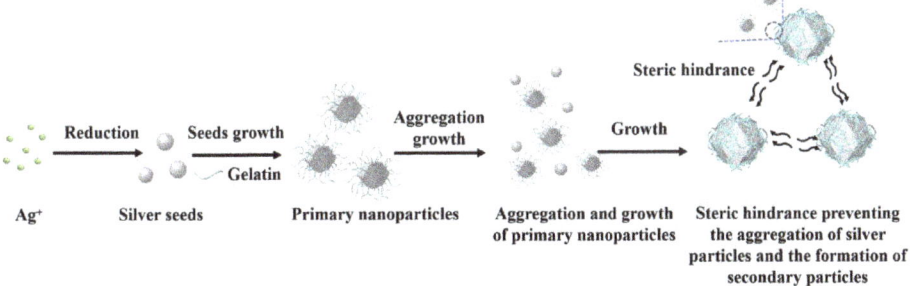

Figure 9. A schematic illustration of the proposed growth process of spherical micrometer silver particles.

3.4. Resistivity of Silver Paste

The silver particles used in the preparation of conductive silver paste were carried out under the following conditions: silver nitrate 0.1 mol/L, ascorbic acid 0.06 mol/L, the content of gelatin was 5% by weight of silver nitrate, and pH = 1.0, and the speed of the reaction solution was adjusted to 50 rpm. The conductive silver paste was prepared according to the description of the test process in 2.3 and printed on polyethylene terephthalate (PET) substrate to form different shapes of conductive films (100 mm × 10 mm,

1500 mm × 0.3 mm and 1500 mm × 0.4 mm). The screen-printing mesh was shown in Figure 10a, and the printed conductive films were shown in Figure 10b.

Figure 10. The screen-printing mesh (**a**) and the printed conductive films (**b**).

Measure the thickness of the conductive film using a micrometer (thickness of the conductive film printed on the PET substrate minus the thickness of the PET substrate). Using the resistance value measured by the four probes and the thickness information measured by the micrometer, the resistivity of the conductive film can be calculated. The thickness of five different points is measured on a conductive film pattern, the average of which is used to calculate resistivity. The average resistivity of three different conductive film patterns is shown in Table 2. The average resistivity of the three conductive films was calculated according to Equation (1) in Section 2.3, and the results are shown in Figure 11. The error bar in Figure 11 was based on the standard.

Table 2. The average resistivity of three different conductive film patterns.

Notations	Length (mm)	Width (mm)	Thickness (μm)	Resistance (Ω)	Resistivity (Ω·cm)
Line 1	100	10	6.7	5.42	3.63×10^{-5}
Line 2	1500	0.3	6.3	2777.78	3.58×10^{-5}
Line 3	1500	0.4	5.6	2 343.75	3.50×10^{-5}

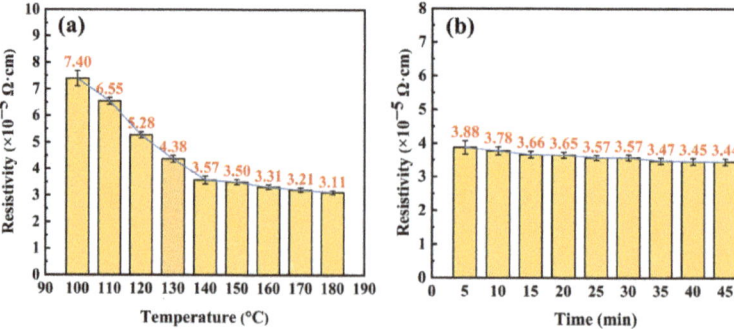

Figure 11. The average resistivity of conductive films: (**a**) Resistivity versus sintering temperature (**b**) Resistivity versus sintering time.

The sintering process of the conductive silver paste has an extremely important impact on its conductivity. The proper sintering temperature and sintering time are conducive to the formation of a low resistivity conductive silver film [30,31]. Figure 11a shows the average resistivity of the three conductive films versus different temperatures (100–180 °C) for 10 min. It can be clearly seen that the resistivity decreased slowly with the increasing of sintering temperature. When the temperature is 100 °C, the average resistivity is 7.40×10^{-5} $\Omega \cdot$cm. Increased the sintering temperature to 140 °C, the average resistivity drops to 3.57×10^{-5} $\Omega \cdot$cm. However, with the sintering temperature increased to 140–180 °C, the average resistivity changes slowly from 3.57×10^{-5} $\Omega \cdot$cm to 3.11×10^{-5} $\Omega \cdot$cm. This is mainly because most of the solvent in the silver film has evaporated and the silver particles were effectively connected with the shrinkage of the film. Only a small portion of the solvent is volatilized by subsequent reheating, which has little effect on the electrical properties.

Figure 11b shows the average resistivity of the three conductive films versus different sintering times (5–45 min) at 140 °C. It shows that the average resistivity changed little with the increased sintering time, and the average resistivity is 3.88×10^{-5} $\Omega \cdot$cm after sintering at 140 °C for 5 min. When the sintering time increased to 30 min, the resistivity dropped a little to 3.57×10^{-5} $\Omega \cdot$cm, and when the sintering time was increased to 35–45 min, the average resistivity dropped slowly from 3.47×10^{-5} $\Omega \cdot$cm to 3.44×10^{-5} $\Omega \cdot$cm. The possible reason is due to the silver film being almost sintered completely when sintered at 140 °C for 5 min and could not be further reduced substantially by prolonging the sintering time.

It had been reported that a silver paste consisting of 80 wt.% Ag nanoparticles, 1.0 wt.% lead-free glass material and 19 wt.% organic carrier have a volume resistivity of 4.11×10^{-6} $\Omega \cdot$cm after sintering at temperatures ranging from 250 °C to 450 °C [32]. However, the conductive silver paste we studied has a lower sintering temperature (140 °C).

4. Conclusions

A simple, efficient and fast method for the preparation of micron-size spherical silver powder was presented in this paper. Silver nitrate as precursor, gelatin as dispersant and ascorbic acid as reducing agent were used to produce silver powder with an average particle size of 5–8 μm. The results indicate that the type of dispersant, the pH value of the solution and the stirring speed have significant effects on the powder's morphology and particle size distribution of silver powder. The conductive silver paste produced by the prepared micron-size spherical silver powder has an excellent electrical conductivity of 3.57×10^{-5} $\Omega \cdot$cm after sintering at 140 °C for 30 min, which indicates that the prepared micron-size spherical silver powders can be used as a conductive silver paste.

Author Contributions: N.L.: writing—original draft preparation, conceptualization, methodology, and investigation; X.W.: visualization, Supervision; Y.N.: resources, Validation; Y.G.: supervision and writing—review and editing; J.L.: formal analysis, supervision and writing—review and editing; G.C.: visualization and funding acquisition; S.J.: investigation, validation, and funding acquisition. All authors have read and agreed to the published version of the manuscript.

Funding: This work was financially supported by the National Natural Science Foundation of China (No. 51964032).

Institutional Review Board Statement: Not applicable.

Informed Consent Statement: Not applicable.

Data Availability Statement: The data presented in this study are available in the article.

Conflicts of Interest: The authors declare no conflict of interest.

References

1. Lee, K.J.; Jun, B.H.; Choi, J.R.; Lee, Y.I.; Joung, J.; Oh, Y. Environmentally friendly synthesis of organic-soluble silver nanoparticles for printed electronics. *Nanotechnology* **2007**, *18*, 335601–335605. [CrossRef]
2. Cano-Raya, C.; Denchev, Z.Z.; Cruz, S.F.; Viana, J.C. Chemistry of solid metal-based inks and pastes for printed electronics—A review. *Appl. Mater. Today* **2019**, *15*, 416–430. [CrossRef]
3. Derakhshankhah, H.; Mohammad-Rezaei, R.; Massoumi, B.; Abbasian, M.; Rezaei, A.; Samadian, H.; Jaymand, M. Conducting polymer-based electrically conductive adhesive materials: Design, fabrication, properties, and applications. *J. Mater. Sci. Mater. Electron.* **2020**, *31*, 10947–10961. [CrossRef]
4. Aradhana, R.; Mohanty, S.; Nayak, S.K. A review on epoxy-based electrically conductive adhesives. *Int. J. Adhes. Adhes.* **2020**, *99*, 102596. [CrossRef]
5. Cheng, Y.; Zhang, J.; Fang, C.; Qiu, W.; Chen, H.; Liu, H.; Wei, Y. Preparation of Low Volatile Organic Compounds Silver Paste Containing Ternary Conductive Fillers and Optimization of Their Performances. *Molecules* **2022**, *27*, 8030. [CrossRef] [PubMed]
6. Tsai, J.T.; Lin, S.T. Silver powder effectiveness and mechanism of silver paste on silicon solar cells. *J. Alloys Compd.* **2013**, *548*, 105–109. [CrossRef]
7. Yang, M.; Chen, X.H.; Wang, Z.D.; Zhu, Y.Z.; Pan, S.W.; Chen, K.X.; Wang, Y.L.; Zheng, J.Q. Zero→Two-Dimensional Metal Nanostructures: An Overview on Methods of Preparation, Characterization, Properties, and Applications. *Nanomaterials* **2021**, *11*, 1895. [CrossRef] [PubMed]
8. Koo, H.Y.; Yi, J.H.; Kim, J.H.; Ko, Y.N.; Jung, Y.N.; Kang, Y.C.; Lee, J.H. Conductive silver films formed from nano-sized silver powders prepared by flame spray pyrolysis. *Mater. Chem. Phys.* **2010**, *124*, 959–963. [CrossRef]
9. Gan, W.; Pan, Q.; Zhang, J. Effect of Silver Powder for Back Silver Paste on Properties Of Crystalline Silicon Solar Cells. *J. Gan Semicond. Optoelectron.* **2014**, *35*, 1016. [CrossRef]
10. Ye, X.Y.; Xiao, X.Q.; Zheng, C.; Hua, N.B.; Huang, Y.Y. Microemulsion-assisted hydrothermal synthesis of mesoporous silver/titania composites with enhanced infrared radiation performance. *Mater. Lett.* **2015**, *152*, 237–239. [CrossRef]
11. Lai, Y.B.; Guo, Z.C.; Huang, H.; Zhang, H.Y.; Wang, S. Synthesis of Monodisperse Micrometer-sized Spherical Silver Particles. *J. Mater. Sci. Technol.* **2014**, *32*, 221. [CrossRef]
12. Prozorova, G.F.; Korzhova, S.A.; Kon'kova, T.; Ermakova, T.G.; Pozdnyakov, A.S.; Sukhov, B.G.; Arsentyev, K.Y.; Likhoshway, K.Y.; Trofimov, B.A. Specific features of formation of silver nanoparticles in the polymer matrix. *Dokl. Chem.* **2011**, *437*, 47–49. [CrossRef]
13. Fadli, A.L.; Hanifah, A.; Fitriani, A.; Rakhmawati, A.; Dwandaru, W. Application of silver-chitosan nanoparticles as a prevention and eradication of nosocomial infections due to *Staphylococcus aureus* sp. In *AIP Conference Proceedings*; AIP Publishing LLC: Long Island, NY, USA, 2018. [CrossRef]
14. Haider, A.; Kang, I.K. Preparation of Silver Nanoparticles and Their Industrial and Biomedical Applications: A Comprehensive Review. *Mater. Sci. Eng. A* **2015**, *2015*, 16. [CrossRef]
15. Venkatesham, M.; Ayodhya, D.; Madhusudhan, A.; Babu, N.V.; Veerabhadram, G. A novel green one-step synthesis of silver nanoparticles using chitosan: Catalytic activity and antimicrobial studies. *Appl. Nanosci.* **2014**, *4*, 113–119. [CrossRef]
16. Guo, G.Q.; Gan, W.P.; Luo, J.A.; Xiang, F.; Zhang, J.L.; Zhou, H.; Liu, H.A. Preparation and dispersive mechanism of highly dispersive ultrafine silver powder. *Appl. Surf. Sci.* **2010**, *256*, 6683–6687. [CrossRef]
17. Gan, W.; Luo, J.; Guo, G.; Xiang, F.; Liu, H. Preparation of ultra-fine silver powder used in electronic paste by chemical reduction. *J. Electron. Mater.* **2010**, *29*, 15–18. [CrossRef]
18. Qin, Y.Q.; Ji, X.H.; Jing, J.; Liu, H.; Wu, H.L.; Yang, W.S. Size control over spherical silver nanoparticles by ascorbic acid reduction. *Colloids Surf. A Physicochem. Eng. Asp.* **2010**, *372*, 172–176. [CrossRef]
19. An, B.; Cai, X.H.; Wu, F.S.; Wu, Y.P. Preparation of micro-sized and uniform spherical Ag powders by novel wet-chemical method. *T. Nonferr. Metal. Soc.* **2020**, *20*, 1550–1554. [CrossRef]
20. Xie, W.; Zheng, Y.Y.; Kuang, J.C.; Wang, Z.; Yi, S.H.; Deng, Y.J. Preparation of Disperse Silver Particles by Chemical Reduction. *Russ. J. Phys. Chem.* **2016**, *90*, 848–855. [CrossRef]
21. Tan, F.; Wang, H.; Wang, W.; Chen, J.; Qiao, X. Morphology and size control of spherical conductive silver particles. *J. Electron. Mater.* **2011**, *30*, 52. [CrossRef]
22. Cai, X.H.; Chen, X.C.; An, B.; Wu, F.S.; Wu, Y.P. The effects of bonding parameters on the reliability performance of flexible RFID tag inlays packaged by anisotropic conductive adhesive. In Proceedings of the 2009 International Conference on Electronic Packaging Technology & High Density Packaging, Beijing, China, 10–13 August 2009. [CrossRef]
23. Sannohe, K.; Ma, T.L.; Hayase, S. Synthesis of monodispersed silver particles: Synthetic techniques to control shapes, particle size distribution and lightness of silver particles. *Adv. Powder Technol.* **2019**, *30*, 3088–3098. [CrossRef]
24. Liu, C.; Fu, Q.; Zou, J.; Huang, Y.; Zeng, X.; Cheng, B. Effects of polyester resin molecular weight on the performance of low temperature curing silver pastes. *J. Mater. Sci. Mater. Electron.* **2016**, *27*, 6511–6516. [CrossRef]
25. Tian, Q.H.; Deng, D.; Li, Y.; Guo, X.Y. Preparation of ultrafine silver powders with controllable size and morphology. *T. Nonferr. Metal Soc.* **2018**, *28*, 524–533. [CrossRef]
26. Ajitha, B.; Ashok, K.; Reddy, P.S. Enhanced antimicrobial activity of silver nanoparticles with controlled particle size by pH variation. *Powder Technol.* **2015**, *269*, 110–117. [CrossRef]

27. Bai, X.H.; Li, W.; Du, X.S.; Zhang, P.; Lin, Z.D. Synthesis of spherical silver particles with micro/nanostructures at room temperature-ScienceDirec. *Compos. Commun.* **2017**, *4*, 54–58. [CrossRef]
28. Gu, S.; Wei, W.; Hui, W.; Tan, F.; Qiao, X. Effect of aqueous ammonia addition on the morphology and size of silver particles reduced by ascorbic acid. *J. Chem. Pow. Technol.* **2013**, *233*, 91–95. [CrossRef]
29. Li, Y.F.; Gan, W.P.; Liu, X.G.; Lin, T.; Huang, B. Dispersion mechanisms of Arabic gum in the preparation of ultrafine silver powder. *Korean. J. Chem. Eng.* **2014**, *31*, 1490–1495. [CrossRef]
30. Deng, D.; Huang, C.; Ma, J.; Bai, S. Reducing resistance and curing temperature of silver pastes containing nanowires. *J. Mater. Sci. Mater. Electron.* **2018**, *29*, 10834–10840. [CrossRef]
31. Huang, W.W.; Qiu, H.J.; Zhang, Y.Q.; Zhang, F.; Gao, L.; Omran, M.; Chen, G. Microstructure and phase transformation behavior of Al_2O_3-ZrO_2 under microwave sintering. *Ceram. Int.* **2022**, *49*, 4855–4862. [CrossRef]
32. Park, K.J.; Seo, D.S.; Lee, J.K. Conductivity of silver paste prepared from nanoparticles. *Colloids Surf. A-Physicochem. Eng. Aspects.* **2008**, *313*, 351–354. [CrossRef]

Disclaimer/Publisher's Note: The statements, opinions and data contained in all publications are solely those of the individual author(s) and contributor(s) and not of MDPI and/or the editor(s). MDPI and/or the editor(s) disclaim responsibility for any injury to people or property resulting from any ideas, methods, instructions or products referred to in the content.

Article

Study on the Roasting Process of Guisha Limonite Pellets

Chuang Zhang, Xiaolei Zhou *, Lei Gao * and Haoyu Fang

Faculty of Metallurgical and Energy Engineering, Kunming University of Science and Technology, Kunming 650093, China
* Correspondence: zhouxiaolei81@163.com (X.Z.); glkust2013@hotmail.com (L.G.)

Abstract: In this paper, a pelletizing method has been researched to enhance the subsequent iron-making process applying Guisha limonite, with advantages including large reserves and low price. The purpose is to provide an alternative for the sinter, thus reducing the greenhouse gas emission during the iron-making process. The response surface method is used to optimize the experimental design of the pelleting process. A multivariate regression model for estimating the compressive strength of pellets was developed using Box–Behnken experimental methodology, where the relevant factors were the roasting temperature, pellet diameter, and bentonite content. The maximum influencing factors of each experimental design response are determined using analysis of variance (ANOVA). Under optimum conditions, the compressive strength of pure limonite pellets is 2705 N, similar to the response goal value of 2570.3 N, with a relative error of 5.20%. Since the high-grade iron ore resources are depleted, the comprehensive utilization of ore resources is becoming increasingly important. The aim of this paper was to provide a valuable technical foundation for lignite pellet-roasting processes in the iron and steel industries, since steel companies is increasing its imports of Guisha limonite.

Keywords: bentonite; pellet ore; pure limonite; response surface; DOE

Citation: Zhang, C.; Zhou, X.; Gao, L.; Fang, H. Study on the Roasting Process of Guisha Limonite Pellets. *Materials* 2022, *15*, 8845. https://doi.org/10.3390/ma15248845

Academic Editor: Pramod Koshy

Received: 21 October 2022
Accepted: 7 December 2022
Published: 11 December 2022

Publisher's Note: MDPI stays neutral with regard to jurisdictional claims in published maps and institutional affiliations.

Copyright: © 2022 by the authors. Licensee MDPI, Basel, Switzerland. This article is an open access article distributed under the terms and conditions of the Creative Commons Attribution (CC BY) license (https://creativecommons.org/licenses/by/4.0/).

1. Introduction

The iron and steel industry is one of the significant contributors to CO_2 emissions and faces the challenge of a deficiency in high-grade ores. Sintering is an important preprocess before iron making in the blast furnace. However, the sintering process has a long flow, with enormous greenhouse gas emissions and high energy consumption [1–3]. Pellets can improve the breathability in the blast furnace, which increases the reduction process's efficiency and reduces greenhouse gas emissions. Additionally, the application of pellets can increase the tolerance of low-grade ores and thus is noticed as a cost-saving technique compared to the application of sinter [4–6].

Guisha limonite is mainly produced in Laojie Province as a typical lean ore, with abundant reserves and affordable pricing. Guisha limonite ($2Fe_2O_3 \cdot H_2O$) is a kind of hydrous iron oxide ore after weathering and has low iron content. Although significant progress was made in applying Guisha limonite in sinter production, research on the production of Guisha limonite pellets is still quite limited [7]. Because of the low sintering strength and high reduction pulverization rate of Guisha limonite, it is not easy to further improve the sintering process with Guisha limonite as raw material. To solve this problem, industries in various countries have increased the development and utilization of limonite pellets. With increased Guisha limonite imports in iron and steel companies, research on Guisha limonite pellets has become a hot topic recently. The consolidation mechanism of Guisha limonite can be generally divided into three steps: (1) Fe_2O_3 grain development, (2) expansion of the grain, and (3) interconnection of the grains into a complete consolidation, which is commonly referred to as solid-phase consolidation. Many experiments have revealed that Fe_2O_3 crystallization is a process that progresses from primary crystals to developed crystals and finally to interconnected crystals, with primary crystals typically forming at about 1150–1200 °C. At 1220–1250 °C, developed crystals form; above 1280 °C, interconnected crystals form [8–10]. The crystallization process is critical to the compressive strength of pellets [11]. As limonite is a highly crystalline water ore, the desorption of crystalline water occurs at 200–500 °C, which increases the internal stress of the pellets and causes pellet blast [12,13]. At the same time, during the hardening process of the pellets, the crystal water converts into steam under high temperature, escaping from the pellets, resulting in the generation of cracks and pores. As a result, the process reduces the compressive strength of the pellets. The motivation behind this study was to optimize the roasting conditions of low-grade limonite to produce pellets with high-quality characteristics. The results may provide suggestions for the utilization of limonite.

In the past, research on the compressive strength of pellets mostly used single-factor analysis or orthogonal testing to qualitatively describe the impact factor of each parameter. However, the influences of the interaction of various aspects have not been considered [14]. The predictability is poor because the number of trials is small [15,16]. The response surface methodology is utilized in this paper to optimize the experimental design of the preparation of limonite pellets [17–19]. The experimental factors affecting the strength of limonite pellets were studied systematically. The optimal roasting process was obtained, including the bentonite content, roasting temperature, and pellet diameter, which improved pellet thermal burst temperature, compressive strength, and so on [20,21]. The attainment of complete resource utilization and pollutant emission reduction is projected to provide a valuable technical foundation for the roasting process of limonite pellets in the iron and steel industries.

2. Materials and Methods

A ball mill was used to grind Guisha limonite for 72 h prior to this experiment, and an automatic screening machine was used to complete the screening. We used 200-mesh Guisha limonite powder (200 g) and combined it with bentonite in a specific proportion (0.8wt%, 1.0wt%, 1.2wt%, 1.4wt%, 1.6wt% of the mineral powder's mass). Distilled water (3.0wt%) was added equally to the mineral powder for wetting. Distilled water (7.0wt%) was sprayed to produce pellets with a diameter of 11–15 mm. The pellets were dried for 1 h in a drying oven at 110 °C to eliminate moisture and volatiles for subsequent use. Figure 1 depicts the experimental procedure, where the experiment was carried out in 75 groups. The details of the experimental design are shown in Table 1. During the experiments, 20 pellets from each group were placed in corundum crucibles and subsequently placed in a resistance furnace for calcination in an air atmosphere. After natural cooling of the samples in air, we tested the compressive strength of the pellets according to GB/T14201-93, namely "Determination of Compressive Strength of Iron Ore Pellets" [14]. The change in appearance of the pellet samples during the roasting process is shown in Figure 2.

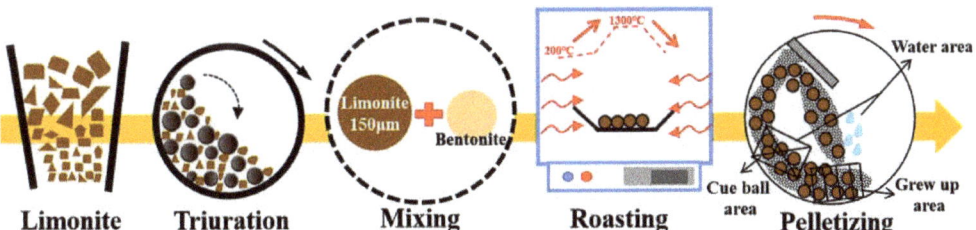

Figure 1. Limonite pellet-roasting process.

Table 1. Experimental design.

Factor	Variable	Level				
		1	2	3	4	5
Roasting temperature/°C	A	1100	1150	1200	1250	1300
Pellet diameter/mm	B	9	10	11	12	13
Bentonite content/%	C	0.8	1.0	1.2	1.4	1.6

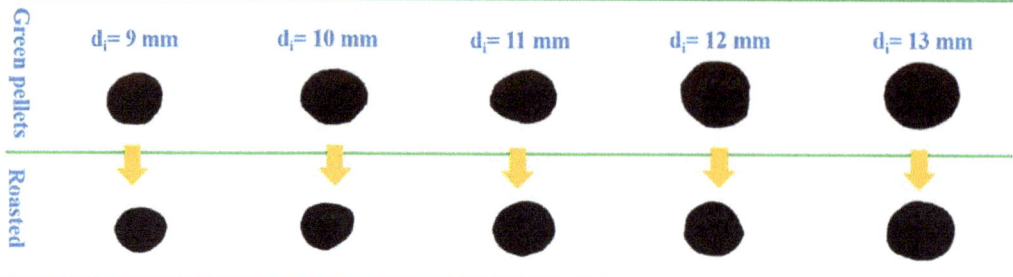

Figure 2. Different sizes of green pellets and roasted pellets.

3. Results and Discussion

3.1. Experimental Materials

The Guisha limonite utilized in this experiment has a rough surface, a loose and porous structure, and small crystals. It is a typical highly crystalline ferrihydrite with a morphology resembling seafloor coral and a typical limonite crystalline condition. This largely determines that it has the characteristics of high porosity, strong water absorption, high wetting capacity, and easy melting. An SEM image of limonite is shown in Figure 3. Guisha limonite has 54.67% total iron concentration, making it a relatively low-grade mineral. The mass fraction of SiO_2 sourced from gangue is 4.04%. It has a large variety of impurities, but the amount is minimal, and 14.82% of it is total moisture content. The different wettability of the raw materials mainly causes the difference in the growth of green pellets. When the raw material has good wettability, the water will form a film that adheres tightly to the pellet's surface. After colliding with another pellet, a liquid bridge is formed, and the two pellets start to bond under the action of surface tension, which promotes growth. It can be found from the SEM image that the limonite particles of Guisha limonite sample are granular, with a rough surface and loose and porous structure. Guisha limonite has strong water absorption, and the wetting capacity of mineral powder is large. This means that limonite has strong surface hydrophilicity in the process of pelletizing, forming a more liquid film, which tends to bond more iron ore particles, which is beneficial to the growth of pellets. The Guisha limonite samples used in this study were from A certain Iron and Steel company in Yunnan Province, the chemical composition of which is shown in Table 2. X-ray diffraction was used to describe the material, and the limonite XRD pattern is displayed in Figure 4. The pattern revealed that FeOOH, which contains a significant amount of crystal water, is the primary mineral form of limonite. The bentonite selected in the experiment contains 85.97% montmorillonite, the montmorillonite content is high, and the physical and chemical properties of the bentonite are also good. The data are shown in Table 3.

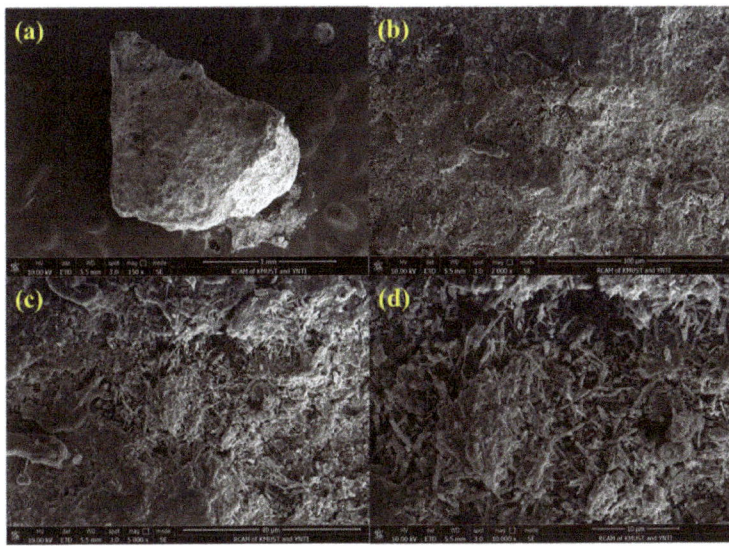

Figure 3. SEM of Guisha limonite: (**a**) 150×, (**b**) 2000×, (**c**) 5000×, (**d**)10,000×.

Table 2. Iron ore chemical composition.

w(TFe)	w(FeO)	w(SiO$_2$)	w(S)	w(MnO)	w(TiO$_2$)	w(Pb)	w(Zn)	w(K$_2$O)	w(Na$_2$O)	w(Cu)	w(V$_2$O$_5$)	Burning Loss
54.67	0.29	4.04	0.048	3.47	0.27	0.009	0.018	0.095	0.001	0.009	0.040	14.82

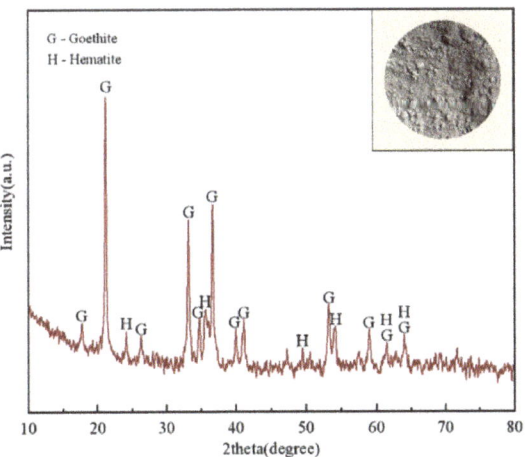

Figure 4. XRD pattern of Guisha limonite.

Table 3. The chemical composition of bentonite used in the experiment.

Moisture/%	Colloid Index/(%·(3g)$^{-1}$)	Swelling Capacity/(mL·g^{-1})	Water Absorption/%	Methylene Blue Index/(g·(100g)$^{-1}$)	Montmorillonite Mass Fraction/%
9.38	20.0	34.5	408	38.0	85.97

Crystal water desorption is a critical step in the limonite roasting process, primarily affected by oxidized pellets' quality. The adsorbed water and crystal water of Guisha limonite will be lost during the roasting process as the material temperature rises. Figure 5 depicts the thermal–thermogravimetric (TG-DTA) curve of Guisha limonite samples, showing that the limonite has a noticeable weight loss before 220 °C and the initial weight loss is 3.33%. The weight loss of the sample is primarily caused by the loss of primarily surface moisture when the temperature is less than 220 °C, based on thermodynamics. When the temperature rises above 220 °C, the crystal water in the minerals begins to evaporate. When the temperature was raised to 220 °C, the limonite decomposed violently, resulting in an endothermic peak in the DTA curve, as shown in Equation (1).

$$nFe_2O_3 \cdot mH_2O \rightarrow n\gamma\text{-}Fe_2O_3 + mH_2O \tag{1}$$

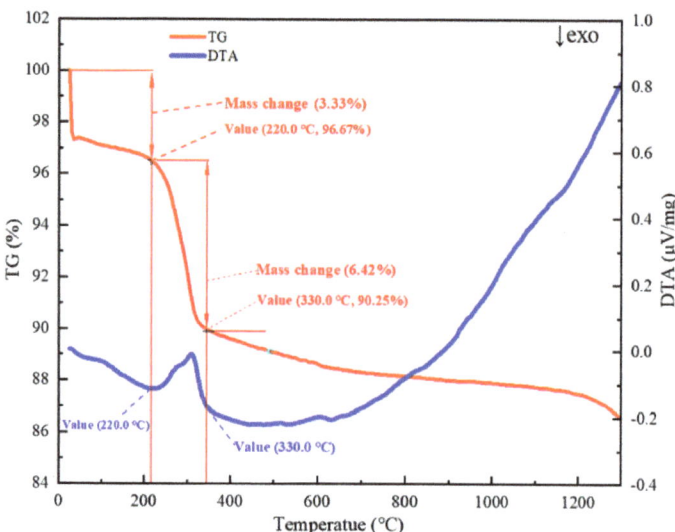

Figure 5. TG-DTA curve of limonite roasting process.

At this stage, only the conversion of $nFe_2O_3 \cdot mH_2O$ to $\gamma\text{-}Fe_2O_3$ takes place; no lattice transformation occurs. However, $\gamma\text{-}Fe_2O_3$ is not stable. When the temperature exceeds 330 °C, the breakdown of limonite ceases. Since the $\gamma\text{-}Fe_2O_3$-phase is unstable, as shown in Equation (2), the crystals will rearrange at high temperatures [22].

$$\gamma\text{-}Fe_2O_3 \rightarrow \alpha\text{-}Fe_2O_3 \tag{2}$$

$\gamma\text{-}Fe_2O_3$ is converted into $\alpha\text{-}Fe_2O_3$, and when the temperature is too high, part of the cations directly diffuse into the second stage, i.e., $nFe_2O_3 \cdot mH_2O$ is transformed into $\alpha\text{-}Fe_2O_3$. The sample's quality gradually deteriorated as the temperature rose, and the crystal water left in the minerals was lost. When the temperature is elevated to 1300 °C, the sample's weight loss can exceed 13%. Figure 6 depicts the microstructure evolution of limonite pellets at various temperatures. It is clear that as the temperature rises, the pores on the surface of the pellets gradually expand due to the removal of crystal water.

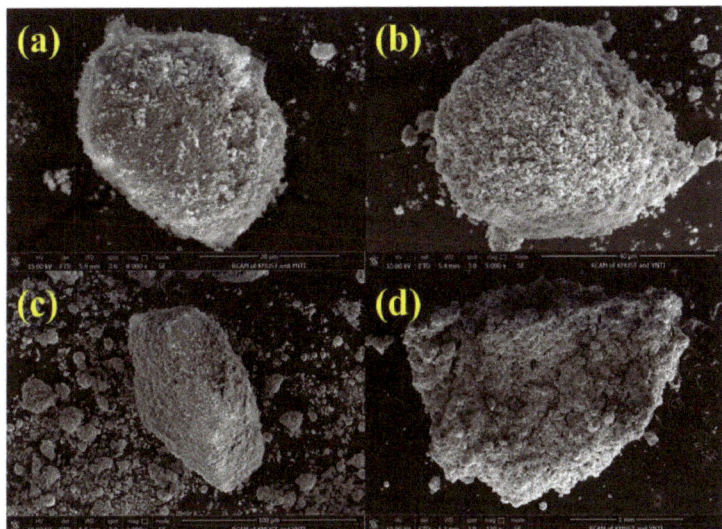

Figure 6. SEM of the evolution of limonite pellets at different temperatures: (**a**) 0 °C, (**b**) 200 °C, (**c**) 800 °C, (**d**) 1250 °C.

3.2. Single-Factor Test

3.2.1. Influence of Roasting Temperature on Compressive Strength of Pellets

The diameter of pellets was controlled to be 9–13 ± 0.5mm, and the addition of bentonite was 0.8wt%, 1.0wt%, 1.2wt%, 1.4wt%, and 1.6wt%. The experiments were carried out in 25 groups to study the effect of roasting temperature on the compressive strength of pellets, with 20 pellets for each group. Figure 7 depicts the experimental outcomes. The normality test and linear analysis were carried out for 25 groups of experimental data. The analysis results are shown in Figure 8. The effect of roasting temperature on pellet compressive strength is substantial. When the temperature is between 1100 °C and 1250 °C, the compressive strength of pellets increases in a positive relationship with the roasting temperature, and the strength is virtually the same between 1250 °C and 1300 °C. As a result, the ideal roasting temperature for pure limonite pellets is around 1250 °C 1300 °C. This finding is because the strength of the pellets is mainly determined by the iron ore's oxidative recrystallization process. The volume of the pellets will alter during the crystallization process. The higher the degree of crystallization, the faster the pellets expand as the temperature rises. It generates fissures in the pellets, diminishing their strength, and the effect on strength becomes more evident as the temperature rises.

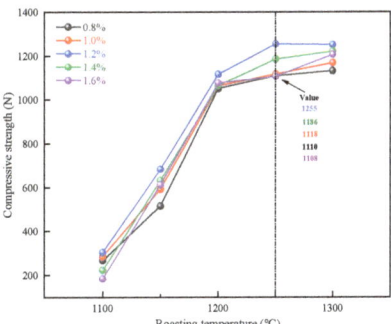

Figure 7. Effect of roasting temperature on the compressive strength of pellets.

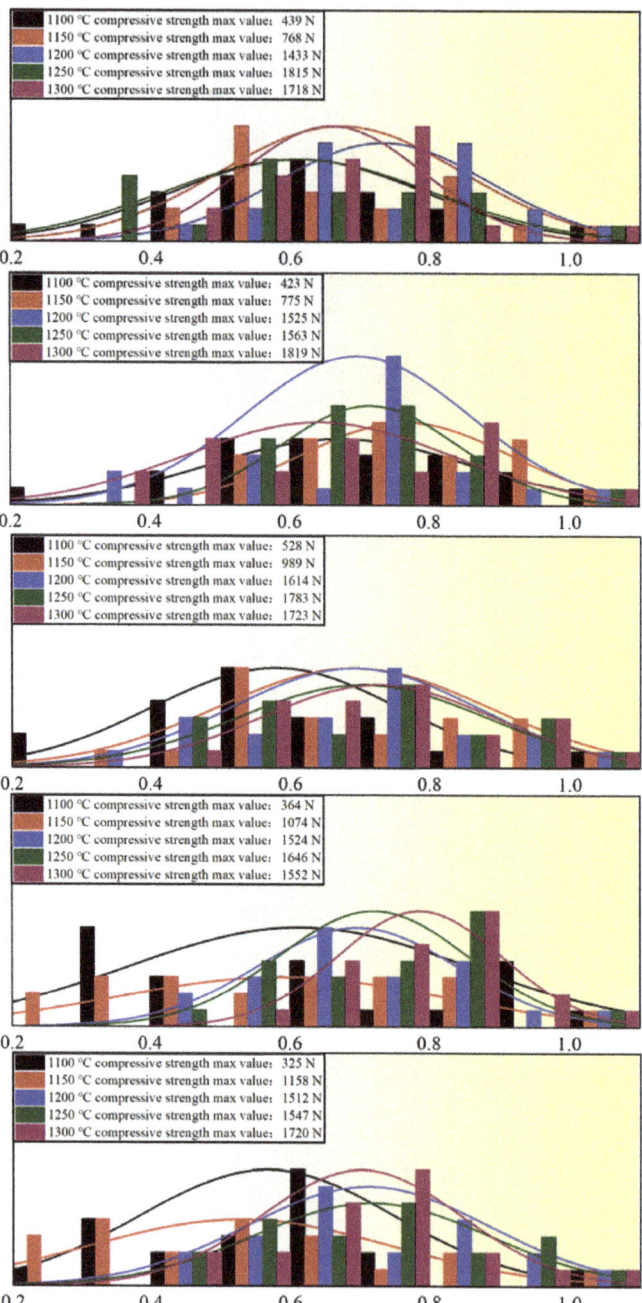

Figure 8. Normality test of pellet compressive strength under the influence of roasting temperature.

3.2.2. Influence of Roasting Temperature on Compressive Strength of Pellets

To investigate the influences of pellet diameter on compressive strength, the bentonite addition amount was regulated at 0.8wt%, 1.0wt%, 1.2wt%, 1.4wt%, and 1.6wt%, and the calcination temperature was 1250 °C. The experiments were carried out in 25 groups, with

20 pellets for each group. Pellets with diameters ranging from 9 mm to 13 mm were used for the compressive strength test. Figure 9 depicts the experimental findings. For 25 groups of experimental data, the normality test and linear analysis were performed. Figure 10 illustrates the analysis results. The compressive strength of the pellets does not vary much within a diameter range of 9–10 mm. It shows a growing tendency with the rise of the pellet diameter within the diameter of 10–13 mm, and the range exceeds 1000 N. This demonstrates that pellet diameter has a significant influence on pellet compressive strength. However, if the pellet diameter is too large, the unit heat consumption will be excessive, the oxidation will be insufficient, and the roasting and chilling time will be extended. As a result, the ideal pellet diameter is around 13 mm.

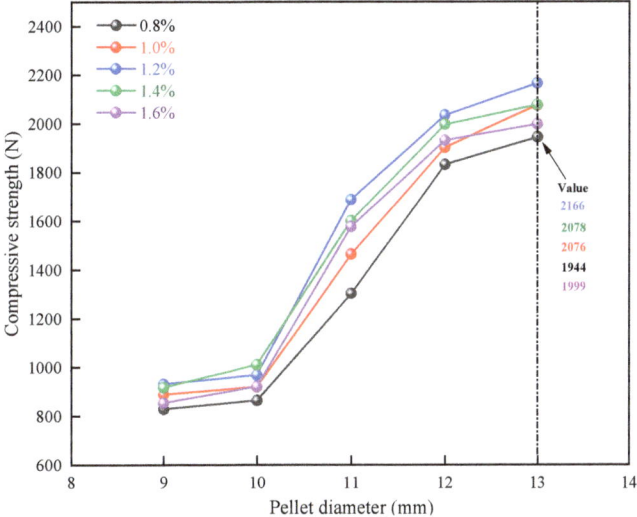

Figure 9. Effect of pellet diameter on the compressive strength of pellets.

3.2.3. Influence of Bentonite Addition on Compressive Strength of Pellets

The diameter of the pellets was controlled to be 9 mm, 10 mm, 11 mm, 12 mm, and 13 mm, and the calcination temperature was 1250 °C. The experiments were carried out in 25 groups to study the effect of the additional amount of bentonite on the compressive strength of the pellets, with 20 pellets for each group. The experimental results are shown in Figure 11. Twenty-five groups of experimental data were examined using the normality test and the linear analysis. Figure 12 displays the analysis findings. With the increased bentonite content, the pellet strength increases and decreases. When the bentonite content is below 1.2wt%, there is a positive correlation, and the effect is very significant; when the bentonite content is between 1.2wt% and 1.6wt%, the strength gradually decreases with the increase in bentonite content. It can be seen that the optimum addition amount of bentonite is about 1.2wt%.

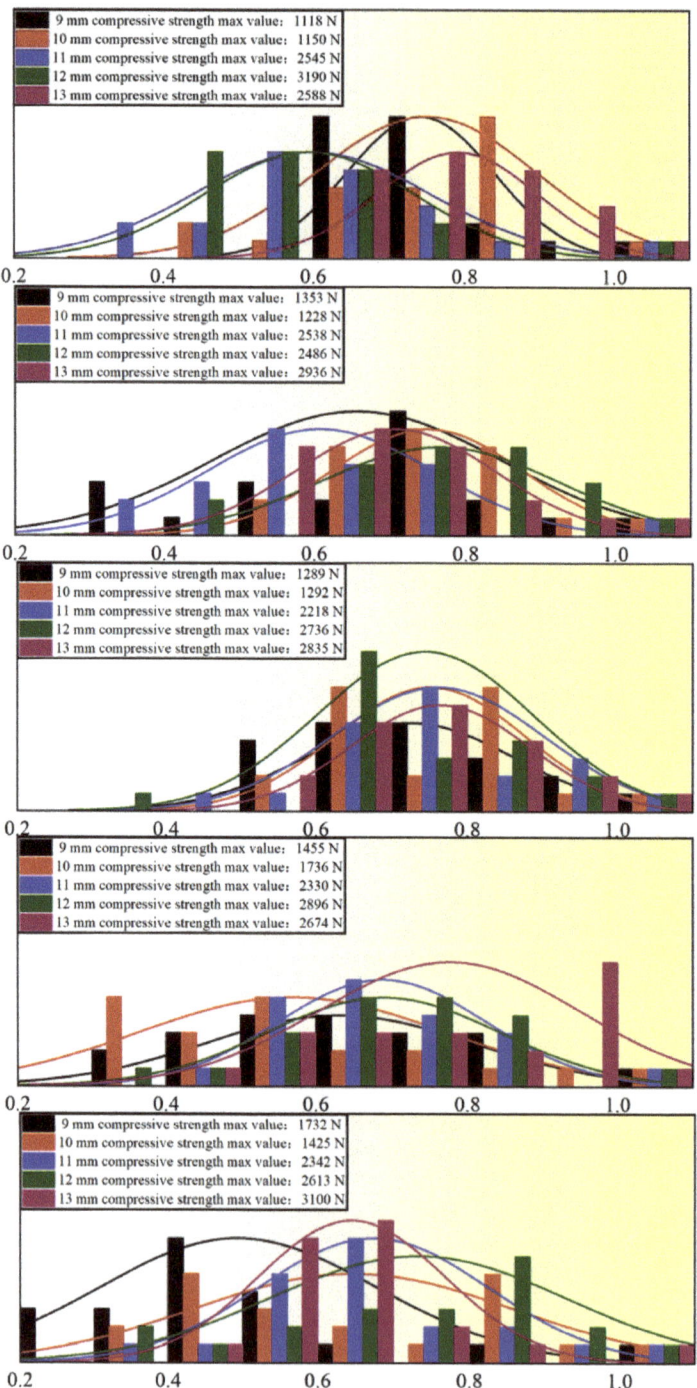

Figure 10. Normality test of pellet compressive strength under the influence of pellet diameter.

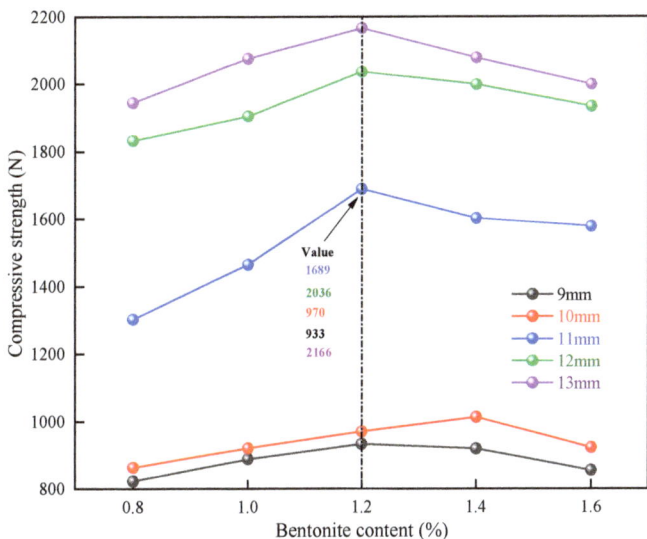

Figure 11. Effect of bentonite content on pellet compressive strength.

The desorption of crystal water from limonite is the main reason for the decrease in pellet strength [23–25]. To investigate the effect of bentonite on the desorption of crystal water, limonite powders with bentonite contents of 0.5wt%, 1wt%, and 2wt% were taken for differential thermal-thermogravimetric (TG-DTA) analysis. In an air atmosphere, the thermogravimetric (TG) and differential thermal (DTA) curves are shown in Figures 13 and 14, respectively. Based on the thermogravimetric curve (Figure 13), it was discovered that the starting and ending temperatures of limonite crystal water desorption are basically the same when different contents of bentonite are added. It shows that the addition of bentonite has no significant effect on the desorption process of limonite crystal water. Analysis of the differential heat (DTA) curve (Figure 14) shows that the endothermic peak intensities of limonite at 200–400 °C are the same when the addition of bentonite is 0–2wt%. It shows that the addition of bentonite has no significant effect on the endothermic or exothermic effect of the desorption process of crystal water. Studies have demonstrated that the desorption performance of crystal water depends only on the degree of its binding to the ore [26]. The function of bentonite is mainly reflected in increasing the nucleation rate, green ball strength, and reducing the growth rate of green balls in ball making. It can be further inferred that the effect of bentonite on the compressive strength of pellets is mainly achieved by changing the pore structure.

3.3. Multifactorial Test

3.3.1. Response Surface Design

The experiment was designed with three factors and three levels. The effects of three experimental parameters, including bentonite content, roasting temperature, and pellet diameter, on the compressive strength (Y) of pure limonite pellets were investigated. The strength of pure limonite was optimized to determine the best roasting process parameters using the response surface design by the design method of the statistical software. The factors and codes used in the experiment are shown in Table 4.

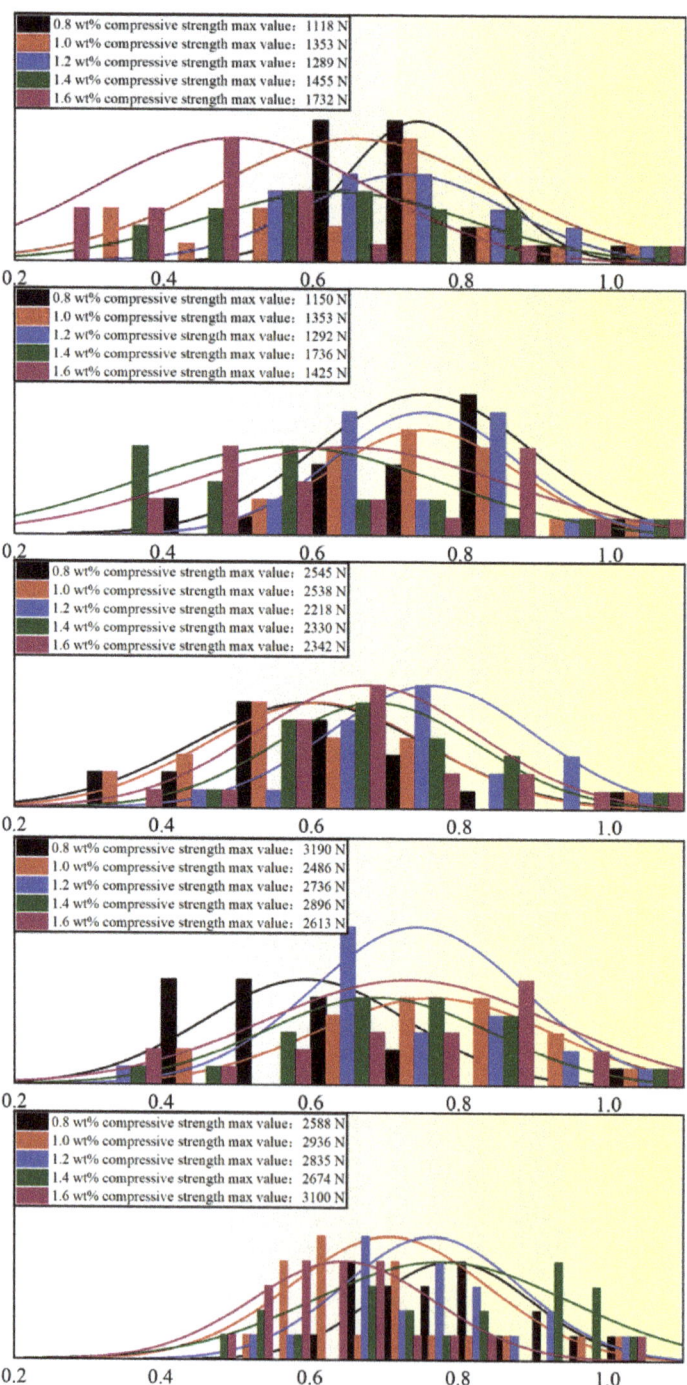

Figure 12. Normality test of pellet compressive strength under the influence of bentonite content.

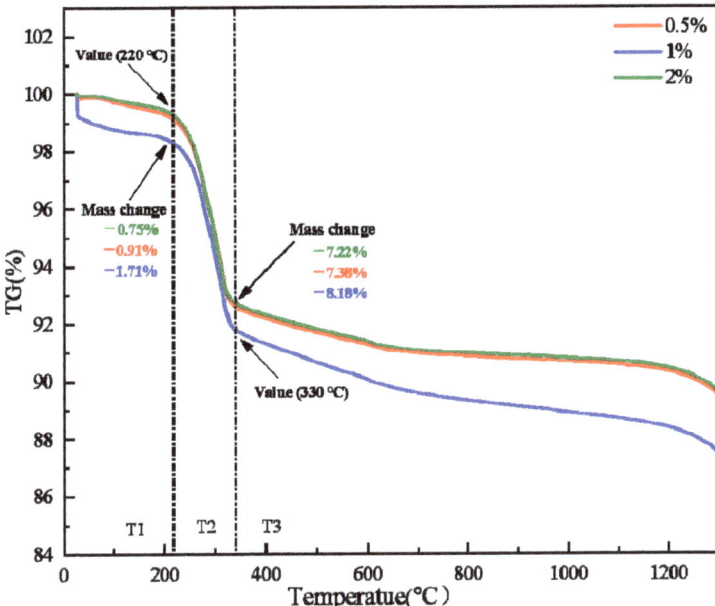

Figure 13. TG curve of limonite with different bentonite ratio.

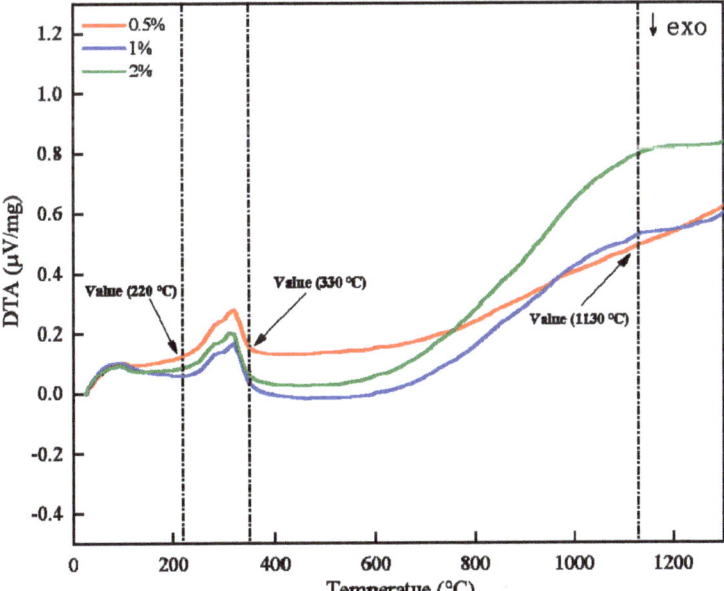

Figure 14. DTA curve of limonite with different bentonite ratio.

Table 4. Experimental factors and coding.

Factor	Variable	Level		
		−1	0	1
Roasting temperature/°C	A	1250	1300	1350
Pellet diameter/mm	B	11	13	15
Bentonite content/%	C	0.5	1.0	1.5

3.3.2. Response Surface Method Design Results

The Box–Behnken experimental design was carried out using Statistical methods, and the experimental results are shown in Table 5. The quadratic model was used to perform regression fitting on the experimental results in Table 5. The mathematical prediction model of each influencing factor of the compressive strength of pure limonite pellets is established, as shown in Equation (3).

$$Y = -242092 + 386.2A - 140B - 11256C - 0.1610A^2 - 56.6B^2 - 415C^2 + 1.490AB + 11.31AC - 176.7BC \quad (3)$$

Table 5. Experimental results.

Run	Roasting Temperature/°C A	Pellet Diameter/mm B	Bentonite Content/% C	Predicted Value/N	Actual Value/N
1	−1	0	−1	2111.05	2059.00
2	0	1	−1	2392.50	2560.00
3	−1	−1	0	1717.20	1881.00
4	0	−1	−1	1444.30	1476.00
5	0	1	1	2359.00	2424.00
6	0	0	0	2408.50	2452.00
7	0	−1	1	2117.60	2047.00
8	0	0	0	2408.50	2366.00
9	1	−1	0	1247.20	1319.00
10	1	0	−1	1373.55	1421.00
11	−1	1	0	2014.00	2039.00
12	0	0	0	2408.50	2553.00
13	1	1	0	2140.00	2073.00
14	−1	0	1	1865.45	1915.00
15	1	0	1	2258.95	2408.00

The statistical methods was used to perform variance analysis on the model, and the results are shown in Table 6. The F value in an ANOVA analysis is the ratio of the mean square between groups to the mean square within each group. The probability value under the corresponding F value is denoted by the p value. The presence of a p value less than 0.05 implies that the model item significantly impacts compressive strength. When the p value is less than 0.01, the factors substantially influence the compressive strength. It can be seen from Table 4 that the p-value of this model is less than 0.01, indicating that the selected experimental model fits well and has statistical significance. For the compressive strength of pure limonite pellets, The p value of A is greater than 0.05, indicating that A has no significant impact on compressive strength; the p value of B is less than 0.01, indicating that B has a very significant effect on compressive strength; and the p value of C is less than 0.05, indicating that C has a significant effect on compressive strength. Similarly, A^2 has a tremendous impact on compressive strength in the square term, B^2 has a significant impact on compressive strength, and C^2 has no major effect on compressive strength. In the interaction term, the influence of AB and BC on compressive strength is negligible; however, the influence of AC on compressive strength is substantial. Among them, for the primary term of a single factor, the order of the influence of the three factors on compressive strength is: pellet diameter > bentonite ratio > calcination temperature.

Table 6. Variance analysis of the model.

Variable	Statistical Analysis				
	Sum of Squares	df	Mean Square	F-Value	p-Value
Model	2,251,335.15	9	250,148.35	13.26	0.0055
A	56,616.13	1	56,616.13	3.00	0.1438
B	703,891.12	1	703,891.12	37.31	0.0017
C	204,160.51	1	204,160.51	10.82	0.0217
AB	88,804.00	1	88,804.00	4.71	0.0822
AC	319,790.25	1	319,790.25	16.95	0.0092
BC	124,962.25	1	124,962.25	6.62	0.0498
A^2	598,176.92	1	598,176.92	31.70	0.0024
B^2	189,423.69	1	189,423.69	10.04	0.0249
C^2	39,744.23	1	39,744.23	2.11	0.2064
Residual	94,341.25	5	18,868.25		
Lack of Fit	76,819.25	3	25,606.42	2.92	0.2652
Pure Error	17,522.00	2	8761.00		
Cor total	12,345,676.40	14			

Figure 15 shows the normal distribution of residuals of the strength of pure limonite pellets. Random scattering patterns that do not follow a specific shape near a straight line indicate the high accuracy of regression modeling. The experimental points are approximately a straight line, indicating that the experimental selection model can be used to predict the experimental process within the normal range of the experimental residual distribution.

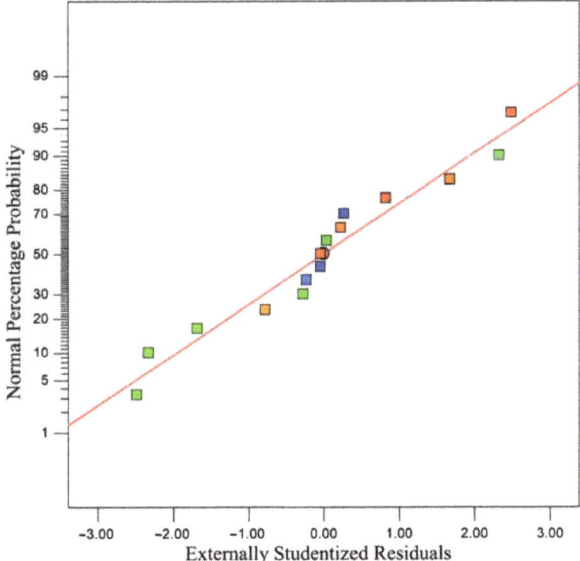

Figure 15. Residual normal probability map.

The comparison between the expected and experimental values of the compressive strength of pure limonite pellets is shown in Figure 16. The diagonal line in the figure means that the expected and experimental values are equal. The dots distributed around the slashes represent the predicted values relative to those experimentally obtained. It can be seen from Figure 16 that the experimental and predicted values are basically on both sides of this straight line, indicating that the experimental and predicted values have a high degree of fit.

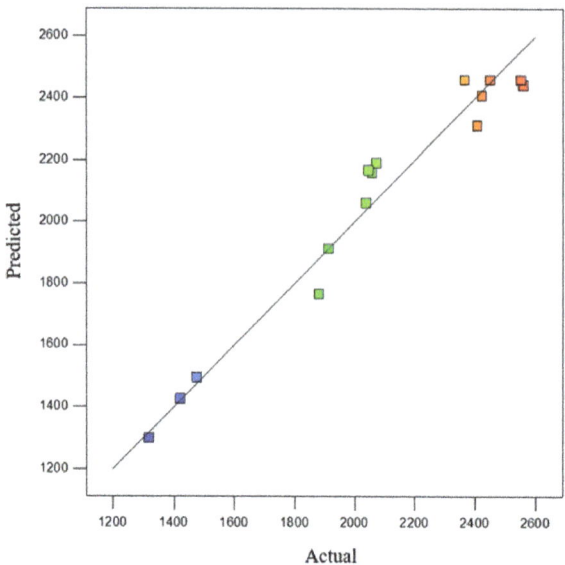

Figure 16. Relationship between experimental values of compressive strength of pellets and predicted values.

3.3.3. Response Surface Method Optimization

A 3D response surface plot shows the effect of model parameters on compressive strength. When the calcination temperature is 1300 °C, the response surface between the pellet diameter, the bentonite content, and the pellet compressive strength is shown in Figure 17a. It can be seen from the response surface in the figure that the experimental range is that the diameter of the pellet is 11–15 mm, and the ratio of bentonite is 0.5–1.5wt%. In the pellet diameter range of 11–13 mm, the compressive strength increases with the increase in pellet diameter; when the pellet diameter increases to 15 mm, the compressive strength decreases. In the bentonite ratio range of 0.5–1.3wt%, the compressive strength increases with the increase of the bentonite ratio, and the effect is significant on compressive strength. When the bentonite ratio is between 1.3wt% and 1.5wt%, the compressive strength declines as the bentonite ratio increases, essentially consistent with the results of the single-component experiment.

The response surface between the calcination temperature, the content of bentonite, and the compressive strength of the pellet when the diameter of the pellet is 13 mm is shown in Figure 18a. It can be seen from the curved surface in the figure that the experimental range is the calcination temperature of 1250 °C–1350 °C, and the ratio of bentonite is 0.5–1.5wt%. In the calcination temperature of 1250 °C–1300 °C, the compressive strength increases with the increase in calcination temperature. When the calcination temperature is between 1300 °C and 1350 °C, the compressive strength decreases with the increase in calcination temperature. In the bentonite ratio range of 0.5–1.3wt%, the compressive strength increases with the increase in bentonite ratio. In the bentonite ratio range of 1.3–1.5wt%, the compressive strength decreases with the rise in bentonite ratio, consistent with the univariate results.

The response surface between pellet diameter, calcination temperature, and compressive strength of pellets when the bentonite content is 1wt% is shown in Figure 19a. It can be seen from the figure that the investigation conditions are 11–15 mm in diameter and 1250–1350 °C in calcination temperature. When the diameter of the pellet is 11–13 mm, with the increase in diameter, the compressive strength of the pellets is continuously improved. When the diameter reaches 15 mm, the compressive strength of the pellets shows a decreasing trend. When the investigated temperature is less than 1300 °C, the effect of temperature on activation energy is insignificant, and the impact on the improvement of compressive strength is weakened. When the temperature is more than 1300 °C, the effect of temperature on the compressive strength of the pellets is also insignificant, which is consistent with the single-factor experimental results.

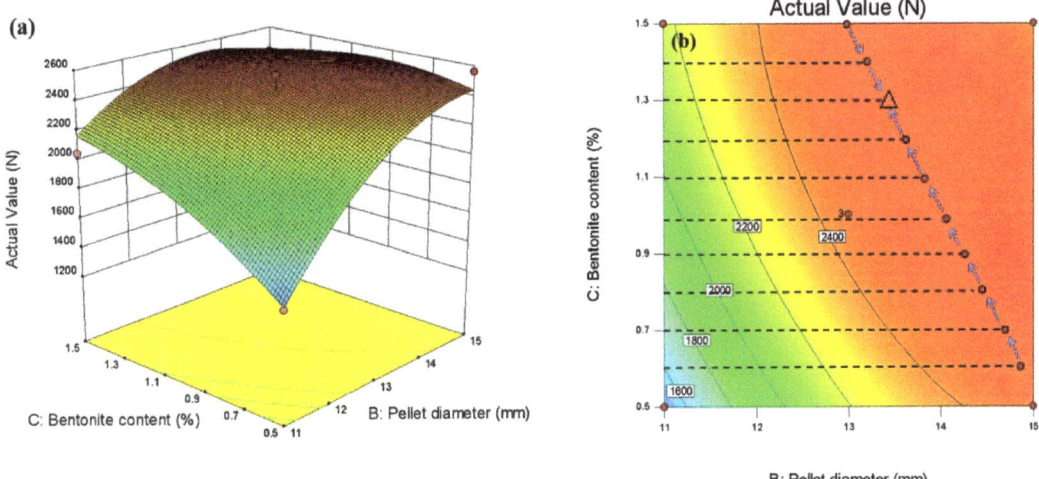

Figure 17. Response surface diagram between pellet diameter, bentonite content, and pellet compressive strength: (**a**) 3D response surface plot, (**b**) optimal parameter contour plot.

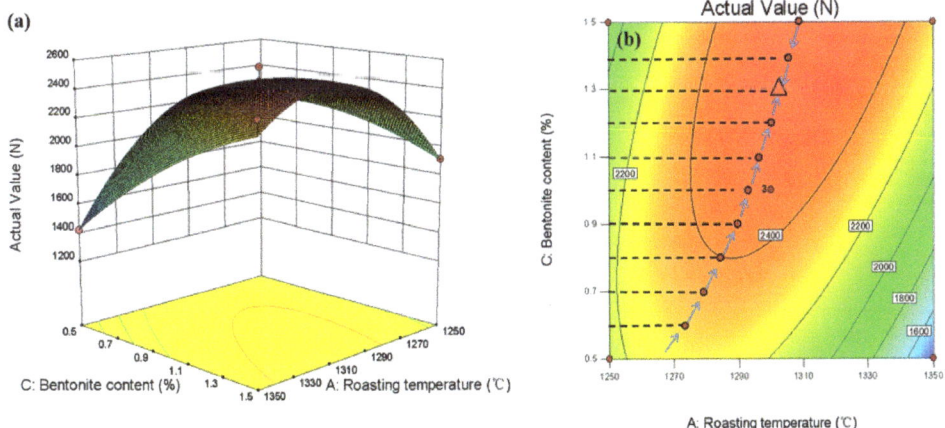

Figure 18. Response surface diagram between roasting temperature, bentonite content, and pellet compressive strength: (**a**) 3D response surface plot, (**b**) optimal parameter contour plot.

Considering the interaction of three factors, the smallest ellipse is the highest point of the response surface in the two-dimensional contour plots. This oval-shaped 2D contour line indicates a significant interaction. Conversely, circular contour plots show weaker interactions. In addition, the higher the strength level of a contour line, the more interaction between the two factors. The contour lines of the optimum compressive strength process parameters are shown in Figures 17b, 18b and 19b. The interaction effect between the two independent variables can be easily understood from the figure. At the same time, the optimal level can be precisely located.

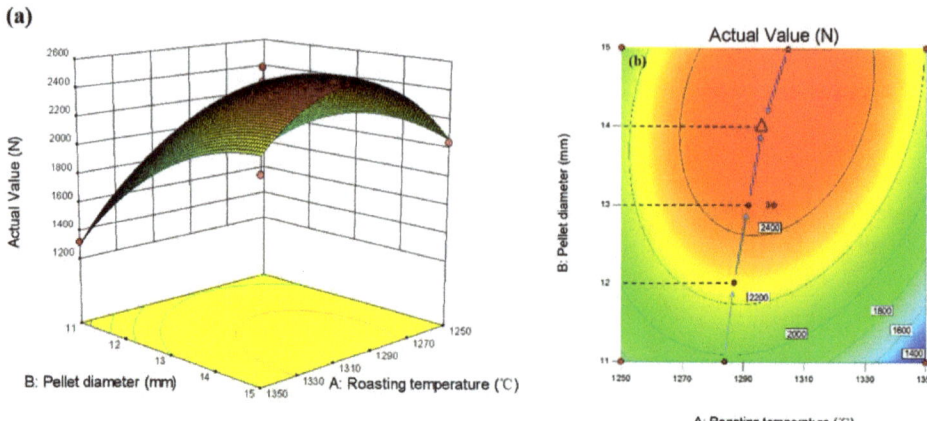

Figure 19. Response surface plot between pellet diameter, roasting temperature, and pellet compressive strength: (**a**) 3D response surface plot, (**b**) optimal parameter contour plot.

3.3.4. Response Surface Method Optimization

The experimental parameters were investigated (calcination temperature 1250–1350 °C, pellet diameter 11–15 mm, bentonite content 0.5–1.5wt%), Comprehensive roasting temperature, roasting time, bentonite content, and other conditions are selected from the optimal experimental conditions optimized by the response surface software: The calcination temperature is 1310.61 °C, the pellet diameter is 13.9495 mm, and the bentonite content is 1.32828%, Its predicted compressive strength is 2570.3 N/piece. According to this optimal parameter condition, the pellet experiment is carried out again, and the experimental value of compressive strength is 2705 N/piece, its value is within the 95% confidence interval, and the relative error between the experimental value and the predicted value is only 5.2%, indicating that the model strength is relatively reliable.

4. Conclusions

1. The compressive strength of pure limonite pellets can reach 2700 N, which can be used in small blast furnaces and rotary hearth furnace processes.
2. Within the set factor range (calcination temperature 1250–1350 °C, bentonite content 0.5–1.5%, pellet diameter 11–15 mm), pellet diameter has the most significant influence on the compressive strength of pure limonite pellets, followed by the bentonite content, and the calcination temperature has no significant impact on the compressive strength of the pellets.
3. The roasting process of pure limonite pellets was optimized by statistical methods. Within the set factor range (calcination temperature 1250–1350 °C, bentonite content 0.5–1.5%, pellet diameter 11–15 mm), the optimized experimental conditions obtained are: bentonite content 1.32828%, calcination temperature 1310.61 °C, pellet diameter 13.9495 mm. The predicted value of pellet compressive strength is 2570.3 N/piece, and the experimentally verified value of pellet compressive strength is 2705 N/piece.
4. The content of bentonite does not affect the desorption process of limonite crystal water.

Author Contributions: The research presented here was carried out with the collaboration of all authors. Conceptualization, X.Z. and L.G.; methodology, C.Z.; validation, C.Z.; resources, X.Z. and L.G.; formal analysis, C.Z.; writing—original draft preparation, C.Z.; writing—review and editing, X.Z., L.G. and H.F; supervision, X.Z., L.G. and H.F. All authors have read and agreed to the published version of the manuscript.

Funding: This work was financially supported by the General Project of Yunnan Science and Technology Department (KKS0202152010 and 202101AT070083).

Data Availability Statement: The data that support the findings of this study are available from the authors upon reasonable request.

Conflicts of Interest: The authors declare no conflict of interest.

References

1. Sahu, S.N.; Sharma, K.; Barma, S.D.; Pradhan, P.; Nayak, B.K.; Biswal, S.K. Utilization of low-grade BHQ iron ore by reduction roasting followed by magnetic separation for the production of magnetite-based pellet feed. *Metall. Res. Technol.* **2019**, *116*, 611. [CrossRef]
2. Nabeel, M.; Karasev, A.; Glaser, B.; Jönsson, P.G. Characterization of Dust Generated during Mechanical Wear of Partially Reduced Iron Ore Pellets. *Steel Res. Int.* **2017**, *88*, 1600442. [CrossRef]
3. Zhao, W.; Chen, J.; Chang, X.; Guo, S.; Srinivasakannan, C.; Chen, G.; Peng, J. Effect of microwave irradiation on selective heating behavior and magnetic separation characteristics of Panzhihua ilmenite. *Appl. Surf. Sci.* **2014**, *300*, 171–177. [CrossRef]
4. Prusti, P.; Barik, K.; Dash, N.; Biswal, S.K.; Meikap, B.C. Effect of limestone and dolomite flux on the quality of pellets using high LOI iron ore. *Powder Technol.* **2021**, *379*, 154–164. [CrossRef]
5. Gao, Q.J.; Shen, F.M.; Wei, G.; Jiang, X.; Zheng, H.Y. Effects of MgO containing additive on low-temperature metallurgical properties of oxidized pellet. *J. Iron Steel Res. Int.* **2013**, *20*, 25–28. [CrossRef]
6. de Morais Oliveira, V.; de Resende, V.G.; Domingues, A.L.A.; Bagatini, M.C.; de Castro, L.F.A. Alternative to deal with high level of fine materials in iron ore sintering process. *J. Mater. Res. Technol.* **2019**, *8*, 4985–4994. [CrossRef]
7. Agrawal, S.; Rayapudi, V.; Dhawan, N. Comparative study of low-grade banded iron ores for iron recovery. *Metall. Res. Technol.* **2020**, *117*, 403. [CrossRef]
8. Iljana, M.; Kemppainen, A.; Paananen, T.; Mattila, O.; Pisilä, E.; Kondrakov, M.; Fabritius, T. Effect of adding limestone on the metallurgical properties of iron ore pellets. *Int. J. Miner. Process.* **2015**, *141*, 34–43. [CrossRef]
9. Zhang, K.; Ge, Y.; Guo, W.; Li, N.; Wang, Z.; Luo, H.; Shang, S. Phase transition and magnetic properties of low-grade limonite during reductive roasting. *Vacuum* **2019**, *167*, 163–174. [CrossRef]
10. Sagar, R.K.; Sah, R.; Maribasappanavar, B.; Desai, S.; Balachandran, G. Optimization of Drying and Preheating Temperatures During Pellet Induration for Utilizing Goethitic Iron Ores. *Min. Metall. Explor.* **2022**, *39*, 1667–1678. [CrossRef]
11. Jiang, X.; Wu, G.S.; Jin, M.F.; Shen, F. Effect of MgO content on softening-melting property of sinter. *J.-Northeast. Univ. Nat. Sci.* **2006**, *27*, 1358.
12. Ghadi, A.Z.; Vahedi, S.M.; Sohn, H.Y. Analysis of the Gaseous Reduction of Porous Wustite Pellets by Response Surface Methodology. *Steel Res. Int.* **2021**, *92*, 2100048. [CrossRef]
13. Aslani, H.; Nabizadeh, R.; Nasseri, S.; Mesdaghinia, A.; Alimohammadi, M.; Mahvi, A.H.; Nazmara, S. Application of response surface methodology for modeling and optimization of trichloroacetic acid and turbidity removal using potassium ferrate (VI). *Desalination Water Treat.* **2016**, *57*, 25317–25328. [CrossRef]
14. Zhang, Z.; Peng, J.; Srinivasakannan, C.; Zhang, Z.; Zhang, L.; Fernández, Y.; Menéndez, J.A. Leaching zinc from spent catalyst: Process optimization using response surface methodology. *J. Hazard. Mater.* **2010**, *176*, 1113–1117. [CrossRef] [PubMed]
15. Kumar, S.; Suman, S.K. Compressive strength of fired pellets using organic binder: Response surface approach for analyzing the performance. *Trans. Indian Inst. Met.* **2018**, *71*, 1629–1634. [CrossRef]
16. Zhang, X.; Sun, C.; Xing, Y.; Kou, J.; Su, M. Thermal decomposition behavior of pyrite in a microwave field and feasibility of gold leaching with generated elemental sulfur from the decomposition of gold-bearing sulfides. *Hydrometallurgy* **2018**, *180*, 210–220. [CrossRef]
17. Chen, Q.; Chen, G.; Tong, L.; Lin, Y.; Yang, H. Optimization of Knelson gravity separation of a quartz vein type gold ore using response surface methodology. *J. Cent. South Univ. Sci. Technol.* **2019**, *50*, 2925–2931.
18. Wu, S.; Yu, X.; Hu, Z.; Zhang, L.; Chen, J. Optimizing aerobic biodegradation of dichloromethane using response surface methodology. *J. Environ. Sci.* **2009**, *21*, 1276–1283. [CrossRef]
19. Lorente, E.; Herguido, J.; Peña, J.A. Steam-iron process: Influence of steam on the kinetics of iron oxide reduction. *Int. J. Hydrog. Energy* **2011**, *36*, 13425–13434. [CrossRef]
20. Fan, J.J.; Qiu, G.Z.; Jiang, T.; Guo, Y.F.; Cai, M.X. Roasting properties of pellets with iron concentrate of complex mineral composition. *J. Iron Steel Res. Int.* **2011**, *18*, 1–7. [CrossRef]
21. Zhang, H.Q.; Lu, M.M.; Fu, J.T. Oxidation and roasting characteristics of artificial magnetite pellets. *J. Cent. South Univ.* **2016**, *23*, 2999–3005. [CrossRef]
22. Wang, G.; Zhang, J.; Hou, X.; Shao, J.; Geng, W. Study on CO_2 gasification properties and kinetics of biomass chars and anthracite char. *Bioresour. Technol.* **2015**, *177*, 66–73. [CrossRef] [PubMed]
23. Wang, G.; Zhang, J.; Shao, J.; Ren, S. Characterisation and model fitting kinetic analysis of coal/biomass co-combustion. *Thermochim. Acta* **2014**, *591*, 68–74. [CrossRef]
24. Wang, L.; Li, T.; Várhegyi, G.; Skreiberg, Ø.; Løvås, T. CO_2 gasification of chars prepared by fast and slow pyrolysis from wood and forest residue: A kinetic study. *Energy Fuels* **2018**, *32*, 588–597. [CrossRef]

25. Rousset, P.; Figueiredo, C.; De Souza, M.; Quirino, W. Pressure effect on the quality of eucalyptus wood charcoal for the steel industry: A statistical analysis approach. *Fuel Process. Technol.* **2011**, *92*, 1890–1897. [CrossRef]
26. Wang, L.; Wang, Z.; Wang, E. Novel hydrogen-bonded three-dimensional network complexes containing copper-pyridine-2, 6-dicarboxylic acid. *J. Coord. Chem.* **2014**, *57*, 1353–1359. [CrossRef]

Article

A Study on the Mechanism and Kinetics of Ultrasound-Enhanced Sulfuric Acid Leaching for Zinc Extraction from Zinc Oxide Dust

Xuemei Zheng [1,2,3], Shiwei Li [1,3], Bingguo Liu [1,3,*], Libo Zhang [1,3,*] and Aiyuan Ma [2,*]

1. Faculty of Metallurgical and Energy Engineering, Kunming University of Science and Technology, Kunming 650093, China
2. School of Chemistry and Materials Engineering, Liupanshui Normal University, Liupanshui 553004, China
3. Key Laboratory of Unconventional Metallurgy, Kunming University of Science and Technology, Kuming 650093, China
* Correspondence: bingoliu@126.com (B.L.); libozhang77@163.com (L.Z.); may_kmust11@163.com (A.M.)

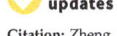

Citation: Zheng, X.; Li, S.; Liu, B.; Zhang, L.; Ma, A. A Study on the Mechanism and Kinetics of Ultrasound-Enhanced Sulfuric Acid Leaching for Zinc Extraction from Zinc Oxide Dust. *Materials* 2022, 15, 5969. https://doi.org/10.3390/ma15175969

Academic Editor: Carlos Manuel Silva

Received: 22 July 2022
Accepted: 22 August 2022
Published: 29 August 2022

Publisher's Note: MDPI stays neutral with regard to jurisdictional claims in published maps and institutional affiliations.

Copyright: © 2022 by the authors. Licensee MDPI, Basel, Switzerland. This article is an open access article distributed under the terms and conditions of the Creative Commons Attribution (CC BY) license (https://creativecommons.org/licenses/by/4.0/).

Abstract: As an important secondary zinc resource, large-scale reserves of zinc oxide dust (ZOD) from a wide range of sources is of high comprehensive recycling value. Therefore, an experimental study on ultrasound-enhanced sulfuric acid leaching for zinc extraction from zinc oxide dust was carried out to investigate the effects of various factors such as ultrasonic power, reaction time, sulfuric acid concentration, and liquid–solid ratio on zinc leaching rate. The results show that the zinc leaching rate under ultrasound reached 91.16% at a temperature of 25 °C, ultrasonic power 500 W, sulfuric acid concentration 140 g/L, liquid–solid ratio 5:1, rotating speed 100 r/min, and leaching time 30 min. Compared with the conventional leaching method (leaching rate: 85.36%), the method under ultrasound increased the zinc leaching rate by 5.8%. In a kinetic analysis of the ultrasound-enhanced sulfuric acid leaching of zinc oxide dust, the initial apparent activation energy of the reaction was 6.90 kJ/mol, indicating that the ultrasound-enhanced leaching process was controlled by the mixed solid product layers. Furthermore, the leached residue was characterized by XRD and SEM-EDS, and the results show that, with ultrasonic waves, the encapsulated mineral particles were dissociated, and the dissolution of ZnO was enhanced. Mostly, the zinc in leached residue existed in the forms of $ZnFe_2O_4$, Zn_2SiO_4, and ZnS.

Keywords: zinc oxide dust (ZOD); ultrasound-enhanced leaching; zinc extraction

1. Introduction

As an extensive metallic element in the world, zinc is broadly used in various fields such as the automobile, construction, shipbuilding, and aerospace industries as plated zinc, zinc-based alloy, and zinc oxide, etc. Mostly, zinc ores are composed of zinc sulfide, zinc oxide, or a mixture thereof. In recent years, zinc resources in China have generally been characterized by less rich ores and more low-grade ores, fewer large mines, and more small or medium mines which are difficult to exploit. At present, 70% of zinc in the world comes from zinc ore resources, while the remaining 30% comes from secondary zinc resources [1–4].

There are various sources of secondary zinc resources, e.g., hot galvanizing slag [5] and zinc ash [6], smelting slag (mud) and zinc-containing dust produced by the copper, lead, and zinc smelting industry [7–9], electric arc furnace (EAF) dust (mud) from the iron and steel industry [10,11], zinc-loaded waste catalysts [12–14], waste zinc manganese batteries [15], and circuit boards [16]. To date, more than 2 million tons of zinc have been recovered from secondary zinc resources, and at the same time, the growth rate for the recycling of zinc metals, alloys, and zinc compounds has been three times higher than that for the production of original zinc, indicating that the recovery of secondary zinc resources plays an important role in the recycling economy at present.

Secondary zinc resources are from a wide range of sources [17–21]. For zinc oxide dust from lead and zinc smelting, which is an important secondary zinc resource, the composition is complicated with many coexisting valuable metals and elementary impurities due to the complex zinc hydrometallurgy process for leached residue. In addition to the dissolution and leaching of ZnO in roasted ore as well as the hydrolysis and purification of Fe^{3+} in leach solution for iron removal, with the addition of neutralizing agents such as zinc calcine and lime milk, the changes in pH value and concentrations of some metal ions (e.g., $Cu^{2+} > 800$ mg/L) may be accompanied by the hydrolysis of copper, cadmium, cobalt, and silicon. The complexity in the composition of slag from the zinc hydrometallurgy process is related to the content of valuable metals and impurities in raw materials (e.g., Zn, Pb, Fe, Cu, Cd, In, Co, Si, As, F, and Cl) and the technical control for the process, and thus, there are often a lot of valuable metals and precious metals in neutral leached residue [22–26].

Mostly, zinc in the leached residue occurs in the forms of ZnO, ZnS, $ZnFe_2O_4$, and $ZnSiO_4$, and for the treatment of zinc-bearing leached residue, usually, a high-temperature reduction volatilization method is used to recover the valuable metals in neutral leached residue in the conventional zinc hydrometallurgy process (e.g., blast furnace smelting and smoke furnace smelting, sulfation roasting, chlorination, sulfation roasting, and rotary kiln roasting). The zinc content in zinc oxide dust produced by high-temperature reduction volatilization may reach 60%, with ZnO as the main zinc phase. However, the high-temperature volatilization process is accompanied by many reactions, in which the polymers tend partially to encase the zinc oxide phase, and thus, the zinc leaching rate for zinc oxide dust leaching by conventional acid method [27–29], alkali method [30,31], or ammonia method [32–35] is low. According to Oustadakis et al. [27], the recovery of Zn from EAFD can reach 80% with diluted sulfuric acid leaching. In this case, iron is partially transferred into the solution, and iron leaching reaches 45%. Sethurajan et al. [28] have examined the sulfuric acid leaching of the three different zinc plant leach residues (ZLR). The results showed that a higher temperature and acid concentration are required to leach the maximum Zn from the ZLRs for sulfates, oxides, and ferrite minerals. Fattahi et al. [29] examined the reductive leaching of zinc, cobalt, and manganese from zinc plant residue with dilute sulfuric acid and citric acid. The maximum Co, Mn, and Zn recoveries were 96.43%, 90.26%, and 64.12%, respectively. According to Ashtari et al. [30], 82.4% of the zinc was recovered from the zinc plant residue by the conventional alkaline leaching under NaOH concentration of 9 M, the temperature of 25 °C, time of 45 min, speed of 400 rpm, and S/L ratio of 1/10. They also found that the unreacted core of ZnO particles can be significantly improved by mechanochemical alkali leaching. Zhang et al. [31] proposed a process of primary normal pressure leaching and secondary alkaline pressure leaching zinc from EAF dust, 66.4%, and 88.7% Zn can be leached, respectively. The optimum conditions were a temperature of 70 °C, NaOH concentration of 6 mol/L, L/S ratio of 20 mL/g, and a reaction time of 2 h. Ma et al. [32] recovered zinc from blast furnace dust in the ammonia leaching system containing different leaching agents: ammonium sulphate, ammonium carbonate, ammonium citrate, and ammonia, from which 75.32%, 72.52%, 65.99%, and 31.92% Zn was recovered, respectively. The study found that Zn_2SiO_4, ZnS, and $ZnFe_2O_4$ did not leach into the ammonia system, which was one of the main causes of the lower zinc extraction rate [33,34]. In addition, the leaching rate of zinc can be effectively improved by microwave calcification pretreatment [9,35].

However, as a new unconventional metallurgical technology [36], ultrasonic metallurgy has been widely used in the comprehensive recycling of valuable metals by many researchers. The ultrasound-enhanced leaching process is mainly reflected in the cavitation effect, mechanical effect, and thermal effect. Under the action of ultrasonic cavitation, cavitation bubbles grow and rupture at a certain sound pressure, forming a local high temperature and high pressure forward flow zone in a tiny space, which promotes the leaching reaction. Under the mechanical action, through agitation and flow in the leaching solution, the ultrasound stirs the liquid intensively, reduces the diffusion resistance, and accelerates the mass and heat transfer to accelerate the diffusion and dissolution process of the medium, Through the thermal effect, the ultrasonic energy is continuously absorbed by the medium and converted into heat energy, which further promotes the reaction. Ultrasound-enhanced leaching provides a very special new physical environment for the difficult or impossible reactions to realize under conventional conditions, and thus, to a certain extent, the leaching conditions are improved, the reaction time is shortened, the recovery rate of valuable metals is increased, and an efficient metallurgical process of energy saving and environmental protection is realized [37–45]. Wang et al. [37] reported enhanced zinc leaching kinetics from zinc residues augmented with ultrasound. Similarly, Brunelli [38] reported an ultrasound-assisted leaching process for the recovery of zinc from electric arc furnace (EAF) dust. The research results show that ultrasonic-assisted leaching is a suitable technique to improve the dissolution of $ZnFe_2O_4$, which represents the main obstacle during conventional leaching. In addition, based on the advantages of using ultrasound in strengthening the leaching process, ultrasonic technology has been widely used in the treatment of metals such as copper (Cu) [39], uranium (U) [40], gold (Au) [41], silver (Ag) [42,43], and germanium (Ge) [44,45], and has achieved relatively significant leaching effects.

With zinc oxide dust volatilized from a rotary kiln for lead and zinc smelting as the object, this study introduced an ultrasound-enhanced method with sulfuric acid as the leaching agent to investigate the effects of ultrasonic power (UP), sulfuric acid concentration (C), leaching time (t), liquid–solid ratio (L/S), rotating speed (r), and temperature (T) on the zinc leaching rate of zinc oxide dust, and explore the kinetics of ultrasound-enhanced leaching of zinc oxide dust. At the same time, the leaching mechanism of zinc oxide dust was analyzed by XRD and SEM-EDS.

2. Experimental Materials and Characterization

2.1. Analysis on the Composition of Raw Materials

The raw materials used in the experiment were from a zinc hydrometallurgy enterprise in Yunnan, China. The chemical composition of the ZOD sample, as shown in Table 1, is complicated, with a zinc content up to 41.37%, coexisting with valuable metal elements, e.g., Pb, Mn, and Cd, a scattered associated metal content as high as 820.8 g/t, as well as large amounts of S, Cl, Si, and Ca. The zinc oxide dust had a high recycling value.

Table 1. The chemical composition of the ZOD sample (mass fraction, %).

Element	Zn	Pb	Cd	Fe	Mn	S
Content/%	41.37	19.77	1.01	2.05	0.20	3.95
Element	Cl	Si	Ca	In	F	
Content/%	0.28	0.19	0.12	820.8 g/t	<0.01	

2.2. Analysis on Mineral Phase

To determine the forms of various metal elements and impurity components existing in zinc oxide dust, the samples were characterized by XRD (TTRIII Multifunctional X-ray Diffractometer, Rigaku, Japan) and SEM-EDS (XL30ESEM-TMP scanning electron microscope, Philips, The Netherlands). The results are shown in Figures 1–3, respectively.

The XRD patterns show that most zinc exists as ZnO, Zn_2SiO_4, ZnS, and $ZnFe_2O_4$, and that most lead exists as PbS and $PbSO_4$. In addition, there are large amounts of gangues in zinc oxide dust, especially SiO_2 and $CaSiO_3$.

The SEM pattern indicates that ore particles in zinc oxide dust are compact with lots of amorphous flocculent inclusions embedded among the particles. The point and surface analysis results of SEM-EDS (see Figures 2 and 3) show that in zinc oxide dust, most of the zinc is distributed in tiny floccule particles as floccules coexisting with O, Fe, Si, and Zn and lumps coexisting with O, S, Pb, and Ca. The zinc oxide dust was different from the original ore in mineral morphology, material existence form, and mineral surface property.

The granular minerals were fused at a high temperature with zinc easy to be enveloped by other valuable metals and gangue components.

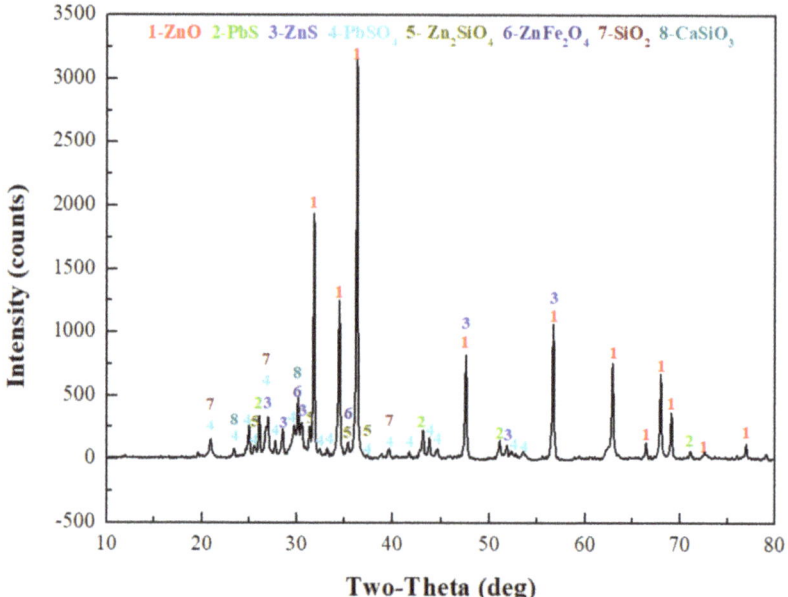

Figure 1. XRD pattern of the raw ZOD sample.

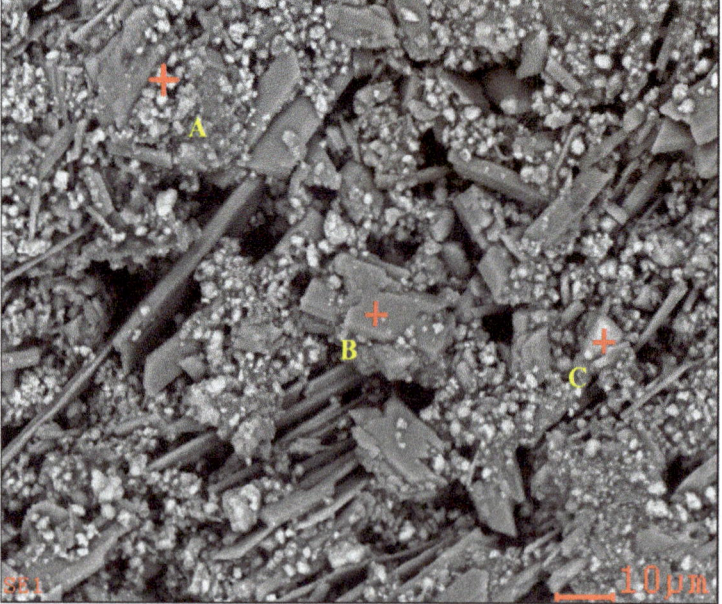

Figure 2. *Cont.*

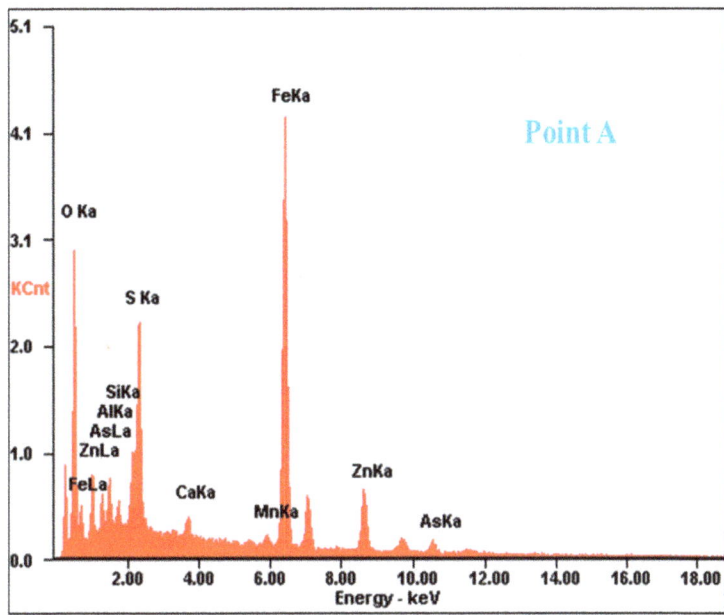
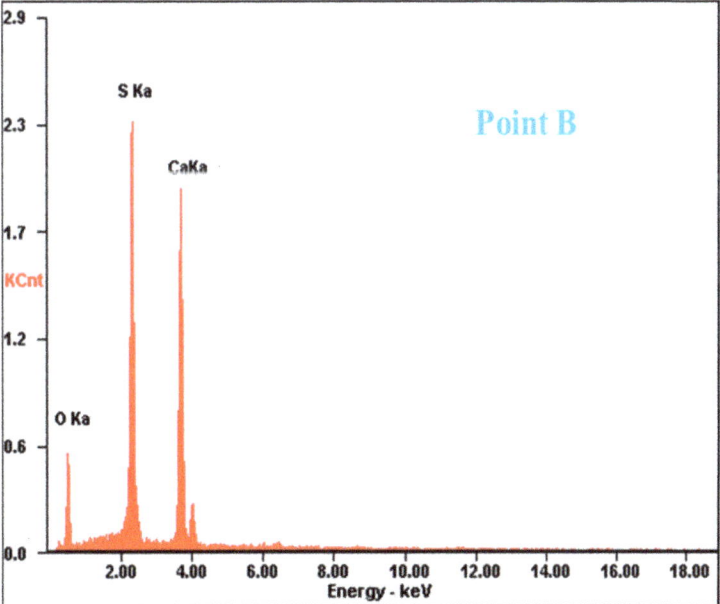

Figure 2. *Cont.*

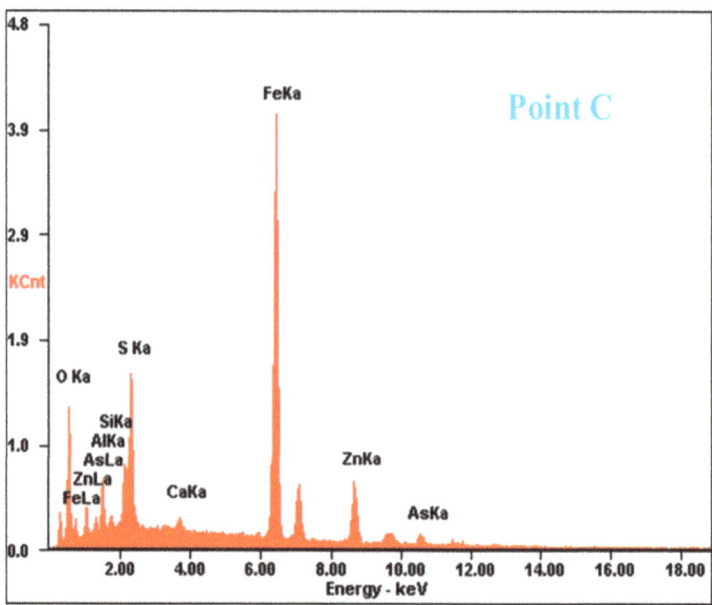

Figure 2. SEM image and EDS point analysis of the raw ZOD sample.

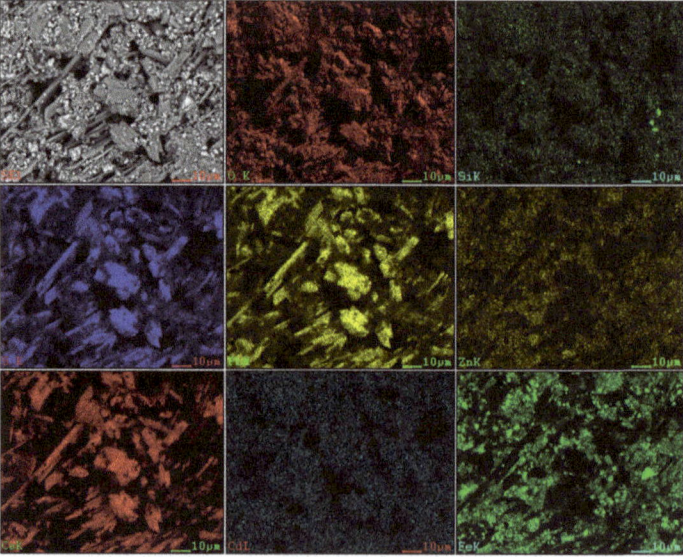

Figure 3. SEM image and EDS mapping analysis of the raw ZOD sample.

2.3. Experimental Methods

A certain amount of zinc oxide dust and a certain concentration of prepared sulfuric acid leaching solvent were added to a 300 mL conical flask in a certain ratio, and then, the conical flask was placed on a thermostatic magnetic agitator with a digital display, and an ultrasonic probe was inserted in the thermostatic water bath at a position equivalent to the level of solution in the conical flask. The experimental apparatus for ultrasonic leaching is presented in Figure 4. The flow diagram of the zinc leaching process is shown in Figure 5. During the ultrasonic leaching process, with the increase in ultrasonic power or the prolongation of the ultrasonic leaching time, the temperature of the water bath will increase to a certain extent. To ensure that the leaching temperature remains constant, the temperature of the water bath was adjusted to 2 °C during the experiment. For the ultrasound system, an ultrasonic probe continuously adjustable within power 0~2000 W and resistant to a certain concentration of acid was adopted to provide an ultrasonic field. The ultrasonic power and stirring speed could be controlled and adjusted as required. Filtered after a certain leaching time, the zinc concentration in the leaching solution was determined by EDTA titration method. In the determination process, there may have been interference from Cu^{2+}, Al^{3+}, and Fe^{2+} on Zn^{2+}. To eliminate the interference, saturated thiourea, ascorbic acid, and potassium fluoride solutions were added before adding the xylenol orange indicator. The zinc extraction rate (η_{Zn}, %) may be calculated by the following formula:

$$\eta_{Zn} = \frac{C_{Zn} \times V}{m \times w_{Zn}^*} \quad (1)$$

where C_{Zn}—Zn concentration in the leaching solution, g/L; V—*the volume of leaching solution*, L; m—*mass of the sample*, g; and w_{Zn}^*—percentage of Zn in the sample.

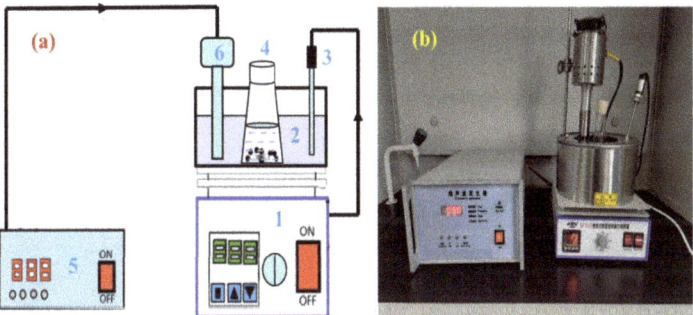

Figure 4. Experimental apparatus for ultrasound leaching: (**a**) schematic diagram; and (**b**) device diagram. In (**a**), 1—heat collection type thermostatic bath; 2—water bath; 3—thermometer, 4—conical flask; 5—ultrasonic generator; 6—ultrasonic probe.

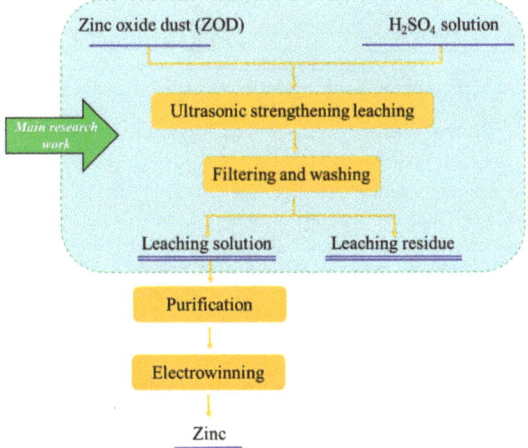

Figure 5. Flow diagram of the leaching process for ZOD.

3. Experimental Results and Relevant Analysis

3.1. Experimental Study on Ultrasound-Enhanced Leaching Conditions

3.1.1. Effect of Ultrasonic Power on Zinc Leaching Rate

The effect of ultrasonic power on the zinc leaching rate over time was investigated under a sulfuric acid concentration of 100 g/L, a liquid–solid ratio of 4:1, a rotating speed of 100 rpm, and a temperature of 65 °C, and the results are as shown in Figure 6. It can be seen that the zinc leaching rate gradually increases with time in direct proportion. At 0–20 min, the zinc leaching rate rapidly increased with time, and after 20 min, the zinc leaching efficiency was obviously lowered. In addition, the zinc leaching rate increased with increasing ultrasonic power, and when the ultrasonic power was 100 W, the zinc leaching rate was only 51.62%; when it was more than 300 W, the zinc leaching rate significantly increased, to 62.31%, 64.6%, 67.25%, and 68.94%, respectively, with the ultrasonic power of 300–900 W, within a time of 30 min. These results indicate that the ultrasonic power had a significant effect on the leaching rate of zinc. Considering that, when the ultrasonic power exceeded 500 W, the zinc leaching rate was obviously not further advanced, and in combination with the relevant energy consumption, the ultrasonic power of leaching was controlled at 500 W.

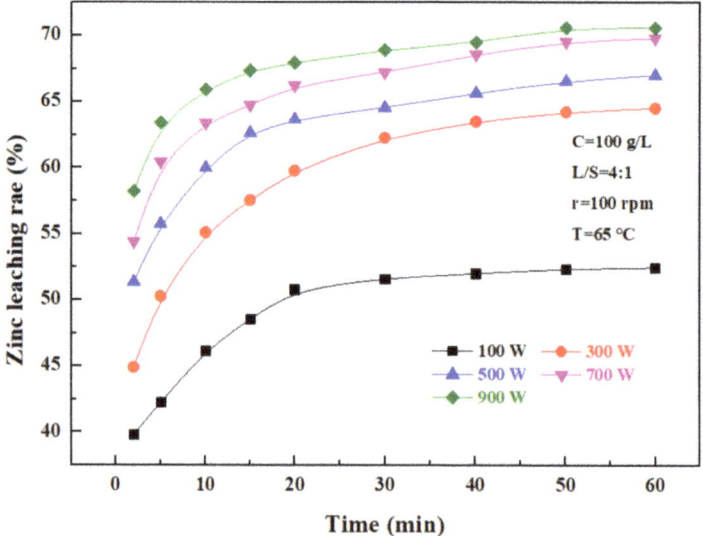

Figure 6. Effect of ultrasonic power on zinc leaching rate.

3.1.2. Effect of Sulfuric Acid Concentration on Zinc Leaching Rate

The effect of the sulfuric acid concentration on the leaching rate of zinc was investigated under an ultrasonic power of 500 W, a liquid–solid ratio of 4:1, a rotating speed of 100 rpm, and a temperature of 65 °C. According to Figure 7, the zinc leaching rate also increased with sulfuric acid concentration. With time, the zinc leaching rate was increasing because the hydrogen ion concentration in the reaction increased with the sulfuric acid concentration. Promoting the contact of sulfuric acid with a granular zinc phase in zinc oxide dust was conducive to the leaching of zinc from the dust, but with time, hydrogen ions in the solution were consumed, and thus, the leaching rate was gradually lowered. Considering the high concentration of sulfuric acid, the dissolved Fe^{3+} ions increased accordingly, making it difficult to perform the subsequent treatment of the leaching solution. Therefore, the optimal concentration of sulfuric acid was determined to be 140 g/L.

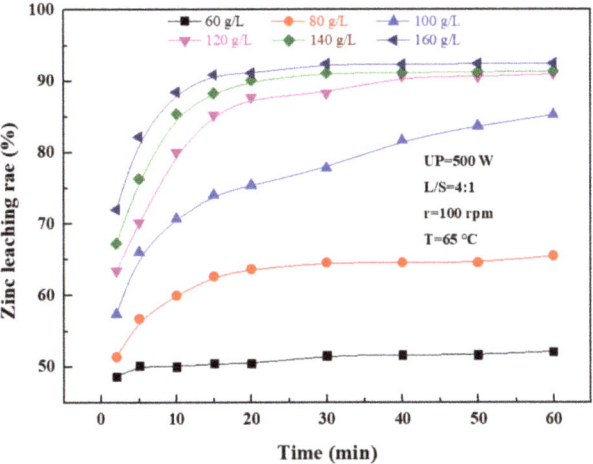

Figure 7. Effect of the sulfuric acid concentration on the zinc leaching rate.

3.1.3. Effect of Liquid–Solid Ratio on Zinc Leaching Rate

The effect of the liquid–solid ratio on the leaching rate of zinc from the zinc oxide dust was investigated under an ultrasonic power of 500 W, a sulfuric acid concentration of 140 g/L, a rotating speed of 100 rpm, and a temperature of 65 °C. The results shown in Figure 8 demonstrate that with the increase in the liquid–solid ratio, the zinc leaching rate first gradually increases, and then reaches a plateau. The main reason for this was that with the increase in the liquid–solid ratio, the fluidity of ions in and out of the system increased, and thus, the movement and collision of fine particles in the zinc oxide dust, as well as relevant reactions, were further intensified. When the liquid–solid ratio was increased from 2:1 to 4:1, the zinc leaching rate was promoted significantly, but when the liquid–solid ratio was further increased, the zinc leaching effect was obviously compromised, although the zinc leaching rate was still increased to some extent. When the liquid–solid ratio was 5:1, the 30 min leaching rate of zinc from the zinc oxide dust was 91.16%. Considering the increased liquid–solid ratio would compromise the subsequent purification and bring difficulties to the follow-up recovery process, and the optimal liquid–solid ratio was determined as 5:1.

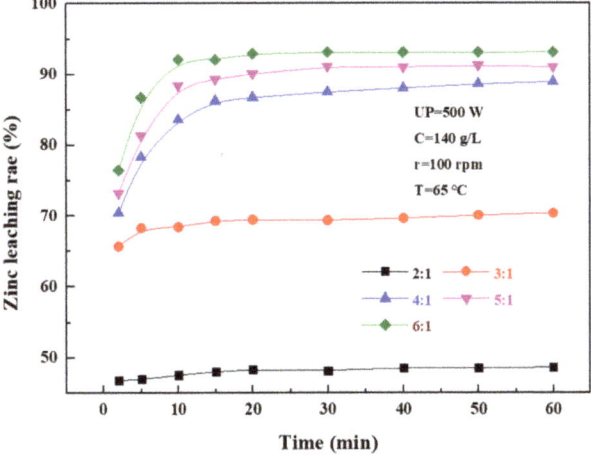

Figure 8. Effect of liquid–solid ratio on zinc leaching rate.

3.1.4. Effect of Rotating Speed on Zinc Leaching Rate

The effect of the rotating speed on the leaching rate of zinc from the zinc oxide dust was investigated under an ultrasonic power of 500 W, a sulfuric acid concentration of 140 g/L, liquid–solid ratio of 5:1, and a temperature of 65 °C. The results were as shown in Figure 9. It can be seen that, as the rotating speed increases, the leaching rate of zinc from the zinc oxide dust gradually increases with time, and ultimately reaches a plateau. If the stirring speed was low, the zinc oxide dust particles dissolved in sulfuric acid solution were easy to settle, which was not conducive to the leaching reaction, while an excessively high stirring speed would increase the energy consumption and the cost of the leaching process. Therefore, it is advisable to control the stirring speed at 100 rpm.

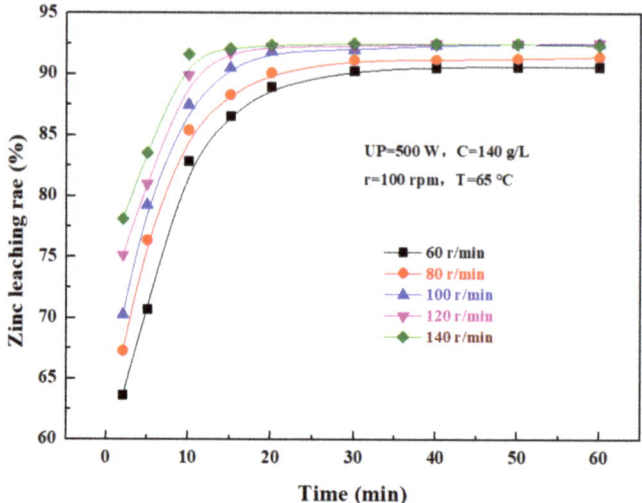

Figure 9. Effect of rotating speed on zinc leaching rate.

3.1.5. Effect of Temperature on Zinc Leaching Rate

The effect of temperature on the leaching rate of zinc from the zinc oxide dust was investigated under an ultrasonic power of 500 W, a sulfuric acid concentration of 140 g/L, a liquid–solid ratio of 5:1, and a rotating speed of 100 rpm. The results are as shown in Figure 10. Figure 10a shows that the change in temperature has a significant effect on the zinc leaching rate when the time was 0 to 20 min, and the zinc leaching rate increases with the increase in temperature. At different temperatures, the zinc leaching rate increased rapidly with time at first, and after 20–30 min, gradually reached a plateau. This is because, with the increase in leaching temperature, the leaching reaction rate was advanced accordingly, and at the same time, the viscosity of the solution decreased, which was conducive to the diffusion of a leaching solvent and product, and thus, the zinc leaching rate was significant at the beginning, while at a later stage, with the continuous consumption of a leaching agent, the leaching efficiency was gradually lowered, and a further increase in leaching temperature did not affect the dissolution of zinc oxide dust. Figure 10b shows that the zinc leaching rate of zinc oxide dust was less affected by temperature after the leaching time reaches 30 min, and the zinc leaching rates were 91.16% and 92.44% at 25 °C and 75 °C, respectively. Considering that high temperatures may increase the volatilization of acid, resulting in a high acid consumption and increased economic cost, the leaching temperature was controlled at 25 °C, and the 30 min zinc leaching rate reached 91.16%.

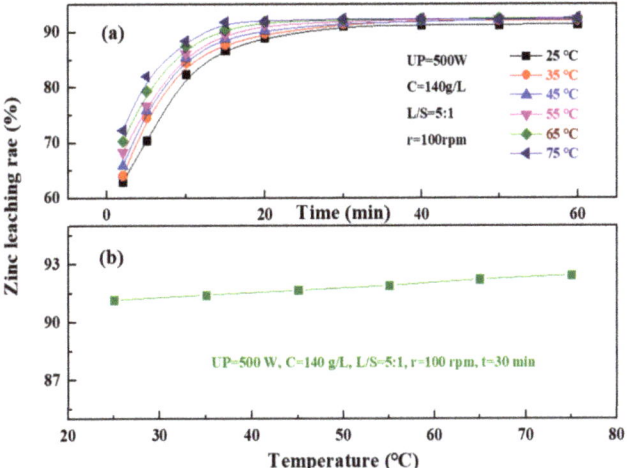

Figure 10. Effect of temperature on zinc leaching rate: (**a**)—different times; and (**b**)—30 min.

3.2. Kinetics of Ultrasound-Enhanced Leaching

The leaching of zinc from zinc oxide dust is a process of liquid–solid reaction, and the leaching reaction process may be controlled by the following steps: (i) the diffusion of a reactant or product for the leaching agent through the liquid boundary layer; (ii) the diffusion of a reactant or product for a leaching agent through the solid product layer; (iii) the chemical reaction of the reactant for a leaching agent with the surface of unreacted nuclear material; and (iv) the mixture of the solid film diffusion and interfacial chemical reaction.

An analysis of the raw material showed that the Zn-bearing dust particles was irregular in morphology, with a relatively complex composition, which included a granular zinc oxide phase and gangue particles, and most of the particles wrapped the Zn-bearing phase. Mostly, the zinc was embedded in the gangue mineral, and in the leaching process, the leaching agent was diffused to the gap or crack of gangue, and reacted with a zinc mineral contained in the zinc oxide dust. With the reaction, the interface for reaction continuously shrank into the center of zinc mineral particles, and the by-products or residual solid layer was thickening to enlarge the path for the diffusion of a reactant or product for leaching agent. In addition, the inert solid residue of the gangue tended to wrap the unreacted shrunk nuclei as a factor controlling the zinc leaching rate of zinc-bearing mineral particles. Therefore, a model of shrinking core was used to explore the kinetic behavior for the leaching of zinc from the zinc-baring metallurgical dust.

Based on the model of a shrinking core, when the solid–liquid phase reaction is controlled by diffusion reaction, the leaching kinetics equation of zinc oxide dust particles may be expressed as follows:

$$k_d \cdot t = 1 - 2/3x - (1-x)^{2/3} \quad (2)$$

when the solid–liquid phase reaction is controlled by interfacial chemical reaction, the leaching kinetics equation of zinc oxide dust particles may be expressed as follows:

$$k_r \cdot t = 1 - (1-x)^{1/3} \quad (3)$$

Furthermore, when the solid–liquid reaction is controlled by both the diffusion reaction and interfacial chemical reaction, the leaching kinetics equation of zinc oxide dust particles may be expressed as follows:

$$k_0 \cdot t = 1/3 \ln(1-x) - [1 - (1-x)^{-1/3}] \quad (4)$$

where k_d is the diffusion rate constant of the solid–liquid phase reaction, k_r, the constant of solid–liquid interfacial chemical reaction; k_0, the reaction rate constant under mixed solid–liquid control, x, the leaching rate of zinc from zinc oxide dust, and t, the leaching time.

To define the procedure for controlling an ultrasound-enhanced leaching process, a kinetic study was conducted for the ultrasound-enhanced leaching of zinc oxide dust. Relevant data from the experiment concerning the effect of a leaching time on the leaching rate of zinc from zinc-bearing metallurgical dust were put into Equations (2)–(4), respectively, for plotting the curves of $1 - 2/3x - (1 - x)^{2/3}$, $1 - (1 - x)^{1/3}$ and $1/3\ln(1 - x) - 1 + (1 - x)^{-1/3}$ vs. time t (0–20 min), and the results are as shown in Figure 11a–c, representing the curves of zinc leaching processes under the control of solid product layer diffusion, control of the interfacial chemical reaction and mixed control, respectively.

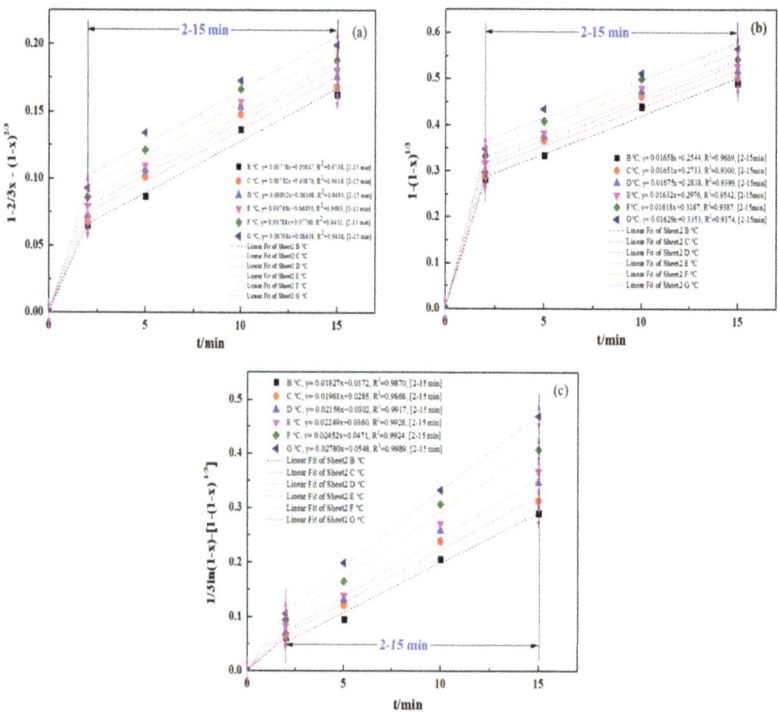

Figure 11. Plot of $1 - 2/3x - (1 - x)^{2/3}$, $1 - (1 - x)^{1/3}$, and $1/3\ln(1 - x) - [1 - (1 - x)^{-1/3}]$ vs. time for various temperatures: (**a**) $1 - 2/3x - (1 - x)^{2/3}$; (**b**) $1 - (1 - x)^{1/3}$; and (**c**) $1/3\ln(1 - x) - [1 - (1 - x)^{-1/3}]$.

It should be mentioned that the experimental data for the initial stage of the process (0–2 min) are ignored in Figure 11a–c. The purpose was to reduce the data disturbance caused by the uncontrolled transfer process in the initial stage. Bringing relevant data into the model will make it difficult for the fitting results to correctly reflect the kinetic conditions of the main reaction process. Therefore, the 2–15 min stage is selected for fitting the experimental data. Similar research methods were applied in the study of Wang et al. [39] and Gui et al. [41].

The apparent reaction rate constants at different leaching temperatures obtained by the fitting (k_d, k_r, and k_0) and the fitting coefficients related to the kinetic equations of reaction rates are as shown in Table 2.

Table 2. Correlation coefficients of fitting results for various models at various temperatures.

T (°C)	$1 - 2/3x - (1-x)^{2/3}$		$1 - (1-x)^{1/3}$		$1/3\ln(1-x) - 1 + (1-x)^{-1/3}$	
	k_d	R^2	k_r	R^2	k_0	R^2
25	0.00778	0.9738	0.01658	0.9689	0.01827	0.9870
35	0.00782	0.9416	0.01651	0.9300	0.01961	0.9868
45	0.00802	0.9495	0.01675	0.9399	0.02156	0.9917
55	0.00788	0.9603	0.01632	0.9542	0.02249	0.9926
65	0.00788	0.9452	0.01618	0.9387	0.02452	0.9924
75	0.00798	0.9436	0.01629	0.9374	0.02780	0.9989

In addition, the reaction rate constant k before the zinc leaching process approaching equilibrium (i.e., a fitting equation) at different temperatures was obtained according to Figure 11a–c and substituted into the Arrhemus empirical equation, respectively, as follows:

$$K = A \cdot \exp(-Ea/RT) \qquad (5)$$

where Ea represents the activation energy of reaction, kJ/mol, A represents the frequency factor as a constant, T represents the temperature (K), and R represents the gas constant, 8.314×10^{-3} kJ/(mol·K).

Take the logarithm for both sides of Equation (4) to obtain the relation between ln$k$ and $1/T$:

$$\mathrm{Ln}k = \ln A - Ea/RT \qquad (6)$$

A curve of ln$k$ vs. $1/T$ is plotted as shown in Figure 12. The degree of fitting for the ln$k$ vs. $1/T$ curve under mixed control (R^2 = 0.9684) was significantly higher than that under diffusion control (R^2 = 0.2315) and interfacial chemical reaction control (R^2 = 0.3929), further indicating that the ultrasound-enhanced leaching of zinc oxide dust was primarily controlled by the mixed control of solid product layers, with an initial apparent activation energy of reaction 6.90 kJ/mol.

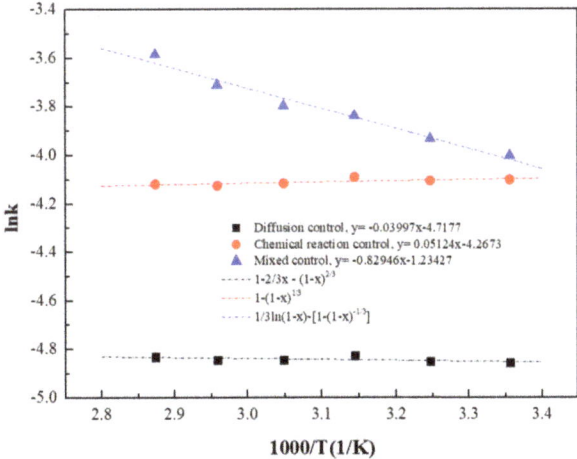

Figure 12. Arrhenius curve obtained for the dissolution of ZOD.

3.3. Comparative Experiment of Conventional-Ultrasonic Leaching

The optimal zinc leaching conditions for zinc oxide dust under an ultrasonic field determined under time conditions are as follows: leaching temperature 25 °C; ultrasonic power 500 W; sulfuric acid concentration 140 g/L; liquid–solid ratio 5:1; rotating speed 100 rpm; and leaching time 30 min, for which the zinc leaching rate can reach 91.16%. Under the optimal process conditions, a comparison between conventional stirring and ultrasound enhancement for an effect on the leaching rate of the zinc from oxide dust was conducted. The results are as shown in Figure 13: the zinc leaching rate under conventional stirring was 85.36%, and was increased by 5.8% under ultrasound enhancement at 500 W.

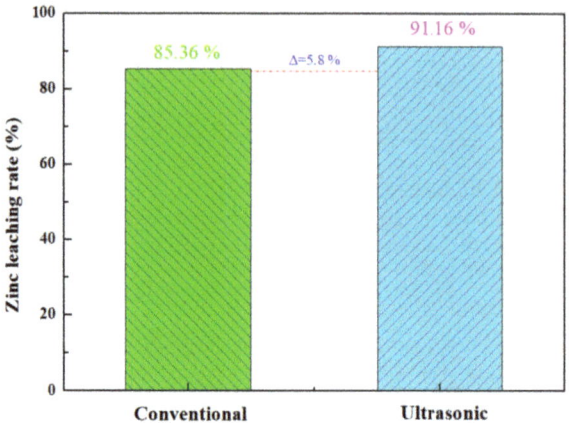

Figure 13. Comparison of zinc leaching rate between conventional and ultrasonic treatment of zinc oxide dust.

3.4. Analysis of Leaching Mechanism

3.4.1. Characterization by XRD

The XRD patterns of the zinc oxide dust and leached residue are as shown in Figure 14. Obviously, the ZnO phase was mostly dissolved after sulfuric acid leaching, and for the conventional leached residue, there was still a ZnO peak. The reason for this might be that there were wraps in the zinc oxide dust, and the ZnO in the wrapped particles was not leached. For both ultrasonic leached residue and conventional leached residue, the ZnO peak intensity was weakened or the peak disappeared as compared with the raw material. Zinc in the residues existed in forms of $ZnFe_2O_4$, Zn_2SiO_4, and ZnS accompanied by large amounts of $PbSO_4$, PbS, and SiO_2.

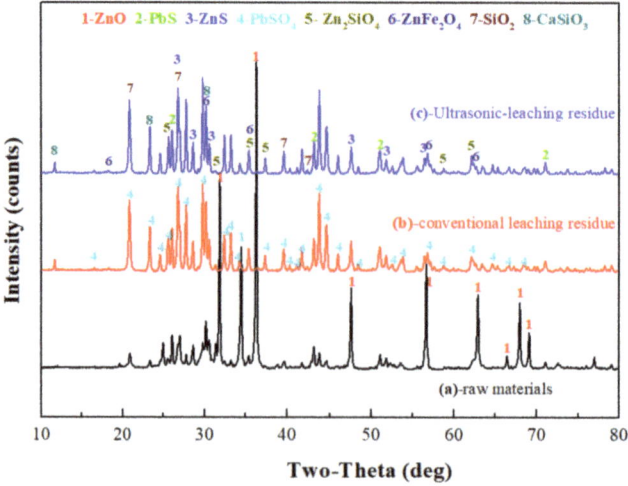

Figure 14. XRD patterns of raw materials (**a**), conventional leaching residue (**b**), and ultrasonic-leaching residue (**c**).

3.4.2. Comparative Analysis of SEM-EDS

For the raw material, the conventional leached residue and ultrasonic leached residue underwent SEM morphology characterization, for which the results are shown in Figure 15. Figure 15 shows that the zinc oxide dust particles were under a state of aggregation, and under the conventional leaching conditions, the wrapped particles had no change in morphology with a corrosion phenomenon on their surfaces. However, under the ultrasound enhancement, the leached residue had particles dispersed evenly, and the wrapped particles and the wraps were opened. This is also an important reason why the zinc leaching rate under ultrasound is higher than that under conventional leaching conditions. An analysis of leached residue by EDS spot scanning was carried out, and the results were as shown in Figures 15 and 16. The leached residue was complicated in composition, with the coexistence of Zn, Pb, Fe, O, S, and Si.

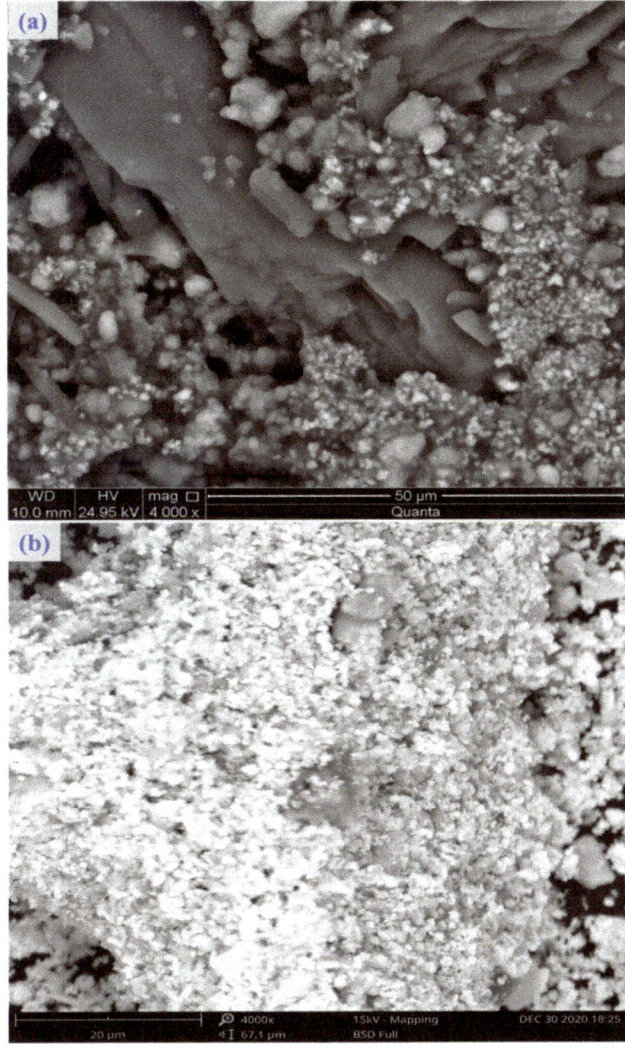

Figure 15. *Cont.*

Figure 15. SEM patterns of raw materials (**a**), conventional leaching residue (**b**), and the ultrasonic leaching residue (**c**).

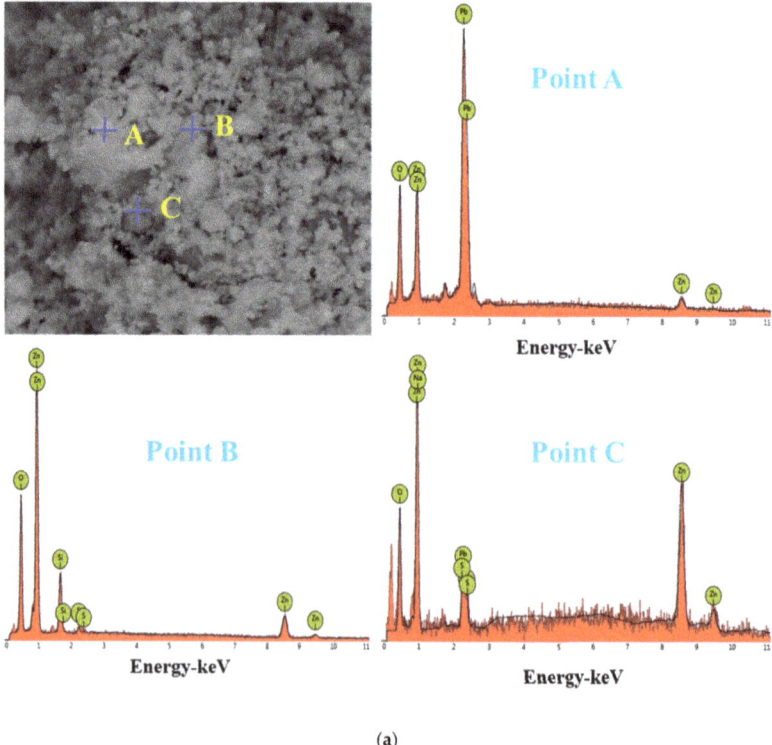

(**a**)

Figure 16. *Cont.*

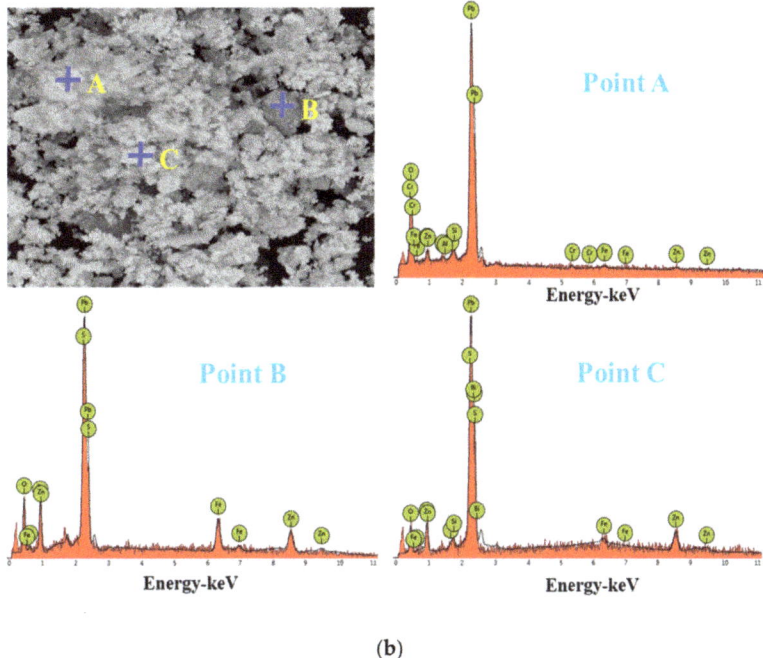

(b)

Figure 16. EDS point analysis of the conventional leaching residue (**a**), and the ultrasonic leaching residue (**b**).

3.4.3. Particle Size Analysis and Mechanism of Leaching ZOD

Table 3 lists the detailed particle size parameters of the ZOD sample (A), the conventional leaching residue (B), and the ultrasonic leaching residue (C). As presented in Table 3, compared with the raw ZOD sample, the parameter values of D_{10}, D_{50}, D_{90}, and D_{av} are increased and the values of the surface area-to-volume ratio are decreased for the conventional leaching residue. The change of parameters may be due to the disappearance of the ZnO phase mainly distributed in the fine particles after leaching, while a large number of encapsulated particles remain in the conventional leaching residue. However, the parameter values of D_{10}, D_{50}, D_{90}, and D_{av} are decreased and the values of surface area-to-volume ratio are increased for the ultrasonic leaching residue compared with the raw ZOD sample. The change in parameters may be due to the generation of a large number of bubbles under the action of ultrasonic cavitation. With the growth and burst of the bubbles, huge energy is released, which promotes the inclusion of the ZnO phase surface to fall off, and realizes the dissociation of the encapsulated particles to generate a large number of tiny particles. The schematic diagram of the leaching mechanism of ultrasonic-enhanced ZOD particles is shown in Figure 17.

Table 3. Particle size parameters of the ZOD sample (A), the conventional leaching residue (B), and the ultrasonic leaching residue (C).

Samples	D_{10} (μm)	D_{50} (μm)	D_{90} (μm)	D_{av} (μm)	Surface Area-to-Volume Ratio (m^2/cm^3)
A	0.821	1.047	1.230	1.031	5.9886
B	1.038	1.355	1.608	1.336	4.6482
C	0.725	0.871	1.007	0.861	7.0916

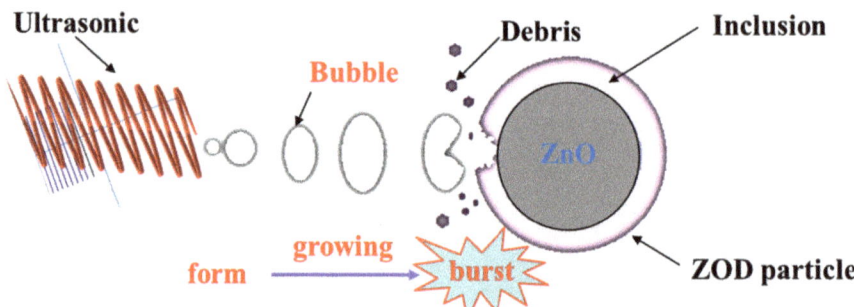

Figure 17. The schematic diagram of the leaching mechanism of ultrasonic-enhanced ZOD particle.

In summary, combined with the above analysis results of zinc leaching efficiency, XRD characterization, SEM-EDS characterization, and particle size analysis, it can be inferred that the leaching efficiency of zinc from zinc oxide dust (ZOD) was improved by ultrasonic strengthening treatment.

4. Conclusions

(1) Through an ultrasound-enhanced sulfuric acid leaching experiment, the optimal zinc leaching conditions for zinc oxide dust were determined as follows: leaching temperature of 25 °C; ultrasonic power of 500 W; sulfuric acid concentration of 140 g/L; liquid–solid ratio of 5:1; rotating speed of 100 rpm; and a leaching time of 30 min—for which the zinc leaching rate could reach up to 91.16%.

(2) In a kinetic analysis of the ultrasound-enhanced sulfuric acid leaching of zinc oxide dust, the initial apparent activation energy of the reaction was 6.90 kJ/mol. indicating that the ultrasound-enhanced leaching of zinc oxide dust was primarily controlled by the mixed control of the solid product layers.

(3) The leached residue was characterized by XRD and SEM-EDS, and the results showed that with ultrasonic waves, the encapsulated mineral particles were dissociated, and the dissolution of ZnO was enhanced. Mostly, the zinc in the leached residue existed in the forms of $ZnFe_2O_4$, Zn_2SiO_4, and ZnS accompanied by large amounts of $PbSO_4$, PbS, and SiO_2.

Author Contributions: X.Z.: writing—original draft, conceptualization, methodology, investigation, and validation; S.L.: visualization and resources; B.L.: supervision and writing—review and editing; L.Z.: supervision, writing—review and editing; A.M.: investigation, validation, and funding acquisition. All authors have read and agreed to the published version of the manuscript.

Funding: This work was supported from the Guizhou Province Science and Technology Project (No. [2019]1444, No. [2020]2001); the Guizhou Provincial Colleges and Universities Science and Technology Talent Project (KY [2020]124); the Science and Technology Innovation Group of Liupanshui Normal University (LPSSYKJTD201801); and the Key Cultivation Disciplines of Liupanshui Normal University (LPSSYZDXK202001).

Institutional Review Board Statement: Not applicable.

Informed Consent Statement: Not applicable.

Data Availability Statement: The data presented in this study are available upon request from the corresponding author. The data are not publicly available due to technical or time limitations.

Acknowledgments: The authors acknowledge the financial support received for carrying out this work.

Conflicts of Interest: The authors declare no conflict of interest.

References

1. Nayak, A.; Jena, M.S.; Mandre, N.R. Beneficiation of lead-zinc ores—A review. *Miner. Process. Extr. Metall. Rev.* **2021**, *43*, 1–20. [CrossRef]
2. Jia, N.N.; Wang, H.G.; Zhang, M.; Guo, M. Selective and Efficient Extraction of Zinc from Mixed Sulfide–oxide Zinc and Lead Ore. *Miner. Process. Extr. Metall. Rev.* **2016**, *37*, 418–426. [CrossRef]
3. Ma, A.Y.; Peng, J.H.; Zhang, L.B.; Li, S.W.; Yang, K.; Zheng, X.M. Leaching Zn from the Low-Grade Zinc Oxide Ore in NH_3-$H_3C_6H_5O_7$-H_2O Media. *Braz. J. Chem. Eng.* **2016**, *33*, 907–917. [CrossRef]
4. Yang, K.; Zhang, L.B.; Zhu, X.C.; Peng, J.H.; Li, S.W.; Ma, A.Y.; Li, H.Y.; Zhu, F. Role of manganese dioxide in the recovery of oxide-sulphide zinc ore. *J. Hazard. Mater.* **2018**, *343*, 315–323. [CrossRef]
5. Dong, M.G.; Xue, X.X.; Kumar, A.; Yang, H.; Sayyed, M.I.; Liu, S.; Bu, E. A novel method of utilization of hot dip galvanizing slag using the heat waste from itself for protection from radiation. *J. Hazard. Mater.* **2018**, *344*, 602–614. [CrossRef]
6. Rudnik, E. Recovery of zinc from zinc ash by leaching in sulphuric acid and electrowinning. *Hydrometallurgy* **2019**, *188*, 256–263. [CrossRef]
7. Zhang, B.K.; Guo, X.Y.; Wang, Q.M.; Tian, Q.H. Thermodynamic analysis and process optimization of zinc and lead recovery from copper smelting slag with chlorination roasting. *Trans. Nonferrous Met. Soc. China* **2021**, *31*, 3905–3917. [CrossRef]
8. Xia, Z.; Zhang, X.; Huang, X.; Yang, S.; Ye, L. Hydrometallurgical stepwise recovery of copper and zinc from smelting slag of waste brass in ammonium chloride solution. *Hydrometallurgy* **2020**, *197*, 105475. [CrossRef]
9. Ma, A.Y.; Zheng, X.M.; Gao, L.; Li, K.Q.; Omran, M.; Chen, G. Enhanced Leaching of Zinc from Zinc-Containing Metallurgical Residues via Microwave Calcium Activation Pretreatment. *Metals* **2021**, *11*, 1922. [CrossRef]
10. Wang, J.; Zhang, Y.Y.; Cui, K.K.; Fu, T.; Gao, J.J.; Hussain, S.; AlGarni, T.S. Pyrometallurgical recovery of zinc and valuable metals from electric arc furnace dust—A review. *J. Clean. Prod.* **2021**, *298*, 126788. [CrossRef]
11. Luo, X.G.; Wang, C.Y.; Shi, X.G.; Li, X.B.; Wei, C.; Li, M.T.; Deng, Z.G. Selective separation of zinc and iron/carbon from blast furnace dust via a hydrometallurgical cooperative leaching method. *Waste Manag.* **2022**, *139*, 116–123. [CrossRef] [PubMed]
12. Zhang, Z.Y.; Peng, J.H.; Srinivasakannan, C.; Zhang, Z.B.; Zhang, L.B.; Fernández, Y.; Menéndez, J.A. Leaching zinc from spent catalyst: Process optimization using response surface methodology. *J. Hazard. Mater.* **2010**, *176*, 1113–1117. [CrossRef]
13. Ma, A.Y.; Zheng, X.M.; Liu, C.H.; Peng, J.H.; Li, S.W.; Zhang, L.B.; Liu, C. Study on regeneration of spent activated carbon by using a clean technology. *Green Process. Synth.* **2017**, *6*, 499–510. [CrossRef]
14. Ma, A.Y.; Zheng, X.M.; Li, K.Q.; Omran, M.; Chen, G. The adsorption removal of tannic acid by regenerated activated carbon from the spent catalyst of vinyl acetate synthesis. *J. Mater. Res. Technol.* **2021**, *10*, 697–708. [CrossRef]
15. Zhan, L.; Li, O.Y.; Xu, Z.M. Preparing nano-zinc oxide with high-added-value from waste zinc manganese battery by vacuum evaporation and oxygen-control oxidation. *J. Clean. Prod.* **2020**, *251*, 119691. [CrossRef]
16. Meng, L.; Zhong, Y.W.; Guo, L.; Wang, Z.; Chen, K.Y.; Guo, Z.C. Recovery of cu and Zn from waste printed circuit boards using super-gravity separation. *Waste Manag.* **2018**, *78*, 559–565. [CrossRef] [PubMed]
17. Song, J.Q.; Peng, C.; Liang, Y.J.; Zhang, D.K.; Lin, Z.; Liao, Y.; Wang, G.J. Efficient extracting germanium and gallium from zinc residue by sulfuric and tartaric complex acid. *Hydrometallurgy* **2021**, *202*, 105599. [CrossRef]
18. Rao, S.; Wang, D.X.; Liu, Z.Q.; Zhang, K.F.; Cao, H.Y.; Tao, J.Z. Selective extraction of zinc, gallium, and germanium from zinc refinery residue using two stage acid and alkaline leaching. *Hydrometallurgy* **2018**, *183*, 38–44. [CrossRef]
19. Kaya, M.; Hussaini, S.; Kursunoglu, S. Critical review on secondary zinc resources and their recycling technologies. *Hydrometallurgy* **2020**, *195*, 105362. [CrossRef]
20. Xue, Y.; Hao, X.S.; Liu, X.M.; Zhang, N. Recovery of Zinc and Iron from Steel Mill Dust—An Overview of Available Technologies. *Materials* **2022**, *15*, 4127. [CrossRef]
21. Ma, A.Y.; Li, Y.Y.; Li, M.X.; Zheng, X.M. Zinc extraction from zinc oxide dust by tartrate. *Nonferr. Met. Eng.* **2021**, *11*, 70–77. [CrossRef]
22. Hu, H.; Deng, Q.; Chao, L.; Yue, X.; Dong, Z.; Wei, Z. The recovery of zn and pb and the manufacture of lightweight bricks from zinc smelting slag and clay. *J. Hazard. Mater.* **2014**, *271*, 220–227. [CrossRef] [PubMed]
23. Wang, J.; Huang, Q.F.; Li, T.; Xin, B.P.; Chen, S.; Guo, X.M.; Liu, C.H.; Li, Y.P. Bioleaching mechanism of Zn, Pb, In, Ag, Cd and As from Pb/Zn smelting slag by autotrophic bacteria. *J. Environ. Manag.* **2015**, *159*, 11–17. [CrossRef] [PubMed]
24. Liu, T.; Li, F.; Jin, Z.; Yang, Y. Acidic leaching of potentially toxic metals cadmium, cobalt, chromium, copper, nickel, lead, and zinc from two Zn smelting slag materials incubated in an acidic soil. *Environ. Pollut.* **2018**, *238*, 359–368. [CrossRef]
25. Ma, A.Y.; Zheng, X.M.; Peng, J.H.; Zhang, L.B.; Srinivasakannan, C.; Li, J.; Wei, C.L. Dechlorination of Zinc Oxide Dust Derived from Zinc Leaching Residue by Microwave Roasting in a Rotary Kiln. *Braz. J. Chem. Eng.* **2017**, *34*, 193–202. [CrossRef]
26. Ma, A.Y.; Zheng, X.M.; Wang, S.X.; Peng, J.H.; Zhang, L.B.; Li, Z.Q. Study on dechlorination kinetics from zinc oxide dust by clean metallurgy technology. *Green Process. Synth.* **2016**, *5*, 49–58. [CrossRef]
27. Oustadakis, P.; Tsakiridis, P.E.; Katsiapi, A.; Agatzini-Leonardou, S. Hydrometallurgical process for zinc recovery from electric arc furnace dust (EAFD): Part I: Characterization and leaching by diluted sulphuric acid. *J. Hazard. Mater.* **2010**, *179*, 1–7. [CrossRef]
28. Sethurajan, M.; Huguenot, D.; Jain, R.; Lens, P.N.L.; Horn, H.A.; Figueiredo, L.H.A.; Hullebusch, E.D.V. Leaching and selective zinc recovery from acidic leachates of zinc metallurgical leach residues. *J. Hazard. Mater.* **2016**, *324*, 71–82. [CrossRef]
29. Fattahi, A.; Rashchi, F.; Abkhoshk, E. Reductive leaching of zinc, cobalt and manganese from zinc plant residue. *Hydrometallurgy* **2016**, *161*, 185–192. [CrossRef]

30. Ashtari, P.; Pourghahramani, P. Selective mechanochemical alkaline leaching of zinc from zinc plant residue. *Hydrometallurgy* **2015**, *156*, 165–172. [CrossRef]
31. Zhang, D.C.; Liu, R.L.; Wang, H.; Liu, W.F.; Yang, T.Z.; Chen, L. Recovery of zinc from electric arc furnace dust by alkaline pressure leaching using iron as a reductant. *J. Cent. South Univ.* **2021**, *28*, 2701–2710. [CrossRef]
32. Ma, A.Y.; Zhang, L.B.; Peng, J.H.; Zheng, X.M.; Li, S.H.; Yang, K.; Chen, W.H. Extraction of zinc from blast furnace dust in ammonia leaching system. *Green Process. Synth.* **2016**, *5*, 23–30. [CrossRef]
33. Ma, A.Y.; Zheng, X.M.; Li, S.; Wang, Y.H.; Zhu, S. Zinc recovery from metallurgical slag and dust by coordination leaching in NH_3–CH_3COONH_4–H_2O system. *R. Soc. Open Sci.* **2018**, *5*, 180660. [CrossRef] [PubMed]
34. Ma, A.Y.; Zheng, X.M.; Zhang, L.B.; Peng, J.H.; Li, Z.; Li, S.; Li, S.W. Clean recycling of zinc from blast furnace dust with ammonium acetate as complexing agents. *Sep. Sci. Technol.* **2018**, *53*, 1327–1341. [CrossRef]
35. Ma, A.Y.; Zheng, X.M.; Gao, L.; Li, K.Q.; Omran, M.; Chen, G. Investigations on the Thermodynamics Characteristics, Thermal and Dielectric Properties of Calcium-ActivatedZinc-Containing Metallurgical Residues. *Materials* **2022**, *15*, 714. [CrossRef] [PubMed]
36. Cravotto, G.; Gaudino, E.C.; Cintas, P. On the mechanochemical activation by ultrasound. *Chem. Soc. Rev.* **2013**, *42*, 7521–7534. [CrossRef]
37. Wang, X.; Srinivasakannan, C.; Duan, X.H.; Peng, J.H.; Yang, D.J.; Ju, S.H. Leaching kinetics of zinc residues augmented with ultrasound. *Sep. Purif. Technol.* **2013**, *115*, 66–72. [CrossRef]
38. Brunelli, K.; Dabalà, M. Ultrasound effects on zinc recovery from EAF dust by sulfuric acid leaching. *Int. J. Miner. Metall. Mater.* **2015**, *22*, 353–362. [CrossRef]
39. Wang, S.X.; Cui, W.; Zhang, G.W.; Zhang, L.B.; Peng, J.H. Ultra fast ultrasound-assisted decopperization from copper anode slime. *Ultrason. Sonochem.* **2017**, *36*, 20–26. [CrossRef]
40. Li, J.; Zhang, L.B.; Peng, J.H.; Hu, J.M.; Yang, L.F.; Ma, A.Y.; Xia, H.Y.; Guo, W.Q.; Yu, X. Removal of Uranium from Uranium Plant Wastewater Using Zero-Valent Iron in an Ultrasonic Field. *Nucl. Eng. Technol.* **2016**, *48*, 744–750. [CrossRef]
41. Zhang, G.W.; Wang, S.X.; Zhang, L.B.; Peng, J.H. Ultrasound-intensified Leaching of Gold from a Refractory Ore. *ISIJ Int.* **2016**, *56*, 714–718. [CrossRef]
42. Chang, J.; Zhang, E.D.; Zhang, L.; Peng, J.H.; Zhou, J.W.; Srinivasakannan, C.; Yang, C.J. A comparison of ultrasound-augmented and conventional leaching of silver from sintering dust using acidic thiourea. *Ultrason. Sonochem.* **2017**, *34*, 222–231. [CrossRef] [PubMed]
43. Li, H.Y.; Zhang, L.B.; Xie, H.M.; Yin, S.H.; Peng, J.H.; Li, S.W.; Yang, K.; Zhu, F. Ultrasound-Assisted Silver Leaching Process for Cleaner Production. *JOM* **2020**, *72*, 766–773. [CrossRef]
44. Zhang, L.B.; Guo, W.Q.; Peng, J.H.; Lin, G.; Yu, X. Comparison of ultrasonic-assisted and regular leaching of germanium from by-product of zinc metallurgy. *Ultrason. Sonochem.* **2016**, *31*, 143–149. [CrossRef] [PubMed]
45. Xin, C.F.; Xia, H.Y.; Zhang, Q.; Zhang, L.B.; Zhang, W. Recovery of Zn and Ge from zinc oxide dust by ultrasonic-H_2O_2 enhanced oxidation leaching. *RSC Adv.* **2021**, *11*, 33788–33797. [CrossRef]

MDPI
St. Alban-Anlage 66
4052 Basel
Switzerland
www.mdpi.com

Materials Editorial Office
E-mail: materials@mdpi.com
www.mdpi.com/journal/materials

Disclaimer/Publisher's Note: The statements, opinions and data contained in all publications are solely those of the individual author(s) and contributor(s) and not of MDPI and/or the editor(s). MDPI and/or the editor(s) disclaim responsibility for any injury to people or property resulting from any ideas, methods, instructions or products referred to in the content.

www.ingramcontent.com/pod-product-compliance
Lightning Source LLC
LaVergne TN
LVHW070505100526
838202LV00014B/1795